REPUBLIC OF SOUTH AFRICA
REPUBLIEK VAN SUID-AFRIKA

DEPARTMENT OF AGRICULTURAL TECHNICAL SERVICES
DEPARTEMENT VAN LANDBOU-TEGNIESE DIENSTE

FLORA OF SOUTHERN AFRICA

VOL. 16, PART I

ISBN 0 621 02263 2

i

FLORA OF SOUTHERN AFRICA

which deals with the territories of

THE REPUBLIC OF SOUTH AFRICA, LESOTHO, SWAZILAND AND SOUTH WEST AFRICA

VOLUME 16, PART I

Edited by

J. H. ROSS

Editorial Committee: B. de Winter, D. J. B. Killick,

O. A. Leistner and J. H. Ross

Botanical Research Institute,

Department of Agricultural Technical Services

1975

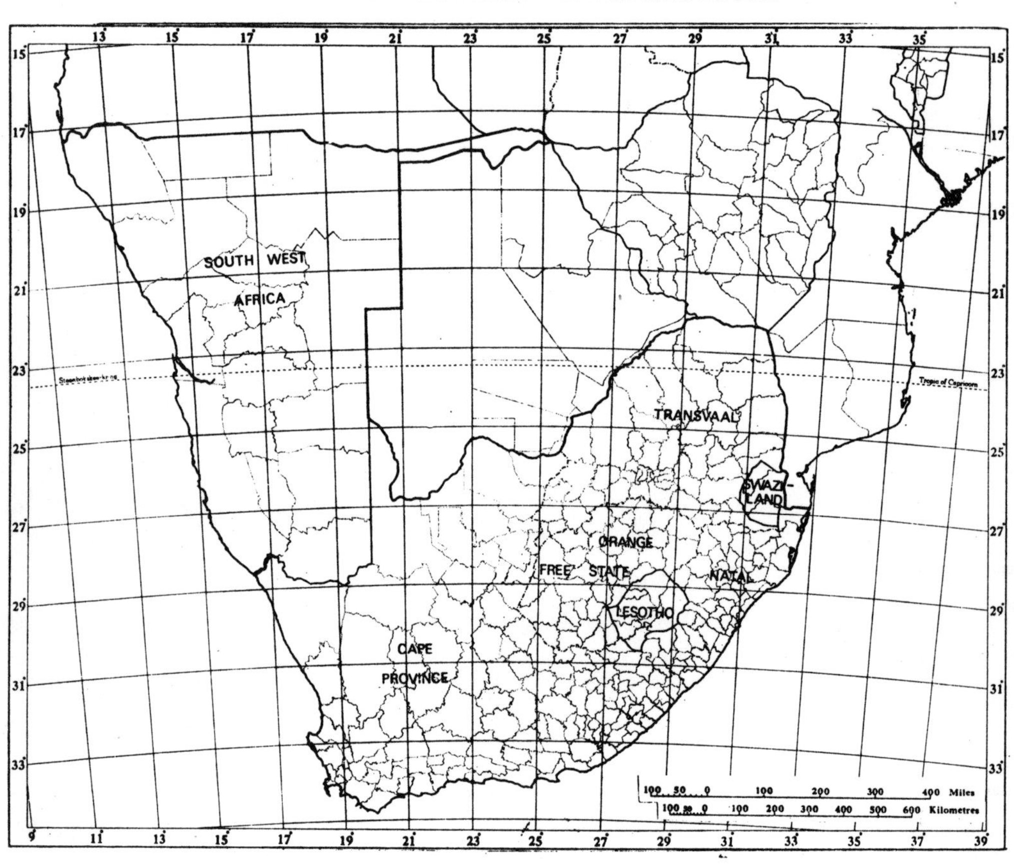
SOUTH WEST
AFRICA
Steenbokkeerkring
Tropic of Capricorn
TRANSVAAL
SWAZI-
LAND
ORANGE
FREE STATE
NATAL
LESOTHO
CAPE
PROVINCE
100 50 0 100 200 300 400 Miles
100 50 0 100 200 300 400 500 600 Kilometres

INTRODUCTION

Volume 16, part 1 is the fourth part of the Flora of Southern Africa to be published, the three which have already appeared being Vol. 1 (1966), Vol. 26 (1963) and Vol. 13 (1970).

For a key to the families, the Flora should be used in conjunction with Phillips's Genera of South African Flowering Plants, which is arranged on the lines of the Engler system. The genera are numbered according to the list published by De Dalla Torre and Harms in order to facilitate reference, though genera in the Flora are not necessarily arranged in this sequence.

As in previous volumes, generally accepted abbreviations are used for literature references, except in the following cases which appear frequently and are, therefore, considerably condensed:

C.F.A................... Conspectus Florae Angolensis.

F.C.................... Flora Capensis.

F.C.B................. Flore du Congo et du Rwanda-Burundi.

F.S.W.A.............. Prodromus einer Flora von Südwestafrika.

F.T.A................. Flora of Tropical Africa.

F.T.E.A.............. Flora of Tropical East Africa.

F.Z.................. Flora Zambesiaca.

Phill., Gen.............. The Genera of South African Flowering Plants by E.P. Phillips, ed. 2 (1951).

Burtt Davy, Fl. Transv..... Manual of the Flowering Plants and Ferns of the Transvaal and Swaziland, Vol. 1 (1926) and Vol. 2 (1932).

As before, the abbreviation "l.c." is used for previously cited references even though "op. cit." or "tom. cit." would in certain cases be more correct.

In citing specimens the grid reference system has been used. The spelling of the names of some localities has been brought into line with the findings of the Committee on Standardisation of Place Names.

In the text, species which show evidence of becoming naturalized are treated in the same way as indigenous species. In the Index, synonyms are in italics while exotic species are signified by an asterisk*.

v

PLAN OF CONTENTS OF VOLUMES

VOL. 1

Stangeriaceae
Zamiaceae
Podocarpaceae
Pinaceae
Cupressaceae
Welwitschiaceae
Typhaceae
Zosteraceae
Potamogetonaceae
Ruppiaceae
Zanichelliaceae
Najadaceae
Aponogetonaceae
Juncaginaceae
Alismataceae
Hydrocharitaceae

VOL. 2

Poaceae

VOL. 3

Cyperaceae
Arecaceae
Araceae
Lemnaceae
Flagellariaceae

VOL. 4

Restionaceae
Mayacaceae
Xyridaceae
Eriocaulaceae
Commelinaceae
Pontederiaceae
Juncaceae

VOL. 5

Liliaceae

VOL. 6

Haemodoraceae
Amaryllidaceae
Velloziaceae
Dioscoreaceae

VOL. 7

Iridaceae

VOL. 8

Musaceae
Zingiberaceae
Burmanniaceae
Orchidaceae

VOL. 9

Piperaceae
Salicaceae
Myricaceae
Ulmaceae
Moraceae
Urticaceae
Proteaceae

VOL. 10

Loranthaceae
Santalaceae
Grubbiaceae
Olacaceae
Balanophoraceae
Aristolochiaceae
Rafflesiaceae
Hydnoraceae
Polygonaceae
Chenopodiaceae
Amaranthaceae
Nyctaginaceae

VOL. 11

Phytolaccaceae
Aizoaceae

VOL. 12

Portulacaceae
Basellaceae
Caryophyllaceae
Nymphaeaceae
Ceratophyllaceae
Ranunculaceae
Menispermaceae
Annonaceae
Monimiaceae
Lauraceae
Hernandiaceae
Papaveraceae
Fumariaceae

VOL. 13

Brassicaceae
Capparaceae
Resedaceae
Moringaceae
Droseraceae
Podostemaceae
Hydrostachyaceae

VOL. 14

Crassulaceae

VOL. 15

Saxifragaceae
Pittosporaceae
Cunoniaceae
Myrothamnaceae
Bruniaceae
Hamamelidaceae
Rosaceae
Connaraceae

VOL. 16

Fabaceae

VOL. 17

Geraniaceae
Oxalidaceae

PLAN OF CONTENTS OF VOLUMES

VOL. 18

Linaceae
Erythroxylaceae
Zygophyllaceae
Rutaceae
Simaroubaceae
Burseraceae
Meliaceae
Malpighiaceae

VOL. 19

Polygalaceae
Dichapetalaceae
Euphorbiaceae
Callitrichaceae
Buxaceae
Anacardiaceae
Aquifoliaceae

VOL. 20

Celastraceae
Hippocrateaceae
Icacinaceae
Sapindaceae
Melianthaceae
Balsaminaceae
Rhamnaceae
Heteropyxidaceae
Vitaceae

VOL. 21

Tiliaceae
Malvaceae
Bombacaceae
Sterculiaceae

VOL. 22

Ochnaceae
Clusiaceae
Elatinaceae
Frankeniaceae
Tamaricaceae
Canellaceae
Violaceae
Flacourtiaceae
Turneraceae
Passifloraceae
Achariaceae
Loasaceae
Begoniaceae
Cactaceae

VOL. 23

Geissolomaceae
Penaeaceae
Oliniaceae
Thymelaeaceae
Lythraceae
Lecythidaceae

VOL. 24

Rhizophoraceae
Combretaceae
Myrtaceae
Melastomataceae
Onagraceae
Hydrocaryaceae
Halorrhagidaceae
Araliaceae
Apiaceae
Cornaceae

VOL. 25

Ericaceae

VOL. 26

Myrsinaceae
Primulaceae
Plumbaginaceae
Sapotaceae
Ebenaceae
Oleaceae
Salvadoraceae
Loganiaceae
Gentianaceae
Apocynaceae

VOL. 27

Asclepiadaceae

VOL. 28

Convolvulaceae
Hydrophyllaceae
Boraginaceae
Verbenaceae
Lamiaceae
Solanaceae

VOL. 29

Scrophulariaceae

VOL. 30

Bignoniaceae
Pedaliaceae
Martyniaceae
Orobanchaceae
Gesneriaceae
Lentibulariaceae
Acanthaceae
Myoporaceae

VOL. 31

Plantaginaceae
Rubiaceae
Valerianaceae
Dipsacaceae
Cucurbitaceae

VOL. 32

Campanulaceae
Goodeniaceae

VOL. 33

Asteraceae

vii

FABACEAE

(alternative name Leguminosae)

Trees, shrubs, climbers or herbs. *Leaves* nearly always alternate, often pinnate or pinnately 3-foliolate, less often bipinnate or digitately 3-foliolate, occasionally 1-foliolate, simple or digitate*; stipules usually present. *Receptacle* usually expanded, flat or saucer- or cup-shaped, occasionally tubular or not expanded. *Sepals* usually united. *Petals* usually 5, occasionally fewer or absent, usually free, less often some or rarely all united. *Stamens* inserted with the petals at the rim of the receptacle, usually 10 but sometimes fewer or more. *Ovary* almost always a single superior 2-several-ovulate carpel, rarely 1-ovulate, the ovules attached to the adaxial suture. *Fruit* nearly always a dry legume (pod), usually dehiscing into 2 valves, less often dehiscing only along the upper suture or splitting transversely into 1-seeded segments or, indehiscent. *Seed* without, or with very little, endosperm.

Key to Subfamilies

Flowers regular, the petals valvate in the bud, often united basally; sepals usually united basally; stamens as many as the petals or twice as many or numerous, free or all united into a tube or to the base of the petals; leaves bipinnate or, rarely, pinnate in indigenous species*; seeds normally with an areole on each face or side..........**Mimosoideae**

Flowers nearly always zygomorphic; petals imbricate in the bud, free or some of them united; seeds usually without an areole:

Adaxial (uppermost) petal overlapped by the adjacent lateral petals when these are present; sepals often free; stamens 10 or fewer or occasionally more, free or, less often, variously united; leaves bipinnate or pinnate, rarely simple or 1-foliolate; seeds with radicle usually straight.............................**Caesalpinioideae**

Adaxial (uppermost) petal outside the adjacent lateral petals; sepals united basally; stamens 10 or rarely fewer, sometimes free, most often the adaxial free or nearly so and the other 9 united, sometimes united into 2 groups of 5 or all united; leaves never bipinnate; seeds with radicle usually curved..................**Papilionoideae**

A cosmopolitan family of about 690 genera and 18 000 species, many of which are of great economic significance.

---

* Some exotic species of *Acacia* grown in our area have phyllodes which may appear to be simple leaves.

Subfamily 1. MIMOSOIDEAE

DC., Prodr. 2 : 424 (1825), as suborder or tribe *Mimoseae;* Benth. in Hook., J. Bot. 3 : 133 (1841), in Trans. Linn. Soc. Lond. 30 : 335 (1875).

by J. H. ROSS

Trees, shrubs or rarely herbs, often armed with prickles or spines. *Leaves* bipinnate or (in exotic species only) simply pinnate or modified to phyllodes. *Stipules* present, sometimes spinescent. *Inflorescences* usually spikes, racemes or heads of sessile or shortly pedicellate, usually small or very small, regular, (3)5(6)-merous flowers. *Sepals* with valvate or rarely imbricate aestivation, often open from an early stage of bud, usually united to form a toothed or lobed calyx, rarely free. *Petals* valvate in bud, free or more often united below into a tube. *Stamens* 4–10 (as many as or twice as many as the petals) or indefinite, free or adnate below to the corolla, or the filaments united below into a tube, usually ±exserted; anthers small, versatile, sometimes with an apical gland; pollen-grains sometimes simple but frequently compound or united. *Pods* and seeds various, the latter usually marked with areoles*; radicle of embryo in seed usually straight.

Genera 56, with about 2 800 species, mainly tropical and subtropical, especially numerous in the southern hemisphere. 13 genera and 106 species occur in our area.

The genera in our area are grouped under a number of tribes. The sequence of the genera, and consequently of the tribes, follows the numerical order adopted by de Dalla Torre and Harms in their Genera Siphonogamarum (1900–1907). A conspectus of the tribes is given below:—

1. Tribe **Ingeae** *Benth.* in Trans. Linn. Soc. Lond. 30 : 359 (1875). Calyx-lobes valvate; stamen-filaments more or less united into a tube; stamens numerous (more than 10); pollen-grains usually collected into 2–6 masses in each loculus. Genus 1.

2. Tribe **Acacieae** *Benth.* in Hook., Lond. J. Bot. 1 : 318 (1842); Benth. in Trans. Linn. Soc. Lond. 30 : 359 (1875). Calyx-lobes valvate; stamen-filaments free from one another or united basally only; stamens numerous (more than 10); pollen-grains usually collected into 2–6 masses in each loculus. Genus 2.

3. Tribe **Mimoseae** *Bronn*, De Formis Pl. Legum. 130 (1822); Benth. in Hook., J. Bot. 2 : 127 (1840); (*Eumimoseae* Benth. in Trans. Linn. Soc. Lond. 30 : 359, 1875). Calyx-lobes valvate; stamen-filaments free from one another or united basally only; stamens 10 or fewer; anthers without a gland at the apex; pollen-grains numerous and separate in each loculus. Genera 3–5.

4. Tribe **Adenanthereae** *Benth.* in Trans. Linn. Soc. Lond. 30 : 358 (1875). Calyx-lobes valvate; stamen-filaments free from one another or united basally only; stamens 10 or fewer; anthers with a (sometimes deciduous) gland at the apex; pollen-grains numerous and separate in each loculus. Genera 6–13.

It is often difficult to identify some of the genera of this subfamily without complete material, including flowers and fruits. As either flowers or fruits are frequently absent, two alternative keys have been constructed, one for flowering and the other for fruiting specimens.

Key to genera based on vegetative and floral characters

Plant armed with prickles or spines:

Inflorescence bicoloured, the upper part yellow, the lower mauve, pink or white; short
 lateral branchlets terminating in spines, plants otherwise unarmed....7. **Dichrostachys**

---

* The seeds of nearly all Mimosoideae show on each face an area, usually circular to elliptic or oblong in shape, bounded (except usually for a gap opposite the micropyle) by a fine line which frequently appears as a fissure in the testa. The size and shape of this area are often taxonomically important, and I have followed Brenan in F.T.E.A. Legum.–Mimos. : 1 (1959) in employing the term "areole" to refer to it. It corresponds to the term "pleurogram" as used by Corner in Phytomorphology 1 : 117–150 (1951).

The seeds of nearly all our Mimosoideae show these areoles. They appear to be absent only in *Elephantorrhiza, Newtonia* and the giant *Entada* species.

While areoles are of common occurrence in Mimosoideae, they do not seem to occur in the other subfamilies of Leguminosae, except in a few genera of Caesalpinioideae, namely, *Burkea, Tamarindus* and in a rather modified way in some species of *Cassia*.

Inflorescence concolorous; plants armed with recurved prickles or stipular spines, very rarely short lateral branchlets terminating in spines:

Plant armed with recurved prickles or stipular spines, spine-tipped lateral branchlets absent:

Flowers in globose or subglobose heads, very rarely the inflorescence reduced to only 2–4 flowers per "head":

Stamens as many as or twice as many as the (3)4–5(6) corolla-lobes:

Small rigid much branched shrub with pale green to olive, cano-puberulous, substriate branches; stipules spinescent, recurved, up to 3,5 mm long; leaves small, each with 1 pinna pair; stamens 10, the 5 opposite the petals shorter than those alternating with the petals; anthers with a minute deciduous apical gland..................................8. **Xerocladia**

Annual or perennial herb or small shrub; stems armed with scattered prickles and usually densely hispid or setulose; stipules not spinescent, persistent; leaves sensitive to touch, with (1)2–14 pinnae pairs; stamens 4 or 8; anthers eglandular apically; flowers pink or mauve..................4. **Mimosa**

Stamens numerous ($\pm$ 35–200); flowers mostly white, yellowish-white or deep yellow..2. **Acacia**

Flowers in spikes or spiciform racemes:

Stamens 10; recurved prickles scattered along the stem; leaflets 2,5–9 mm wide..13. **Entada**

Stamens numerous ($\pm$ 35–200); prickles or stipular spines usually confined to the nodes and in pairs, sometimes solitary or in threes at the nodes, seldom prickles scattered but then leaflets less than 1,75 mm wide.........2. **Acacia**

Plant armed with spine-tipped lateral shoots, recurved prickles and stipular spines absent; inflorescence globose; stamen-filaments united basally into a tube 1. **Albizia**

Plant unarmed:

Inflorescence capitate, globose or subglobose:

Aquatic herb with creeping, usually floating and swollen stems; flowers of two sorts, hermaphrodite in upper part of head, neuter with elongate staminodes in lower part...6. **Neptunia**

Trees, shrubs or suffrutices, rarely herbaceous and then not aquatic:

Leaves reduced to simple entire phyllodes...............................2. **Acacia**

Leaves bipinnate:

Stamens 10:

Leaves with 1 pinna pair; leaflets large, 2–6 cm wide; large trees up to 15 m high..11. **Xylia**

Leaves with (2)3–8 pinnae pairs; leaflets small, less than 0,5 cm wide; perennial herb, suffrutex, shrub or small tree:

Anthers with conspicuous scattered hairs; flowers in heads up to 1,8 cm in diameter; shrub or small tree with densely grey-puberulous branchlets; leaflets 1,5–4 mm wide............................3. **Leucaena**

Anthers glabrous; flowers in heads up to 0,8 cm in diameter; perennial herb or suffrutex with subglabrous to sparingly puberulous $\pm$ angular stems; leaflets 0,8–1,5 mm wide........................5. **Desmanthus**

Stamens many to numerous ($\pm$ 19–200):

Stamen-filaments free; central flower of inflorescence not different from the rest...2. **Acacia**

Stamen-filaments united basally into a tube, tube included or exserted beyond the corolla; central flower of inflorescence usually differing from and often larger than the rest....................................1. **Albizia**

Inflorescence elongate, spicate or racemose:

Leaflets alternate to subopposite, 0,7–1,9 cm wide, with petiolules 1,5–3 mm long.........................9. **Amblygonocarpus**

Leaflets opposite, very rarely alternate but then sessile and less than 2 mm wide:

Stamens 10:

Leaf-rhachis with a gland at the junction of each pair of pinnae or the upper pairs; large tree..............................10. **Newtonia**

Leaf-rhachis eglandular; liane, climber, scandent shrub, suffrutex or small tree:

Liane, climber, scandent shrub or erect suffrutex, the latter with leaflets 7–20 mm wide.............................13. **Entada**

Suffrutex, shrub or small tree, leaflets less than 6,5 mm wide...12. **Elephantorrhiza**

Stamens many to numerous ($\pm$ 19–200):

Stamen-filaments free...2. **Acacia**

Stamen-filaments united basally into a tube, tube included or exserted beyond the corolla...1. **Albizia**

Key to genera based on vegetative and pod characters

Plant armed with prickles or spines:

Pod at maturity splitting transversely into segments each containing a seed:

Valves of pod falling away at maturity, leaving the persistent margins; stem and often leaves with scattered prickles:

Pod without bristles or prickles on the surface or margins (very rarely a few prickles occur along one margin), 2,8–4 cm wide; leaflets 2,5–9 mm wide; climber or scandent shrub...13. **Entada**

Pod $\pm$ bristly or prickly on the surface or on the margins only, up to 1,2 cm wide; leaflets up to 3 mm wide; annual or perennial herb or small shrub......4. **Mimosa**

Valves of pod not falling away from the margins; stems armed with paired stipular spines...2. **Acacia**

Pod not splitting transversely into segments:

Plant armed with stipular spines or with recurved prickles; spine-tipped lateral branchlets absent:

Pod indehiscent, broadly falcate-ovate to semi-orbicular, up to $\pm$ 1,5 cm long and 1,5 cm broad, compressed, the lower suture arched and winged; small rigid much branched shrub with pale green to olive, cano-puberulous, sub-striate branches; armed with paired, short, recurved stipular spines; leaves small, each with 1 pinna pair.............................8. **Xerocladia**

Pod dehiscent or indehiscent, if indehiscent then not as above; small shrubs to large trees; armed with paired stipular spines or with prickles which are in pairs, solitary or in threes at the nodes, or scattered along the stem; leaves with 1–60 pinnae pairs.............................2. **Acacia**

Plant armed with spine-tipped lateral shoots; recurved prickles and stipular spines absent:

Pods densely clustered, usually contorted or spirally twisted; leaflets less than 3 mm wide.............................7. **Dichrostachys**

Pods not densely clustered, ±straight or slightly curved, not contorted or twisted; leaflets more than 3 mm wide.....................................1. **Albizia**

Plant unarmed (except for a single minute very inconspicuous prickle below the node which is sometimes present in *Albizia harveyi*):

Aquatic herb with creeping, usually floating and swollen stems; pods clustered, bent almost at right angles to the short basal stipe, 1,3–2,8 cm long..........6. **Neptunia**

Trees, shrubs or suffrutices, rarely herbaceous and then not aquatic; pods usually longer than 3,5 cm:

Leaves reduced to simple entire phyllodes................................2. **Acacia**
Leaves bipinnate:

Leaflets alternate to subopposite, distinctly petiolulate; pod bluntly tetragonal or subcylindric in section, woody, indehiscent.................9. **Amblygonocarpus**

Leaflets opposite, very rarely alternate but then sessile and very narrow; pod usually flattened, rarely ±turgid, never tetragonal:

Pod at maturity splitting transversely into segments each containing a seed
....13. **Entada**

Pod not splitting transversely into segments:

Valves of pod woody, separating along both margins, recurving; leaves with only 1 pinna pair..11. **Xylia**

Valves of pod membranous to rigidly coriaceous but not woody or recurving:

Pod dehiscent:

Valves of pod separating along one margin only:

Pod 0,3–1(1,3) cm wide, usually ±moniliform; seeds small, black to brown, unwinged.....................................2. **Acacia**

Pod 1,3–2 cm wide, margins entire, bright red when young becoming brown with age; seeds large, brown, conspicuously winged
....10.**Newtonia**

Valves of pod separating along both margins:

Pod up to 4 mm wide; perennial herb or small suffrutex....5. **Desmanthus**
Pod over 1 cm wide, shrub or tree:

Seeds 6–13 mm wide; leaflets variable, often exceeding 4 mm in width; large shrub or tree; widespread.................1. **Albizia**

Seeds 3,5–5 mm wide; leaflets up to 4 mm wide; shrub or small tree to 4 m high, naturalized exotic confined to Natal coast..3. **Leucaena**

Pod indehiscent:

Valves at maturity separating from the persistent margins, the outer layer often peeling off the inner layer, the layers remaining intact or breaking up irregularly; suffrutex, shrub or small tree..12. **Elephantorrhiza**

Valves not separating from the margins and splitting into layers; large shrub or tree...1. **Albizia**

Conspectus of the pod differences

In all genera except *Acacia* the pod is usually rather constant in form and structure.

1. Pod dehiscing into two separate valves:

Valves woody:
Acacia
Xylia

Valves papery to rigidly coriaceous:

Acacia
Albizia
Desmanthus
Leucaena
Neptunia

2. Pod dehiscing into two valves which remain attached to one another along one margin:

Acacia
Newtonia

3. Pod splitting transversely into segments each containing a seed:

Acacia
Entada
Mimosa

4. Pod indehiscent, not splitting transversely:

Valves separating from the margins, usually splitting into two layers:

Elephantorrhiza

Valves spiral or contorted, but neither separating from the margins nor splitting:

Acacia
Dichrostachys

Valves flattened, but neither spiral nor contorted nor separating from the margins nor splitting:

Acacia
Albizia
Xerocladia

Valves not flattened but convex or angled, neither spiral nor contorted nor separating from the margins nor splitting:

Acacia
Amblygonocarpus

Exotic species

Several exotic species of Mimosoideae are planted in our area and most of them are mentioned under their appropriate genera. In addition to these, however, the genus *Prosopis* occurs only as an exotic and is dealt with briefly below. Some species of *Prosopis* bear a superficial resemblance to certain of the indigenous *Acacia* species but may be readily distinguished from the latter as there are only 10 stamens per flower.

1. **Prosopis pubescens** *Benth.* in Hook., Lond. J. Bot. 5 : 82 (1846); Benson in Am. J. Bot. 28 : 753 (1941); Isely in Madrono 21 : 294 (1972); in Mem. N.Y. Bot. Gdn. 25 (1) : 120 (1973).

Large shrub or small tree up to 4 m high armed with paired straight spines 0,5–1,5 cm long. *Leaves* bipinnate, with 1 pinna pair; leaflets 5–8 pairs per pinna, 5–10 × 1,75–4 mm, finely and rather sparsely appressed-pubescent on both surfaces. *Inflorescence* an axillary spike; flowers sessile, yellow. *Ovary* densely pilose. *Pod* very tightly spirally coiled into a± straight, brownish cylinder 2,5–5 cm long overall, 4–6 mm in diameter, tomentellous when young but becoming glabrescent with age.

The Screw-Bean, a native of the Western United States of America, is found in the northern Cape Province. Recorded from Kimberley, *Wilman sub KMG* 3385, *Acocks* 50; Carnarvon Distr., along River at Van Wyk's Vlei, *Acocks* 1788.

2. **Prosopis velutina** *Wooton* in Bull. Torrey Bot. Club 25 : 456 (1898); Johnston in Brittonia 14 : 86 (1962); Isely in Madrono 21 : 295 (1972); Isely in Mem. N.Y. Bot. Gdn. 25 (1) : 120 (1973).

Shrub or small tree up to 3 m high armed with paired straight spines 1–2 cm long. *Leaves* bipinnate; petioles and rhachillae sparingly to densely pubescent; with 1 pinna pair; leaflets 13–20 pairs per pinna, 6–13 × 2–4 mm, venation usually fairly conspicuous on the lower surface, with marginal cilia. *Inflorescence* an axillary spike, axes densely pubescent; flowers sessile, yellow. *Ovary* densely pilose. *Pod* straight or slightly falcate, 10–20 cm long, 0,8–1 cm wide, somewhat compressed, margin slightly constricted between the seeds, segments oblique, finely longitudinally venose, beaked apically.

The Velvet Mesquite, a native of the Western United States of America. Recorded from the Cape Province: Kenhardt Distr., near Putsonderwater on Marydale–Upington Road, *Merxmuller* 728; Hay Distr., Witsand, *Acocks* 2287.

3. **Prosopis glandulosa** *Torrey* in Ann. Lyc. N.Y. 2 : 192 (1827); Johnston in Brittonia 14 : 82 (1962); Isely in Madrono 21 : 290 (1972); in Mem. N.Y. Bot. Gdn. 25 (1) : 118 (1973).

P. juliflora auct., non (Swartz) DC.
P. chilensis auct., non (Mol.) Stuntz

Shrub or tree to 10 m high armed at the nodes with paired or solitary straight spines 0,5–3 cm long. *Leaves* bipinnate; with 1–2 pinnae pairs; leaflets 7–22 pairs per pinna, 1–6 cm long, 1–4 mm wide, widely spaced on the rhachilla, the intervals wider than the width of the leaflets, venation fairly conspicuous, glabrous. *Inflorescence* an axillary spike; flowers sessile, yellow. *Ovary* densely pilose. *Pod* straight or almost so, 10–22 cm long, ± 8 mm in diameter, margins not or scarcely constricted, valves becoming woody, with a fibrous exocarp, beaked apically.

P. glandulosa, the Mesquite, is one of the most successful and characteristic plants of the southwestern United States of America. *P. glandulosa* forms part of a complex of closely related and taxonomically difficult species (see M. C. Johnston in Brittonia 14 : 72–90, 1962). Leaflet size, shape and number in *P. glandulosa* are extremely variable and two varieties are recognized chiefly on the basis of leaflet size, namely, var. *glandulosa* with leaflets 2,5–6 cm long and var. *torreyana* (Benson) M. C. Johnston with leaflets 1–2,5 cm long. The latter occurs in our area.

Recorded from South West Africa, without precise locality, *Cellier sub PRE* 1603; Transvaal, Wolmaransstad, *Liebenberg* 3422.

3443   1. ALBIZIA

Albizia *Durazz.*, Magazz. Tosc. 3(4) (vol. 12) : 10, 13, illustr. (1772)*; Benth. in Hook., Lond. J. Bot. 3 : 84 (1844); Harv. in F.C. 2 : 284 (1862); Benth. & Hook.f., Gen. Pl. 1 : 596 (1865); Harv., Gen. Pl. ed.2 : 92 (1868); Oliv. in F.T.A. 2 : 355 (1871); Benth. in Trans. Linn. Soc. Lond. 30 : 557 (1875); Taub. in Pflanzenfam. 3, 3 : 106 (1891); Sim, For. Fl. P.E.Afr. 58 (1909); Harms in Engl., Pflanzenw. Afr. 3, 1 : 337 (1915); Bak.f., Leg. Trop. Afr. 3 : 855 (1930); Burtt Davy, Fl. Transv. 2 : 347 (1932); Little in Amer. Midl. Nat. 33 : 510 (1945); Phill, Gen. 390 (1951); Gilbert & Boutique in F.C.B. 3 : 170 (1952); Torre in C.F.A. 2 : 288 (1956); Codd in Bothalia 7 : 68 (1958); Keay in F.W.T.A. ed.2, 1 : 501 (1958); Brenan in F.T.E.A. Legum.–Mimos. : 136 (1959); Dale & Greenway, Kenya Trees & Shrubs 298 (1961); F. White, For. Fl. N.Rhod. 88 (1962); Hutch., Gen. Fl. Pl. 1 : 294 (1964); Schreiber in F.S.W.A 58 : 13 (1967); Brenan in F.Z. 3, 1 : 113 (1970). Type species: *A. julibrissin* Durazz.

Zygia sensu E. Mey., Comm. 164 (1836); Benth. in Hook., Lond. J. Bot. 3 : 92 (1844); Bolle in Peters, Reise Mossamb. Bot. 1 : 1 (1861); Harv. in F.C. 2 : 284 (1862); Gen. Pl. ed.2 : 92 (1868), non P. Br.

Besenna A. Rich., Tent. Fl. Abyss. 1 : 253 (1847)

---

*I have not seen this work. The reference is from Little in Amer. Midl. Nat. 33 : 510 (1945).

Trees or shrubs, very rarely climbing (not in Africa); prickles or stipular spines absent in African species (except for a very small prickle beneath the node in *A. harveyi* sometimes, while in *A. anthelmintica* some branchlets are sharp and spinescent apically). *Leaves* bipinnate; petiole usually glandular; pinnae each with one to many pairs of leaflets. *Stipules* herbaceous, usually soon deciduous. *Inflorescences* globose or (not in indigenous African species) spicate, pedunculate, axillary and solitary or more often fascicled, often aggregated near the ends of the branchlets which may be lateral and much shortened, sometimes paniculate. *Flowers* hermaphrodite or occasionally male and hermaphrodite; 1–2 central flowers in each head frequently larger, different in form from the others and apparently male. *Calyx* gamosepalous, normally with 5 teeth or lobes (rarely 4, 6 or 7). *Corolla* gamopetalous, infundibuliform or campanulate, normally with 5 lobes (rarely 4 or 6). *Stamens* numerous (19–50), fertile, their filaments united basally into a slender tube, tube shorter than or exceeding the corolla. *Ovary* subsessile or shortly stipitate. *Pods* oblong, straight, flattened, dehiscent or not, not septate within, the valves papery to rigidly coriaceous. *Seeds* usually±flattened.

A genus of 100–150 species occurring throughout the tropics, with fewer in the subtropics. 11 indigenous species occur in our area, while 3 introduced species have become naturalized. *A. chinensis* (Osbeck) Merr. is cultivated in the Transvaal but does not appear to have become naturalized yet.

The generic name was previously often spelt *Albizzia*; the reasons for rejecting this spelling are given by Little in Amer. Midl. Nat. 33 : 510 (1945).

The genus *Albizia* is named in honour of Filippo del Albizzi, a Florentine nobleman who in 1749 introduced *A. julibrissin* into cultivation.

In the following key the descriptions of the floral parts do not apply to the 1–2 larger modified flowers usually found in the centre of each of the globose heads; in these the staminal tube is not or scarcely exserted beyond the corolla even when it is long-exserted in the other flowers.

Key based mainly on vegetative and floral characters

Inflorescences spicate..14. *A. lophantha*
Inflorescences globose:
  Staminal tube not or scarcely exserted beyond the corolla:
    Leaflets small or very small, 0,5–4 mm wide:
      Calyx and corolla glabrous or with few whitish hairs; mature leaves glabrous or sparsely (rarely±
        densely) appressed grey-pubescent; pinnae (3)6–10(15) pairs; apex of leaflets symmetric, not
        falcate...1. *A. brevifolia*
      Calyx and corolla fulvous- to golden-pubescent or tomentose; mature leaves clothed with a grey to
        golden spreading indumentum; pinnae 2–44 pairs:
        Leaflets 0,5–1,5 mm wide; pinnae 2–44 pairs:
          Apex of leaflets acute, asymmetric, falcate, the point turned towards the apex of the pinna;
            lateral nerves of leaflets±raised and visible beneath; pinnae 6–22 pairs; bracteoles persistent
            until the flowers open; stamen-filaments ± 1,5–2 cm long; pods dehiscent; glabrous or
            almost so...3. *A. harveyi*
          Apex of leaflets obtuse or subacute, symmetric, not falcate; lateral nerves of leaflets indistinct
            beneath; pinnae 12–44 pairs; bracteoles already fallen when flowers open; stamen-filaments
            1–1,5 cm long; pods apparently indehiscent, puberulous....2. *A. amara* subsp. *sericocephala*
        Leaflets 1,5–4 mm wide, apex rounded to subacute or mucronate, turned towards the apex of the
          pinna, lateral nerves usually±raised and visible beneath; pinnae 2–7 pairs; pods indehiscent,
          very prominently transversely venose, the veins±parallel and 2–4 mm apart......4. *A. forbesii*
    Leaflets exceeding 4 mm in width:
      Young branchlets often forming abbreviated spine-tipped lateral shoots; rhachides and rhachillae
        of all or most leaves projecting at the ends as a short rigid persistent deflexed hook; a single
        similarly bent stipel often near the base of each rhachilla; leaflets (1)2–4(6) pairs; flowers usually
        precocious on leafless or almost leafless branchlets; calyces and corollas glabrous or sparsely
        puberulous...9. *A. anthelmintica*

Young branchlets not forming spine-tipped shoots; rhachides and rhachillae not projecting, or else projections±straight and not hooked (except rarely and casually in *A. versicolor*) and usually deciduous; calyces and corollas usually±densely puberulous to tomentose outside, if glabrous or almost so then flowers not precocious:

Mature leaflets densely pubescent to tomentose beneath, mostly broadly and obliquely obovate to subcircular, in 2–5(6) pairs; young branchlets densely rusty-tomentose; indumentum on corolla-lobes usually rusty-tomentose....................................5. *A. versicolor*

Mature leaflets thinly appressed-pubescent or glabrous beneath, in 3–17 pairs; young branchlets not densely rusty-tomentose:

Indumentum on corolla-lobes conspicuously rusty:

Leaflet margins crisped; stamens 1–1,6 cm long; pods 1,4–2,6 cm wide........8. *A. suluensis*

Leaflet margins not crisped; stamens 1,5–3 cm long; pods 2,5–5 cm wide:

Leaflets markedly discolorous, glaucous to grey or whitish beneath, glabrous on both surfaces; pinnae 1–4 pairs; leaflets 4–9 pairs; bark grey, not peeling off in papery flakes..6. *A. antunesiana*

Leaflets not markedly discolorous, glabrous or sparingly pubescent on both surfaces; pinnae (2)3–6(7) pairs; leaflets 4–13(17) pairs; bark smooth, creamy-white to ochre-yellow, peeling off in large papery pieces.........................7. *A. tanganyicensis*

Indumentum on corolla-lobes grey to whitish or lobes glabrous:

Inflorescences in fascicles; flowers on pedicels 1,5–4,5 mm long; leaflets not discolorous; pods straw-coloured to light brown, 3–4,5 cm wide......................12. *A. lebbeck*

Inflorescences in axillary or terminal panicles; flowers sessile or almost so; leaflets markedly discolorous; pods light to dark brown, 1,5–2 cm wide....................13. *A. procera*

Staminal tube exserted beyond the corolla for a length of ± 0,7–2,5 cm (usually more than 1 cm):

Leaflets 2–5 pairs; pinnae (1)2–3(4) pairs; bracteoles± 1 mm long, rapidly deciduous and falling while the flowers are still in bud; shrub or tree usually branching freely near the ground with many ascending branches...10. *A. petersiana* subsp. *evansii*

Leaflets (4)6–15 pairs; pinnae 4–8 pairs; bracteoles 5–8 mm long, linear-spathulate to oblanceolate, exceeding the flower buds, variably persistent but sometimes deciduous before the flowers open; stipules and bracts at base of peduncles 5–12 × 3–11 mm, ovate; tree usually with a single stem and a flattened spreading crown...11. *A. adianthifolia*

Key based on vegetative and fruit characters

Leaflets small or very small, 0,5–4 mm wide:

Mature leaves glabrous or sparsely (rarely ± densely) appressed grey-pubescent; pinnae (3)6–10(15) pairs; apex of leaflets symmetric, not falcate...................................1. *A. brevifolia*

Mature leaves clothed with a grey to golden spreading indumentum; pinnae 2–44 pairs:

Leaflets 0,5–1,5 mm wide:

Apex of leaflets acute, asymmetric, falcate, the point turned towards the apex of the pinna; lateral nerves of leaflets ± raised and visible beneath; pinnae 6–22 pairs; pods dehiscent, glabrous or nearly so except for slight pubescence along the margins.........................3. *A. harveyi*

Apex of leaflets obtuse or subacute, symmetric, not falcate; lateral nerves of leaflets indistinct beneath; pinnae 12–44 pairs; pods apparently indehiscent, puberulous..................
....2. *A. amara* subsp. *sericocephala*

Leaflets 1,5–4 mm wide:

Pinnae 2–7 pairs; leaflets 6–16 pairs per pinna; pods oblong, 3,2–5 cm wide, indehiscent, very prominently transversely venose, the veins ± parallel and 2–4 mm apart........4. *A. forbesii*

Pinnae 7–12 pairs; leaflets 20–35 pairs per pinna; pods linear-oblong, 1,4–1,7 cm wide, obscurely venose...14. *A. lophantha*

Leaflets exceeding 4 mm in width:

Young branchlets often forming abbreviated spine-tipped lateral shoots; rhachides and rhachillae of all or most leaves projecting at the ends as a short rigid persistent deflexed hook; a single similarly bent stipel often near the base of each rhachilla; leaflets (1)2–4(6) pairs per pinna.......9. *A. anthelmintica*

Young branchlets not forming spine-tipped shoots; rhachides and rhachillae not projecting, or else projections ± straight and not hooked (except rarely and casually in *A. versicolor*) and usually deciduous:

Mature leaflets broadly and obliquely obovate to subcircular, 15–67 × (10)15–45 mm, densely pubescent to tomentose beneath, in 2–5(6) pairs; pods glabrous or almost so, ± glossy, 3–5,5 cm wide; young branchlets densely rusty-tomentose..............................5. *A. versicolor*

Mature leaflets not broadly and obliquely obovate to suborbicular, thinly appressed-pubescent or glabrous beneath, if densely pubescent beneath then smaller than above and usually more numerous; pods glabrous to pubescent, not ± glossy:

Bark smooth, creamy-white to ochre-yellow, peeling off in large papery pieces....7. *A. tanganyicensis*

Bark not peeling off in large papery pieces:

Leaflet margins crisped; pods 1,4–2,6 cm wide................................8. *A. suluensis*

Leaflet margins not crisped:

Pods 1,4–2 cm wide; young branchlets not densely and coarsely rusty- to fulvous-pubescent; stipules <3 mm wide:

Young branchlets densely pubescent; leaflets 2–5 pairs, (4,5)8–22 × (3,5)5–13 mm; rhachillae 0,5–3 cm long; shrub or tree, usually branching freely near the base with many ascending branches...10. *A. petersiana* subsp. *evansii*

Young branchlets glabrous or subglabrous; leaflets 5–11 pairs, 15–60 × 8–21 mm; rhachillae 7–19 cm long; tree, usually single-stemmed..............................13. *A. procera*

Pods 2,2–4,6 cm wide, if less than 2,2 cm wide then young branchlets densely rusty- to fulvous-pubescent and stipules ovate, 5–12 × 3–8(11) mm:

Young branchlets densely and rather coarsely and persistently rusty- to fulvous-pubescent, indumentum sometimes becoming grey with age; pinnae 4–8 pairs; leaflets 4–8 (11) mm wide, obliquely rhombic-quadrate or -oblong, midrib diagonal; stipules ovate, 5–12 × 3–8(11) mm; pods 1,9–3,4 cm wide, densely and persistently pubescent; crown typically flattened and spreading...11. *A. adianthifolia*

Young branchlets very shortly pubescent, often becoming glabrous with age; pinnae 1–4(5) pairs; leaflets 8–28 mm wide, not as above; stipules much smaller than above; pods 2,5–4,6 cm wide:

Leaflets markedly discolorous, glaucous to grey or whitish beneath, glabrous on both surfaces; pods brown or reddish-brown, slightly umbonate over the seeds, longitudinally dehiscent; confined to South West Africa.....................6. *A. antunesiana*

Leaflets not markedly discolorous, glabrous to pubescent beneath; pods straw-coloured to light brown, very conspicuously raised over the seeds, tardily dehiscent and often only after falling to the ground; introduced into Natal..................12. *A. lebbeck*

---

1. **Albizia brevifolia** *Schinz* in Bull. Herb. Boiss., Sér. 2, 2 : 945 (1902); Bak.f., Leg. Trop. Afr. 3 : 864 (1930); Codd in Bothalia 7 : 69 (1958); Von Breitenbach, Indig. Trees S. Afr. 2 : 254 (1965); Schreiber in F.S.W.A. 58 : 14 (1967); Brenan in F.Z. 3, 1 : 125 (1970); Van Wyk, Trees Kruger Nat. Park 1 : 112 (1972); Palmer & Pitman, Trees S. Afr. 2 : 713 (1973). Type: Mozambique, Boruma, on the Nhasinde, *Menyharth* 994 (Z, holo.!; K!).

A. parvifolia Burtt Davy, Fl. Transv. 2 : xvii, 348 (1932). Syntypes: Transvaal, Soutpansberg district, near Messina, *Rogers* 19247a (PRE!); *Rogers* 22118 (K!). *A. rogersii* Burtt Davy, Fl. Transv. 2 : xviii, 348 (1932); Codd, Trees & Shrubs Kruger Nat. Park 56 (1951); O. B. Miller in J. S. Afr. Bot. 18 : 27 (1952); F. White, For. Fl. N. Rhod. 89, fig. 16 I (1962). Type: Transvaal, Soutpansberg district, Messina, *Moss & Rogers* 66 (PRE holo!; BM!; K!).

Shrub or tree 3–15 m high, trunk often branching near ground level into several to many ascending branches. *Bark* grey to grey-black, smooth or shallowly fissured; young branchlets very sparingly to ± densely appressed grey-puberulous when young, becoming glabrous with age; occasionally a few lateral shoots abbreviated and spinescent apically. *Leaves* sparingly to ± densely pubescent or glabrous: petiole 0,8–2,5 cm long, adaxial gland squat, sessile, ± 0,25 mm high; rhachis (1)2,5–7(14) cm long; pinnae (3)6–10(15) pairs; rhachillae 1–4,4 (6,5) cm long; leaflets 9–25 pairs, 3–7,5(9) × 0,75–1,75(2) mm, narrowly oblong to linear-oblong, symmetric, obtuse to subacute apically, not falcate, midrib nearly central (except basally), lateral nerves indistinct beneath, glabrous or margins ± appressed-ciliate. *Inflorescences* globose; peduncles 1,5–3 cm long, sparingly appressed-pubescent. *Flowers* white to creamy-yellow, on pedicels 1–1,5 mm long; bracteoles rapidly deciduous. *Calyx* 1–1,5 mm long, glabrous to ± puberulous outside. *Corolla* 4–5 mm long, tube 2,5–3 mm long, glabrous, lobes 1,5–2 mm long, pubescent apically. *Stamens* 1–1,8 cm long, united basally for ± 3 mm, tube not or scarcely exserted beyond the corolla. *Ovary* ± 1,5 mm long, shortly stipitate, glabrous. *Pods* (9)12–25 × (1,8)2,4–4,4 cm, linear to linear-oblong, valves thin, transversely venose, glabrous to finely puberulous, umbonate over the seeds, apparently indehiscent. *Seeds* 8–10 × ± 6,5 mm, flattened.

Found in South West Africa, Botswana, Rhodesia, Mozambique and the Transvaal. Occurs usually on hot, dry, rocky hillsides, often on granite, quartzite or sandstone formations.

S.W.A.—1720 (Posto Velho): S.E. slope of mountain at Ombepera, *De Winter & Leistner 5510*. 1813 (Ohopoho): Giraffenbergen bei Otjue, *Merxmuller & Giess 1476* (M); 14,4 km W. of Oruwanje, *Giess & Leippert 7387* (M). 1814 (Otjitundua): Otjitoko, *Merxmuller & Giess 1364* (M).

TRANSVAAL.—2229 (Waterpoort): Dongola area, farm Little Muck, *Codd 4459*. 2230 (Messina): Messina, *Pole Evans 2527*. 2231 (Pafuri): Kruger National Park, near Punda Milia, *Lamont 5*. 2327 (Ellisras): 17,6 km S. of Ellisras, *Codd 8493*. 2331 (Phalaborwa): Kruger National Park, The Gorge, *Codd 6188*. 2527 (Rustenburg): 15,4 km N. of Assen, *Codd 6566*.

Vegetative specimens of *A. harveyi* Fourn. are sometimes confused with specimens of *A. brevifolia* but the two species may be distinguished by the shape of the apex of the leaflet which, in *A. harveyi*, is acute and distinctly falcate, the apex turned towards the pinna-apex, while in *A. brevifolia* it is obtuse to subacute and almost straight.

2. **Albizia amara** (*Roxb.*) *Boiv.* in Encycl. XIX-me Siècle 2 : 34 (?1834); Benth. in Trans. Linn. Soc. Lond. 30 : 567 (1875); Brenan in Kew Bull. 10 : 189 (1955); in F.T.E.A. Legum.–Mimos. : 151 (1959); in F.Z. 3, 1 : 123 (1970). Type: India, *Roxburgh* (K, painting of holotype material, No. 486!).

Mimosa amara Roxb., Pl. Corom. 2 : 13, t.122 (1799). Type as above.

subsp. **sericocephala** (*Benth.*) *Brenan* in Kew Bull. 10 : 190 (1955); Palgrave, Trees Cent. Afr. 258 (1956); Codd in Bothalia 7 : 70 (1958); Brenan in F.T.E.A. Legum.–Mimos. : 152 (1959); F. White, For. Fl. N. Rhod. 89,

fig. 16H (1962); Von Breitenbach, Indig. Trees S. Afr. 2 : 255 (1965); Brenan in F.Z. 3, 1 : 125, t.23A (1970); Van Wyk, Trees Kruger Nat. Park 1 : 108 (1972); Palmer & Pitman, Trees S. Afr. 2 : 715 (1973). Syntypes: Sudan, Sennar, *Kotschy 244* (K!; P!) & Kordofan, Milbes, *Kotschy 294* (BM!; E!; K!; OXF!; P!; W!); Ethiopia, Gapdia, *Schimper 818* (BM!; K!; OXF!; P!) & Dscheladscheranne [Jelajeranne], *Schimper 883* (BM!; K!; OXF!; P!).

A. sericocephala Benth. in Hook., Lond. J. Bot. 3 : 91 (1844); Milne-Redhead in Kew Bull. 1934 : 301 (1934); Brenan, Checklist Tang. Terr. 341 (1949); Pardy in Rhod. Agric. J. 49 : 14 (1952). Types as above. *A. amara* sensu Oliv. in F.T.A. 2 : 356 (1871); Bak.f., Leg. Trop. Afr. 3 : 865 (1930); Gilbert & Boutique in F.C.B. 3 : 172 (1952). *A. struthiophylla* Milne-Redhead in Kew Bull. 1933 : 144 (1933); O. B. Miller in J. S.Afr. Bot. 18 : 27 (1952) ("struthio-folia"). Type: Zambia, Mazabuka, *Milne-Redhead 1207* (K, holo.!, PRE!).

Acacia sericocephala Fenzl in Flora 27 : 312 (1844), nomen nudum.

Inga sericocephala A. Rich., Tent. Fl. Abyss. 1 : 236 (1847). Types as for *Albizia sericocephala*.

Tree 2–12 m high, crown rounded or somewhat flattened. *Bark* fissured, rough; young branchlets with rather short dense spreading golden to grey pubescence. *Leaves* golden-tomentose when young, rusty to grey-pubescent when older: petiole 0,6–1,6 cm long, adaxial gland low, sessile, up to ± 0,25 mm high; rhachis 5,8–23 cm long; pinnae 12–44 pairs; rhachillae 1–3 cm long; leaflets (12)20–45 pairs, 2–4 × 0,5–1 mm, linear to linear-oblong, symmetric, obtuse to subacute apically, not falcate, midrib nearly central (except basally), lateral nerves indistinct beneath, rarely slightly raised, ± appressed-pubescent on one or both surfaces or on the margins only, glabrescent or not later. *Inflorescences* globose; peduncles 1–2,4 cm long, golden-tomentose when young, becoming greyish when older. *Flowers* whitish or tinged with pink, subsessile, buds golden-tomentose; bracteoles rapidly deciduous. *Calyx* 1–2 mm long, golden-puberulous or pubescent. *Corolla* 3,5–5 mm long, golden-puberulous or pubescent, especially towards apices of lobes, lobes 2–2,5 mm long. *Stamens* 1–1,5 cm long, united basally for ± 2,5 mm, tube not or scarcely exserted beyond the corolla. *Ovary* 2–2,5 mm long, shortly stipitate, glabrous. *Pods* 12–28 × 2,5–4 cm, linear-oblong, valves thin, transversely venose, puberulous, umbonate over the seeds, apparently indehiscent. *Seeds* 8–13 × 7–8 mm, flattened.

Found from the Sudan and Ethiopia southwards to Botswana, Rhodesia, Mozambique and the north eastern Transvaal. In the Transvaal it is restricted to the area between Sibasa and Punda Milia where it occurs on sandy flats in woodland and lowveld bush.

TRANSVAAL.—2230 (Messina): Sibasa area, *Van Warmelo 5115/24*; 27,2 km E. of Sibasa, *Codd & Dyer 4494*; 38,4 km N.E. of Sibasa near Sambandou, *Codd 6901*; Minga, *Gerstner 6219*. 2231 (Pafuri): Kruger National Park, near Punda Milia, *Van der Schijff 986*; *Van der Schijff 2894*.

A. amara subsp. *sericocephala* may be readily distinguished from *A. harveyi* by its longer leaves, more numerous pinnae and leaflet pairs, and by the apex of the leaflets not being acute and distinctly falcate. Subsp. *sericocephala* differs from *A. brevifolia* in the longer leaves, more numerous pinnae and leaflet pairs, and in the dense spreading golden indumentum on the young branchlets.

Typical subsp. *amara*, which differs in having fewer pinnae and leaflet pairs, is found in India, Kenya, Tanzania and Mozambique.

3. **Albizia harveyi** *Fourn.* in Bull. Soc. Bot. Fr. 12 : 399 (1865); Bak.f., Leg. Trop. Afr. 3 : 865 (1930); Burtt Davy, Fl. Transv. 2 : 348 (1932); Steedman, Trees etc. S. Rhod. 16 (1933); Brenan, Checklist Tang. Terr. 341 (1949); in Mem. N.Y. Bot. Gdn. 8 : 430 (1954); Gomes e Sousa, Dendrol Moçamb. 4 : 46 (1949); Codd, Trees & Shrubs Kruger Nat. Park 56, figs. 49–51 (1951); Gilbert & Boutique in F.C.B. 3 : 173 (1952); O. B. Miller in J.S. Afr. Bot. 18 : 27 (1952); Codd in Bothalia 7 : 71 (1958); Brenan in F.T.E.A. Legum.–Mimos. : 149, fig. 20 (1959); F. White, For. Fl. N. Rhod. 89, fig. 16G (1962); Von Breitenbach, Indig. Trees S. Afr. 2 : 256 (1965); Gomes e Sousa, Dendrol. Moçamb. 1 : 235, t.39 (1966); Compton in J.S. Afr. Bot. Suppl. 6 : 45 (1966); Schreiber in F.S.W.A. 58 : 14 (1967); Brenan in F.Z. 3, 1 : 122, t.22 (1970); Van Wyk, Trees Kruger Nat. Park 1 : 118 (1972); Palmer & Pitman, Trees S. Afr. 2 : 713 (1973). Type: Botswana, near Lake Ngami, *McCabe* (K, holo.!).

A. pallida Harv. in F.C. 2 : 284 (1862) nom. illegit., non *A. pallida* Fourn. in Ann. Sci. Nat. Ser. IV, 14 : 375 (1860). Type as for *A. harveyi*. *A. hypoleuca* Oliv. in F.T.A. 2 : 356 (1871); Benth. in Trans. Linn. Soc. Lond. 30 : 567 (1875). Type as for *A. harveyi*.

Tree up to 15 m high, crown narrowly rounded or rounded. *Bark* grey-brown to blackish, rough, fissured; young branchlets with grey to brown or sometimes slightly golden spreading pubescence, a very small prickle-like outgrowth often present beneath each node. *Leaves* often golden-tomentose when very young, rusty to grey-pubescent when older: petiole 0,8–3 cm long, adaxial gland often prominent and sometimes shortly stalked and up to 1 mm high, often absent; rhachis 2,4–9(15) cm long; pinnae 6–22 pairs; rhachillae 1–3,4 cm long; leaflets (6)12–28 pairs, 2–6 × 0,5–1,5 mm, ± falcate, apex asymmetric, acute, turned towards the apex of the pinna, midrib nearer the distal margin, lateral nerves ± raised and visible beneath, lower surface of leaflet paler than upper, ± appressed-pubescent on both surfaces or the lower only when young, becoming sparingly pubescent to glabrescent with age. *Inflorescences* globose; peduncles 1,5–4 cm long, ± golden-pubescent when young, becoming greyish when older. *Flowers* creamy-white, sessile or up to 0,5 mm pedicellate, buds golden-pubescent; bracteoles persistent during flowering period. *Calyx* 1,5–2,5 mm long, golden-pubescent, lobes ± 0,5 mm long. *Corolla* 4–5,5 mm long, pubescent, tube to 3 mm long, lobes to 2,5 mm long. *Stamens* 1,3–1,8 cm long, united basally for ± 2,5 mm, tube not or scarcely exserted beyond the corolla. *Ovary* ± 2 mm long, very shortly stipitate, puberulous. *Pods* yellowish-brown to brown, 6–15 × 1,8–3,5 cm, oblong, valves thin, glabrous or nearly so except for a slight pubescence along the margins and near the base, umbonate over the seeds, longitudinally dehiscent. *Seeds* 8–12 × 6–9 mm, flattened.

Widespread in eastern and southern tropical Africa from southern Kenya in the north to South West Africa, Botswana, the Transvaal and Swaziland. Occurs in woodland and dry bushveld, sometimes near rivers.

S.W.A.—1719 (Runtu): Runtu, *Volk 2030* (M). 1724 (Katima Mulilo): Katima Mulilo, *West 3249*. 1821 (Andara): Andara, *Volk 2090* (M). 1920 (Tsumkwe): 9,6 km S. of Tsumkwe, *Giess 9898* (M).

TRANSVAAL.—2229 (Waterpoort): Dongola reserve, Cream of Tartarfontein, *Pole Evans 3526*. 2230 (Messina): Messina, *Pole Evans 1440*. 2231 (Pafuri): Kruger National Park, Punda Milia, *Codd 4223*. 2328 (Baltimore): 32 km E. of Ellisras, *Acocks 8820*. 2331 (Phalaborwa): Kruger National Park, Letaba area, *Lang sub TRV 30357*. 2431 (Acornhoek): Kruger National Park, near Skukuza camp, *Letty* 57. 2531 (Komatipoort): Kruger National Park, 20 km N.E. of Malelane on road to Skukuza, *Codd 5218*.

SWAZILAND.—2631 (Mbabane): near Makombo, Mtendekwa stream, *Miller S/15*.

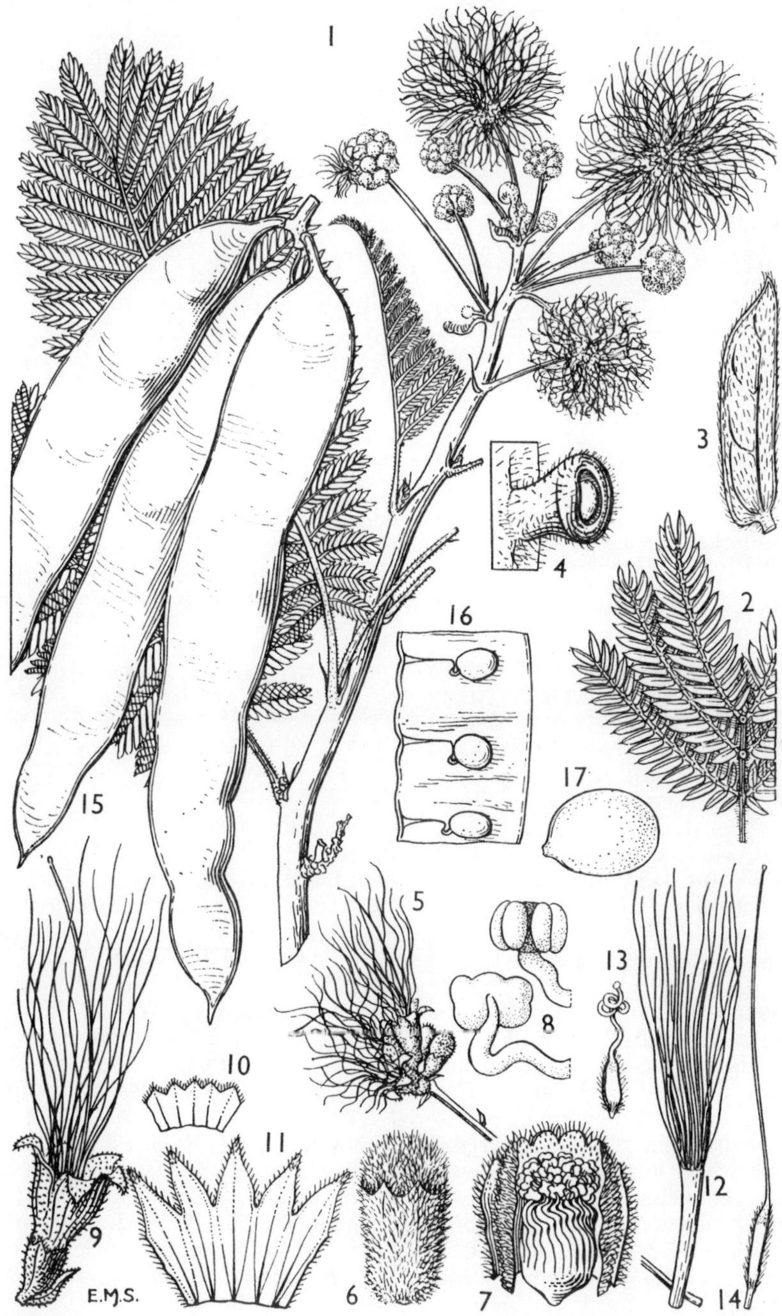

FIG. 1.—**Albizia harveyi.** 1, flowering branch, × 1; 2, part of leaf showing glands on rhachis, × 2; 3, leaflet × 8; all from *Burtt* 3809; 4, gland on rhachis, × 24; 5, young flower head, × 2; 6, flower bud, × 8; 7, flower bud opened, showing arrangement of stamens, × 8; 8, anthers from bud, front and back views, × 40; 9, open flower, × 4; 10, calyx opened out, × 4; 11, corolla opened out, × 4; 12, stamen filaments and tube, × 4; 13, ovary from bud, × 8; 14, ovary from mature flower, × 4; all from *Burtt* 5037; 15, pods, × ⅔, *Burtt* 1661; 16, part of valve of pod, seen from inside, × ⅔; 17, seed, × 2, both from *Legat* 65. Reproduced by permission of the Editor of Flora of Tropical East Africa.

Vegetative specimens of *A. harveyi* are sometimes confused with specimens of *A. brevifolia*. The differences between the two species are given under *A. brevifolia*.

4. **Albizia forbesii** *Benth.* in Hook., Lond. J. Bot. 3 : 92 (1844); Harv. in F.C. 2 : 284 (1862); Benth. in Trans. Linn. Soc. Lond. 30 : 568 (1875); Sim, For. Fl. P.E. Afr. 58, t.39A (1909); Burtt Davy, Fl. Transv. 2 : 348 (1932); Codd, Trees & Shrubs Kruger Nat. Park 54, fig. 48a (1951); in Bothalia 7 : 72 (1958); Brenan in F.T.E.A. Legum.–Mimos. : 151 (1959); Von Breitenbach, Indig. Trees S. Afr. 2 : 256 (1965); Gomes e Sousa, Dendrol. Mocamb. 1 : 236, t.40 (1966); Brenan in F.Z. 3, 1 : 122 (1970); Van Wyk, Trees Kruger Nat. Park 1 : 116 (1972); Palmer & Pitman, Trees S. Afr. 2 : 717 (1973); Ross, Fl. Natal 192 (1973). Type: Mozambique, Delagoa Bay, *Forbes s.n.* (K, holo!; P!; TCD!).

Tree to 20 m high, with a single trunk or sometimes branching near the base, crown typically rounded and spreading. *Bark* grey to blackish, rough or sometimes smooth; young branchlets densely grey-pubescent. *Leaves* densely grey-pubescent, often golden when very young: petiole 0,6–2 (2,8) cm long, adaxial gland low, up to ± 0,25 mm high; rhachis 0,5–3,8 cm long, with a fairly conspicuous gland at the junction of each pinna pair or the top 1–3 pairs only; pinnae 2–7 pairs; rhachillae 1,6–4,8 cm long; leaflets 6–16 pairs, 3,5–8 × 1,5–4 mm, obliquely oblong to oblong-elliptic (the terminal pair obovate), apex rounded to subacute or mucronate, turned towards the apex of the pinna, the midrib nearer the distal margin, lateral nerves usually ± raised and visible beneath, usually glabrous above, glabrous beneath except for spreading pubescence on the midrib and recurved margins, sometimes pubescent throughout. *Inflorescences* globose; peduncles 2–5,5 cm long, ± golden-pubescent when young, becoming greyish when older. *Flowers* creamy-white, sessile or almost so, buds golden-pubescent; bracteoles 1,5–2 mm long, linear or oblanceolate, usually deciduous before flowers open. *Calyx* 3–4 mm long, densely and shortly appressed-pubescent. *Corolla* 5–8(9) mm long, densely and shortly appressed-pubescent, tube 4–5(7) mm long, lobes up to 2,5 mm long. *Stamens* 1–1,5 cm long, united

basally for ± 3 mm, tube not or scarcely exserted beyond the corolla. *Ovary* 2–3 mm long, shortly stipitate, glabrous. *Pods* dark brown or reddish-brown, 9–18 cm (including 1–2 cm long stipe) × 3,2–5 cm, oblong, valves fibrous, puberulous or almost glabrous except on margins and stipe, very prominently transversely venose, the veins±parallel and 2–4 mm apart, margins thickened. *Seeds* 11–12 × 4,5–6,5 mm, oblong-ellipsoid or ellipsoid, slightly flattened.

Found in southern Tanzania, Mozambique, the eastern areas of Rhodesia, the Transvaal and Natal (Tongaland). Occurs in woodland, bushveld and sand-forest.

TRANSVAAL.—2231 (Pafuri): Kruger National Park, near Punda Milia, *Rowland Jones 29*. 2431 (Acornhoek): Kruger National Park, 12,8 km E. of Skukuza, S. side of Sabie River, *Codd 5704*. 2531 (Komatipoort): Kruger National Park, near Malelane, *Van der Schijff 1417*.

NATAL.—2632 (Bela Vista): Ndumu Game Reserve, E. of Pongola floodplain, *Pooley 1403* (K, NU). 2732 (Ubombo): 12,8 km from Makanes bridge on road to Sihangwane, *Ross 2425*.

In Tongaland *A. forbesii* occurs commonly on white sandy soils and is one of the dominant species in sand-forest.

The prominently transversely venose pods are most distinctive, and the seeds are rather narrow in proportion to their length.

5. **Albizia versicolor** *Welw. ex Oliv.* in F.T.A. 2 : 359 (1871); Benth. in Trans. Linn. Soc. Lond. 30 : 562 (1875); Hiern, Cat. Afr. Pl. Welw. 1 : 315 (1896); Eyles in Trans. Roy. Soc. S. Afr. 5 : 361 (1916); Bak.f., Leg. Trop. Afr. 3 : 863 (1930); Burtt Davy, Fl. Transv. 2 : 348 (1932); Steedman, Trees etc. S. Rhod. 16 (1933); Henkel, Woody Pl. Natal 236 (1934); Gomes e Sousa, Dendrol. Mocamb. 1 : 94 (1948); Brenan, Checklist Tang. Terr. 343 (1949); Codd, Trees & Shrubs Kruger Nat. Park 57, figs. 52, 53 (1951); Gilbert & Boutique in F.C.B. 3 : 182, fig. 7 (1952); O. B. Miller in J. S. Afr. Bot. 18 : 27 (1952); Torre in C.F.A. 2 : 293 (1956); Codd in Bothalia 7 : 73 (1958); Brenan in F.T.E.A. Legum.–Mimos. : 146 (1959); F. White, For. Fl. N. Rhod. 89, fig. 16J (1962); Von Breitenbach, Indig. Trees S. Afr. 2 : 258 (1965); Compton in J. S. Afr. Bot., Suppl. 6 : 45 (1966); Schreiber in F.S.W.A. 58 : 14 (1967); Brenan in F.Z. 3, 1 : 117, t.23 fig. D (1970); Van Wyk, Trees Kruger Nat. Park 1 : 122 (1972); Palmer & Pitman, Trees S. Afr. 2 : 717 (1973); Ross, Fl. Natal 192

(1973). Type: Angola, Cuanza Norte : Golungo Alto, Candombo e Trombeta, *Welwitsch* 1760 (LISU, lecto; BM!; COI; K!; P!; iso.).

A. versicolor var. *mossambicensis* Schinz in Bull. Herb. Boiss., Sér. 2, 2 : 946 (1902); Bak.f., Leg. Trop. Afr. 3 : 863 (1930). Type: Mozambique, Baruma, Nhaondue, *Menyharth 77b* (Z, holo.!). *A. mossambicensis* Sim, For. Fl. P.E. Afr. 59, t.60 (1909); Eyles in Trans. Roy. Soc. S.Afr. 5 : 361 (1916). Type: Mozambique, *Sim 6392* (NU, holo.!).

Tree to 18 m high with a spreading rounded crown. *Bark* greyish-brown, usually rough; young branchlets densely rusty-tomentose, older branchlets grey-tomentose. *Leaves* densely rusty-tomentose when young, becoming grey-pubescent when older: petiole 3–7,5 cm long, adaxial gland usually present and situated a short distance above the pulvinus, usually flattened, ±discoid, up to 3 × 2 mm; rhachis 0–18 cm long; pinnae 1–3(5) pairs; rhachillae 2–15 cm long, usually with a small gland below the junction of the top or top 1–2 pinnae pairs; leaflets 2–5(6) pairs, 15–67 × (10)15–45 mm, broadly and obliquely obovate or obovate-oblong to subcircular, apex rounded and often mucronate, sometimes subacute or emarginate, becoming coriaceous, pubescent above, densely tomentose or pubescent beneath, midrib and lateral nerves raised and very prominent beneath. *Inflorescences* globose; peduncles 2–6 cm long, rusty-tomentose. *Flowers* creamy-white, sessile or subsessile; bracteoles present or already deciduous at flowering time. *Calyx* 4,5–7 mm long, densely rusty-pubescent or -tomentose. *Corolla* 7–12 mm long, clothed like the calyx. *Stamens* 2,5–4 cm long, united basally for ± 3 mm, tube not exceeding the corolla. *Ovary* ± 3 mm long, subsessile, puberulous. *Pods* chestnut- to reddish-brown or purplish, 8–22 × (2,2)3–5,5 cm, oblong, valves thin, glabrous or with few hairs on stipe and margins, ±glossy, obscurely transversely venose, margins thickened, longitudinally dehiscent. *Seeds* 9–13 × 8–11 mm, flattened.

Found from Uganda and Kenya southwards to South West Africa, Botswana, Rhodesia, Mozambique, the Transvaal, Swaziland and Natal (Zululand). Occurs in mixed woodland.

S.W.A.—1715 (Ondangua): near Oshikango, *Rodin 2609* (K). 1719 (Runtu): Ndwaki camp, 70,4 km W. of Runtu, *De Winter 3817*. 1721 (Mbambi): Kangongo camp, 76,8 km E. of Nyangana Mission station, *De Winter 4215*. 1821 (Andara): Shitangadimba camp at Andara Mission station, *De Winter 4237*; Andara, *Watt 36* (M).

TRANSVAAL.—2230 (Messina): 6,4 km S.E. of Sibasa, *Codd & Dyer 4491*. 2231 (Pafuri): Kruger National Park, Punda Milia, *Codd & Dyer 4615*. 2330 (Tzaneen): Elim, *Obermeyer 564*. 2430 (Pilgrim's Rest): 1,6 km N. of P.O. Buffelsvlei, *Codd 6674*. 2431 (Acornhoek): Acornhoek, *Keet 1488*. 2531 (Komatipoort): Kruger National Park, Shabin Kop, near Pretorius Kop, *Codd 5694*.

SWAZILAND.—2531 (Komatipoort): near Ngonini, *Compton 29361*. 2631 (Mbabane): Mposi, *I'Ons 59/6*. 2731 (Louwsburg): Ingwavuma River, *West 2105*.

NATAL.—2632 (Bela Vista): Ndumu Game Reserve, Ndumu Hill, *Pooley 710* (NU). 2732 (Ubombo): 38,4 km W. of Maputa on road to Makanes Pont, *De Winter & Vahrmeijer 8631*. 2831 (Nkandla): between Eshowe and Nkandla, *Pole Evans 3622*. 2832 (Mtubatuba): Hluhluwe Game Reserve, *Ward 1861*(NH). 2931 (Stanger): Amatikulu, *Wylie sub Wood 7589* (K).

A very distinct and easily recognized species on account of the combination of the ± rust-coloured indumentum over the vegetative parts, and the few pairs of broad leaflets.

The type specimen of *A. mossambicensis* Sim is cited as *Sim 6392*. The holotype in the University of Natal Herbarium is actually annotated by Sim as *A. umbelusiana*, but the specimen clearly agrees with his description and plate of *A. mossambicensis* and not with his published description and plate of *A. umbalusiana* (see notes under *A. anthelmintica* and *A. petersiana* subsp. *evansii*).

A. versicolor grows into an attractive large tree and yields a useful timber. Brenan l.c. (1949), in addition to mentioning that the timber is used, records that the natives make a soapy substance from the roots and that the roots are also employed as an anthelmintic and as a purgative and enema.

The unripe pods and seeds of *A. versicolor* have proved to be toxic to stock (Needham & Lawrence in Rhod. Agric. J. 63 : 137, 1966).

6. **Albizia antunesiana** *Harms* in Bot. Jahrb. 30 : 317 (1901); R.E. Fr., Schwed. Rhod.–Kongo–Exped. 1 : 63 (1914); Eyles in Trans. Roy. Soc. S. Afr. 5 : 361 (1916); Bak.f., Leg. Trop. Afr. 3 : 861 (1930); Steedman, Trees etc. S. Rhod. 15 (1933); Brenan, Checklist Tang. Terr. 342 (1949); Pardy in Rhod. Agric. J. 48 : 398 (1951); Gilbert & Boutique in F.C.B. 3 : 189, fig. 10C, D (1952); O. B. Miller in J. S. Afr. Bot. 18 : 27 (1952); Torre in C.F.A. 2 : 291 (1956); Palgrave, Trees Cent. Afr. 261 (1956); Codd in Bothalia 7 : 74 (1958); Brenan in F.T.E.A. Legum.–Mimos. : 148 (1959); F. White, For. Fl. N. Rhod. 90, fig. 16M (1962); Von Breitenbach, Indig. Trees S. Afr. 2 : 260

(1965); Schreiber in F.S.W.A. 58 : 14 (1967); Brenan in F.Z. 3, 1 : 119 (1970); Palmer & Pitman, Trees S. Afr. 2 : 720 (1973); Schreiber in Mitt. Bot. Staatssamml. Munchen 11 : 126 (1973). Syntypes: Tanzania, Mbeya District, Unyika, Iyunga village, *Goetze* 1372 (B, †; BM!; P!); Angola, Huila, *Antunes* 330 (B, †).

Tree up to 12 m high, branches usually spreading somewhat. *Bark* grey, rough or sometimes smooth; young branchlets very shortly pubescent, becoming glabrous with age. *Leaves*: petiole 3–8,5 cm long, adaxial gland usually immediately above or a short distance above the pulvinus, humped or flattened and ±discoid and up to 4 × 4 mm; rhachis 0–13 cm long, sparingly pubescent when young but soon becoming glabrous or almost so; pinnae 1–4 pairs; rhachillae 5–16 cm long; leaflets 4–9 pairs, (16)23–50 × 8–28 mm (in our area), oblique, ovate to rhombic-ovate or elliptic-oblong, usually rounded or slightly emarginate apically, papery to sub-coriaceous, glabrous, venose, glaucous, paler beneath. *Inflorescences* globose; peduncles 3,6–8 cm long, sparingly rusty-pubescent. *Flowers* creamy-white, sessile or up to 2 mm pedicellate; bracteoles small, rapidly deciduous and shed before flowering. *Calyx* (3)3,5–5,5 mm long, rusty-puberulous or -pubescent. *Corolla* 5,5–11 mm long, densely minutely appressed-pubescent, lobes up to 4 mm long. *Stamens* 1,5–3 cm long, united basally for up to 5 mm, tube not or scarcely exserted beyond the corolla. *Ovary* ± 2 mm long, shortly stipitate, glabrous. *Pods* 11–16 (23) × 2,5–4(4,6) cm, oblong, valves thin, glabrous or with few hairs near the base and margins, transversely venose, umbonate over the seeds, longitudinally dehiscent. *Seeds* ± 7–9 mm in diameter, flattened.

Found in Tanzania, Zaire, Angola, Zambia, Malawi, South West Africa, Botswana, Rhodesia and Mozambique. Occurs in mixed woodland.

S.W.A.—1821 (Andara): between Bagani camp and Mahango, *De Winter & Wiss 4394*; opposite Andara, *De Winter & Marais 4816*.

Vegetative specimens of *A. antunesiana* are easily recognized by the glabrous, discolorous leaflets, but the flowers are often produced when the tree is leafless.

A. antunesiana is closely related to *A. coriaria* Welw. ex Oliv., the latter differing in having puberulous or pubescent leaf-rhachides and usually smaller, less oblique leaflets.

More material of *A. antunesiana* from our area, particularly fertile material, is needed.

7. **Albizia tanganyicensis** *Bak.f.* in J. Bot. 67 : 199 (1929); Leg. Trop. Afr. 3 : 862 (1930); Brenan, Checklist Tang. Terr. 342 (1949); Torre in C.F.A. 2 : 293 (1956); Codd in Bothalia 7 : 75 (1958); Brenan in F.T.E.A. Legum.–Mimos. : 144 (1959); F. White, For. Fl. N. Rhod. 90 (1962); Von Breitenbach, Indig. Trees S. Afr. 2 : 262 (1965); Brenan in F.Z. 3, 1 : 116 (1970); Van Wyk, Trees Kruger Nat. Park 1 : 120 (1972); Palmer & Pitman, Trees S. Afr. 2 : 721 (1973). Type: Tanzania, Kondoa District, Simbo Hills, *B. D. Burtt* 716 (BM, holo.!; EA; K!).

subsp. **tanganyicensis**.

Brenan in Kew Bull. 29 : 717 (1975). Type as above.

A. rhodesica Burtt Davy, Fl. Transv. 2 : xviii, 348 (1932); Codd, Trees & Shrubs Kruger Nat. Park 56 (1951); O. B. Miller in J. S. Afr. Bot. 18 : 27 (1952); Pardy in Rhod. Agric. J. 51 : 4 (1954); Palgrave, Trees Cent. Afr. 269 (1956). Syntypes: Rhodesia, Matopos, *Galpin* 7082 (PRE!); Victoria Falls, *Allen* 174 (K!); *Rogers* 5319 (K!). *A. lebbeck* var. *australis* Burtt Davy in Burtt Davy & Hoyle, Checklist Nyasaland 53 (1936) nomen nudum; Burtt Davy & Hoyle, rev. Topham, Checklist Nyasaland, ed.2 : 65 (1958) nomen nudum.

Tree up to 12 m high, sometimes sparingly branched near the base and with few ascending branches, crown rounded or somewhat flattened, canopy often sparse. *Bark* smooth (except at base of trunk where burned), creamy-white to light ochre-yellow or yellow-green when young, the older bark peeling off in large papery pieces; young branchlets pubescent to glabrous. *Leaves* sparingly pubescent especially when young but often becoming glabrous: petiole 2–5(8) cm long, adaxial gland usually just above the pulvinus or ± midway along length of petiole, fairly conspicuous and up to 3 mm long; rhachis 6–14(23) cm long; pinnae (2)3–6(7) pairs; rhachillae (2,8)3,5–19 cm long, sometimes with a gland at the junction of the top or top 1–5 leaflet pairs; leaflets 4–13(17) pairs, (9)13–45(55) × 5–25(32) mm, somewhat asymmetric, ovate- to obovate-elliptic or ovate-oblong, rounded to subacute apically, papery to subcoriaceous, pubescent on both surfaces or glabrous, venose. *Inflorescences* globose; peduncles 3–5 cm long, rusty-pubescent. *Flowers* creamy-white, usually produced before the young leaves, sessile or up to 1 mm pedicellate; bracteoles rapidly deciduous and shed before the flowers open. *Calyx* 4–6 mm long, densely ± rusty-tomentellous on the lobes, lobes up to 2 mm long. *Corolla* 7–11 mm long, lobes up to 4 mm long, ± rusty-tomentellous, especially apically. *Stamens* 1,5–3 cm long, united basally for ± 4 mm, tube not or scarcely exserted beyond the corolla. *Ovary* ± 3 mm long, shortly stipitate, glabrous. *Pods* brown, 10–25 × 2,5–5 cm, oblong, valves slightly thickened, glabrous, not or obscurely venose, longitudinally dehiscent. *Seeds* ± 10–17 × 8–13 mm, flattened.

Found in Tanzania, Zambia, Malawi, Angola, South West Africa, Botswana, Rhodesia, Mozambique and the Transvaal. Occurs usually on rocky outcrops and ridges, often on quartzite and granite formations.

S.W.A.—1813 (Ohopoho): Oruwanjei on Orupembe–Ohopoho Road, *Joubert 297* (M); 4,8 km E. of Oruwanjei, near Kaoko–Otavi, *Giess 10533* (M).

TRANSVAAL.—2231 (Pafuri): Kruger National Park, near Punda Milia, *Codd & Dyer 4549*. 2427 (Thabazimbi): Rooiberg, *Pole Evans s.n.* 2428 (Nylstroom): 30,4 km E. of Vaalwater, *Codd 986*; 8 km N.E. of Nylstroom, *Codd 5601*; 16 km N. of Warmbaths on road to Nylstroom, *De Winter 8681*; hills near Warmbaths, *Burtt Davy 2183*; Olifants Spruit, *Repton 3472*.

A very distinct species which is readily distinguished by its thin smooth papery-peeling bark. The flowers are usually produced when the tree is leafless, and specimens collected in this state, without notes about the bark, are liable to be confused with specimens of *A. antunesiana* in the same state. Apart from having an entirely different distribution in our area, *A. antunesiana* may be distinguished by its smaller flowers and (in dried specimens) by the more prominent and raised nerves on the calyx-tube.

A. tanganyicensis has in the past been confused with the introduced *A. lebbeck* (L.) Benth., but the two are quite distinct. Vegetative specimens of the two species are similar but *A. tanganyicensis* differs in its papery peeling bark and in its more closely spaced leaflets. In *A. tanganyicensis* the flowers are sessile or up to 1 mm pedicellate and the corolla-lobes are rusty-tomentellous especially apically, while in *A. lebbeck* the flowers are on pedicels 1,5–4,5 mm long and the corolla-lobes have few whitish hairs. The pods of the two species also differ, those of *A. lebbeck* being straw-coloured and the positions of the seeds marked by distinct bumps in the valves, while in *A. tanganyicensis* they are brown and there are no conspicuous bumps in the valves indicating the positions of the seeds within.

The wood of *A. tanganyicensis* is of no commercial value and Codd l.c. : 76 reports that when the wood is worked the dust irritates the nose and throat, so that in the Transvaal it is locally known as "sneezewood".

8. Albizia suluensis *Gerstn.* in J. S. Afr. Bot. 13 : 62, fig. 6 (1947); Codd in Bothalia 7 : 76 (1958); Von Breitenbach, Indig. Trees S. Afr. 2 : 259 (1965); Ross, Fl. Natal 192 (1973); Palmer & Pitman, Trees S. Afr. 2 : 723 (1973). Type: Natal, Melmoth Distr., 3,2 km west of Dundulu Store, *Gerstner 4337* (PRE, lecto.!).

Tree to 15 m high with a rounded or spreading crown. *Bark* greyish, smooth or fissured; young branchlets sparingly pubescent when young but soon becoming glabrous. *Leaves* glabrous or subglabrous: petiole 2–7 cm long, a small rounded slightly raised gland usually present at or just below the junction of the lowest pinna pair; rhachis (0)2–8 cm long, sulcate above, a small rounded gland usually present at the junction of the top pinna pair, at the junction of each pinna pair or absent from some; pinnae (1)2–4 pairs; rhachillae 5,5–17 cm long, sulcate above, pulvinule sometimes sparsely pubescent; leaflets 4–9 pairs, 18–28 × 8–15 (18) mm, oblique basally, oblong to broadly elliptic or obovate, rounded to truncate apically, mucronate, margins crisped, glabrous to sparsely puberulous on both surfaces, petiolules often sparsely pubescent. *Inflorescences* globose; peduncles 1,5–4 cm long, sparingly rusty-pubescent. *Flowers* whitish, subsessile; bracteoles rapidly deciduous. *Calyx* 2,5–4 mm long, densely ± rusty-tomentellous. *Corolla* 4–6 mm long, lobes densely ± rusty-tomentellous, especially apically. *Stamens* 1–1,6 cm long, united basally for up to 6 mm, tube not or scarcely exserted beyond the corolla. *Ovary* 2–3 mm long, shortly stipitate, glabrous. *Pods* light brown, 8–21 × 1,4–2,6 cm, linear-oblong, valves thin, glabrous, not or scarcely venose, slightly umbonate over the seeds, longitudinally dehiscent.

Restricted to the Ntonjaneni and Hlabisa districts of Zululand. Occurs in forest, often on stream banks or near streams.

NATAL.—2732 (Ubombo): Dukunbane, *Gerstner 714*. 2831 (Nkandla): Hlabisa, *Gerstner 6440*; Gwegwede River, *Gerstner 6261* (BOL, K); Ndhlwati, *Gerstner 730*; Inhlwati, *F. Bayer s.n.* (K, NU); 6,4 km N. of Hlabisa, *Codd 9611*. 2832 (Mtubatuba): Hluhluwe Game Reserve, *Ward 2829* (K, NH).

A very distinct species which is distinguished by the crisped leaflet margins. It appears to be endemic in the Ntonjaneni and Hlabisa districts of Zululand and is nowhere very common.

Gerstner l.c. records that the Zulu's pound the bark in water, producing a foaming mixture which is used as a powerful enema. The timber is said to be hard and durable with an attractive grain, suitable for furniture.

9. Albizia anthelmintica (*A. Rich.*) *Brongn.* in Bull. Soc. Bot. Fr. 7 : 902 (1860); Oliv. in F.T.A. 2 : 357 (1871); Benth. in Trans. Linn. Soc. Lond. 30 : 564 (1875); Sim, For. Fl. P.E. Afr. 60 (1909); Eyles in Trans. Roy. Soc. S. Afr. 5 : 361 (1916) pro parte excl. specim. *Rogers* 5343; Marloth, Fl. S. Afr. 2, 1 : fig. 29 (1925); Bak.f., Leg. Trop. Afr. 3 : 859 (1930); Gomes e Sousa, Pl. Menyharth 68 (1936); Brenan, Checklist Tang. Terr. 341 (1949); Codd, Trees & Shrubs Kruger Nat. Park 53 (1951); Torre in C.F.A. 2 : 289 (1956); Pardy in Rhod. Agric. J. 53 : 952 (1956); Codd in Bothalia 7 : 77 (1958); Brenan in F.T.E.A. Legum.–Mimos. : 148 (1959); F. White, For. Fl. N. Rhod. 90, fig. 16L (1962); Von Breitenbach, Indig. Trees S. Afr. 2 : 259 (1965); Compton in J. S. Afr. Bot., Suppl. 6 : 45 (1966); Leistner in Mem. Bot. Surv. S. Afr. 38 : 123, t.8, 12 (1967); Schreiber in F.S.W.A. 58 : 14 (1967); Brenan in F.Z. 3, 1 : 120, t.23 fig. C (1970); Van Wyk, Trees Kruger Nat. Park 1 : 110 (1972); Palmer & Pitman, Trees S. Afr. 2 : 725 (1973); Ross, Fl. Natal 192 (1973). Type: Ethiopia, near Add'erbati, *Quartin Dillon* (P, holo.!; K.!).

Besenna anthelmintica A. Rich., Tent. Fl. Abyss. 1 : 253 (1847).

Acacia inermis Marloth in Trans. S.Afr. Phil. Soc. 5 : 269 (1889); Wordsworth et al. in Ann. Bolus Herb. 3 : 21 (1920); Ross in Bothalia 10 : 548 (1972) nomen nudum. *A. marlothii* Engl. in Bot. Jahrb. 10 : 19 (1889); Wordsworth et al, l.c. : 21 (1920). Type: South West Africa, near Otjimbingwe, *Marloth* 1317 (B, holo. †; BOL!; GRA!; PRE!).

Albizia umbalusiana Sim, For. Fl. P.E. Afr. 59, t.55A (1909). Type: Mozambique, "Lourenzo Marques, Maputa, Lebombos", *Sim* 6200 (whereabouts unknown, perhaps no longer extant). *A. anthelmintica* var. *australis* Bak.f., Leg. Trop. Afr. 3 : 859 (1930); Torre in C.F.A. 2 : 290 (1956). Type: South West Africa, Okahandja Distr., Okahandja, *Dinter* 269 (BM, holo.!; GRA!; K!; P!; PRE!; SAM!). *A. anthelmintica* var. *pubescens* Burtt Davy, Fl. Transv. 2 : xvii, 348 (1932); O. B. Miller in J. S.Afr. Bot. 18 : 27 (1952). Syntypes: Transvaal, Soutpansberg Distr., Waterpoort, *Rogers* 19347 (K!; PRE!); Messina, *Rogers* 21504 (PRE!).

Shrub or tree to 10 m high, crown often somewhat rounded. *Bark* grey to reddish-brown, smooth; young branchlets shortly pubescent or glabrous, lenticellate, often forming abbreviated divaricate spine-tipped lateral shoots. *Leaves* glabrous to shortly pubescent: petiole 0,5–2 cm long, a small gland usually present ± midway along the petiole or just below the junction of the lowest pinna pair; rhachis 0–3,5 cm long, usually terminating in a short rigid persistent deflexed hook, a small gland often present near the junction of the top pinna pair; pinnae 1–3(4) pairs; rhachillae 1,5–5 cm long, usually terminating in a short rigid persistent deflexed hook, often a single stipel similarly bent near the base; leaflets (1)2–4(6) pairs, (5)8–25(30) × 4–18(24) mm (in our area), obliquely obovate or elliptic to subrotund, mucronate apically, venose, glabrous or sparingly pubescent beneath, especially on the midrib and lateral nerves. *Inflorescences* globose; peduncles 1–3 cm long, glabrous or sparingly pubescent. *Flowers* usually borne on leafless branchlets, whitish, on pedicels 0,5–2,5 mm long; bracteoles rapidly deciduous. *Calyx* greenish, 3–5 mm long, usually deeply slit unilaterally, glabrous throughout or with a tuft of hairs at the apex of each lobe, sometimes sparingly pubescent throughout. *Corolla* greenish, 6–12 mm long, glabrous throughout or puberulous near the apices and margins of the lobes, lobes up to 5 mm long. *Stamens* 1,5–2,8 cm long, united basally for up to 4 mm, tube not exserted beyond the corolla. *Ovary* ± 3 mm long, glabrous, shortly stipitate. *Pods* straw-coloured to light brown when mature, (6)8–18 × 1,5–2,8 cm, oblong, valves thin, glabrous or occasionally puberulous all over, venose, umbonate over the seeds, longitudinally dehiscent. *Seeds* 9–13 mm in diam., rounded, flattened.

Found from the Sudan and Ethiopia southwards to Angola, South West Africa, Botswana, Rhodesia, Mozambique, the Transvaal, Swaziland and Zululand. Occurs in dry bushveld, scrub, woodland and sand forest.

S.W.A.—1715 (Ondangua): near Oshikango, *Rodin 2673.* 1719 (Runtu): 6,4 km E. of Runtu, *De Winter 3782.* 1820 (Tarikora): 56–64 km W. of Andara, *Merxmuller & Giess 2087* (M). 1917 (Tsumeb): Tsumeb, *Basson 15.* 2017 (Waterberg): Okosongomingo, *Volk 232* (M). 2115 (Karibib): road from Karibib to Omaruru, *Kinges 3602* (M). 2116 (Okahandja): Okahandja, *Dinter 269.* 2215 (Trekkopje): farm Nudis, *Seydel s.n.* (M). 2216 (Otjimbingwe): farm Homusas, 128 km W. of Windhoek, *De Winter 2627.* 2217 (Windhoek): 4,8 km N. of Windhoek, *Codd 5793.* 2416 (Maltahöhe): Gamis farmhouse, Great Fish River, *Pearson 8976* (K). 2516 (Helmeringhausen): farm Kleinfontein, *Marloth 5052.* 2618 (Keetmanshoop): Gellap Ost, 19,2 km N.W. of Keetmanshoop, *Acocks 15608.*

TRANSVAAL.—2229 (Waterpoort): Dongola area, farm Little Muck 604, *Codd 4331.* 2230 (Messina): near Messina, *Rogers 19347.* 2231 (Pafuri):

Kruger National Park, Pafuri, *Van der Schijff 642*. 2328 (Baltimore): Leipzig Mission, Blouberg, *Leipolt 3*. 2427 (Thabazimbi): 48 km N.W. of Vaalwater, *Smuts 361*. 2430 (Pilgrim's Rest): 6,4 km N. of P.O. Buffelsvlei, *Codd 6675*. 2431 (Acornhoek): Kruger National Park, 6,4 km E. of Skukuza, *Codd 5701*. 2531 (Komatipoort): near Louws Creek, *Acocks 12879*.

SWAZILAND.—2631 (Mbabane): Sipofaneni, *Compton 29112*; Ranches, *Compton 27013*.

NATAL.—2632 (Bela Vista): Ndumu Game Reserve, *Moll 4257*. 2732 (Ubombo): Mkuzi Game Reserve, *Ward 3519*.

In our area *A. anthelmintica* is usually ± pubescent, whereas in the northern part of its distributional range it is typically glabrous. These pubescent specimens in the southern part of the species range were separated as var. *pubescens* by Burtt Davy, but there are so many gradations connecting the glabrous and pubescent specimens that var. *pubescens* is not considered worthy of formal taxonomic recognition.

Codd 1.c. : 78 (1958) drew attention to the possibility that Sim based his *A. umbalusiana* on a mixture of *A. anthelmintica* and *A. petersiana* subsp. *evansii*. The flowering twig with spine-tipped branchlets illustrated by Sim in Plate 55A of his For. Fl. P.E. Afr. (1909) is definitely *A. anthelmintica*; in *A. petersiana* subsp. *evansii* spine-tipped branchlets are absent and the staminal tube is long and exserted far beyond the corolla. The pods and two leaves illustrated are more difficult to identify with certainty, but the possibility does exist that they may be those of *A. petersiana* subsp. *evansii*. However, in the absence of Sim's type specimen, it is unlikely that the leaves and pods illustrated will ever be identified with certainty. It is, of course, possible that they are indeed those of *A. anthelmintica* after all.

The bark and roots of *A. anthelmintica* are used as an anthelmintic, whence the specific epithet.

10. **Albizia petersiana** (*Bolle*) *Oliv.* in F.T.A. 2 : 362 (1871); Benth. in Trans. Linn. Soc. Lond. 30 : 569 (1875); Gilg in Pflanzen. Ost Afr. B : 299 (1895); Sim, For. Fl. P.E. Afr. 60 (1909); Bak.f., Leg. Trop. Afr. 3 : 867 (1930); Brenan, Checklist Tang. Terr. 340 (1949); in F.T.E.A. Legum.–Mimos. : 162 (1959); in F.Z. 3, 1 : 132, t.23 fig.E (1970). Type: Mozambique, Boror and Sena, 16–18°S. lat., *Peters* (B, holo.†; BM!).

Zygia petersiana Bolle in Peters, Reise Mossamb. Bot. 1 : 1, t.1 (1861). Type as above.

subsp. **evansii** (*Burtt Davy*) *Brenan* in Kew Bull. 21 : 482 (1968); in F.Z. 3, 1 : 133 (1970); Van Wyk, Trees Kruger Nat. Park 1 : 114 (1972); Palmer & Pitman, Trees S. Afr. 2 : 727 (1973); Ross, Fl. Natal 192 (1973). Type: Transvaal, Nelspruit District, Sabie Game Reserve, *Pole Evans H16921* (K, holo.!).

Albizia evansii Burtt Davy, Fl. Transv. 2 : xvii, 348 (1932); Gomes e Sousa, Dendrol. Mocamb. 2 : 54 (1951); Dendrol. Mocamb. 1 : 238, t.42 (1966); Codd, Trees & Shrubs Kruger Nat. Park 54, fig.47, 48b (1951); in Bothalia 7 : 79 (1958); Von Breitenbach, Indig. Trees S.Afr. 2 : 262 (1965). Type as above.

Shrub or tree up to 10 m high, usually branching freely near the base with many ascending branches, sometimes single-stemmed, crown rounded or somewhat flattened. *Bark* grey to brown, usually smooth; young branchlets densely pubescent, not forming abbreviated spine-tipped shoots. *Leaves* ± densely pubescent: petiole 0,5–2,4 cm long, a small gland usually present ± halfway along the petiole; rhachis (0)0,8–3 cm long, a small gland often present at the junction of the top or top two pinnae pairs; pinnae (1)2–3(4) pairs; rhachillae 0,5–3 cm long, a small gland often present just below the junction of the top pair of leaflets; leaflets 2–5 pairs, (4,5)8–22 × (3,5)5–13 mm, obliquely obovate or obovate- to oblong-rhombic, obtuse to mucronate apically, dark green above, glabrous or sparsely pubescent, lower surface paler, sparingly to ± densely pubescent, especially on midrib and lateral nerves. *Stipules* rapidly deciduous, 1,75–3,5 × 0,6–1 mm, oblanceolate or triangular-acute. *Inflorescences* globose; peduncles 0,8–1,5(2) cm long, ± densely pubescent. *Flowers* whitish, usually tinged with red, on pedicels up to 2 mm long; bracteoles ± 1 mm long, rapidly deciduous. *Calyx* 1–1,75 mm long, very shallowly lobed apically, glabrous except for apices of lobes or puberulous throughout. *Corolla* 6–9 mm long, glabrous throughout or sparingly pubescent near the apices of the lobes. *Stamens* 1,6–2,5 cm long, united into a narrow tube for most of their length, tube reddish, exserted beyond the corolla for 1–1,5 cm. *Ovary* ± 2,5 mm long, glabrous, shortly stipitate. *Pods* brown, 6–16 × 1,4–2 cm, oblong, valves semi-woody, shortly pubescent or puberulous, obscurely venose, longitudinally dehiscent. *Seeds* 9–12 mm long or in diam., flattened.

Found in Rhodesia, southern Mozambique, the Transvaal and Zululand. Occurs in woodland and scrub.

TRANSVAAL.—2231 (Pafuri): Kruger National Park, 3,2 km S. of Punda Milia, *Codd 5990*. 2431 (Acornhoek): Kruger National Park, 32 km N.E. of Skukuza, *Codd 5592*. 2531 (Komatipoort): Kruger National Park, 9 km S.W. of Lower Sabie Camp, *Codd 5708*.

NATAL.—2632 (Bela Vista): Ndumu Game Reserve, *Gerstner 3440*; *Ward 3179*. 2732 (Ubombo): 24 km from Ingwavuma on Ndumu road, *Moll 3708*; 1,6 km E. of Makanes Drift, *Ross 1966*.

Subsp. *petersiana* is found in Uganda, Kenya, Tanzania, Malawi and northern Mozambique. It differs from subsp. *evansii* in having only shortly pubescent or glabrous young branchlets, typically more numerous leaflet pairs, sparingly pubescent or glabrous peduncles and ± glabrous pods. Although subsp. *petersiana* is not clearly separable from subsp. *evansii* by any single character except perhaps the ± glabrous pods, the above characters in combination usually enable subsp. *petersiana* to be distinguished.

Attention has already been drawn (see note under *A. anthelmintica*) to the possibility that the description and illustration of *A. umbalusiana* Sim are drawn from *A. anthelmintica* and from *A. petersiana* subsp. *evansii*.

The leaves of *A. anthelmintica* bear a superficial resemblance to those of *A. petersiana* subsp. *evansii*. However, the rhachides and rhachillae of the latter are not terminated by a short rigid persistent deflexed hook, the base of the rhachillae lack a stipel, and the leaflets are sparingly to ± densely pubescent beneath, while those of *A. anthelmintica* are glabrous or with few scattered hairs on the midrib and nerves beneath except in some specimens from South West Africa where the leaflets are also ± pubescent.

11. Albizia adianthifolia (*Schumach.*) *W. F. Wight* in U.S. Dept. Agric. Bur. Pl. Industry Bull. 137 : 12 (1909); Gilbert & Boutique in F.C.B. 3 : 178 (1952); Brenan in Kew Bull. 7 : 520 (1953); in Mem. N.Y. Bot. Gdn. 8 : 430 (1954); Torre in C.F.A. 2 : 295 (1956); Codd in Bothalia 7 : 79 (1958); Brenan in F.T.E.A. Legum.–Mimos. : 160, fig.21/6–9, 22/2 (1959); F. White, For. Fl. N. Rhod. 88, fig. 16E-F (1962); Von Breitenbach, Indig. Trees S.Afr. 2 : 264 (1965); Gomes e Sousa, Dendrol. Mocamb. 1 : 239, t.43 (1966); Compton in J. S. Afr. Bot., Suppl. 6 : 45 (1966); Brenan in F.Z. 3, 1 : 131, t.24 fig.B (1970); Van Wyk, Trees Kruger Nat. Park 1 : 106 (1972); Palmer & Pitman, Trees S.Afr. 2 : 729 (1973); Ross, Fl. Natal 192 (1973). Type: probably from Ghana, Bligusso *Thonning* (C, holo., K, photo.!).

Mimosa adianthifolia Schumach., Beskr. Guin. Pl. 322 (1827). Type as above.

Zygia fastigiata E. Mey., Comm. 1 : 165 (1836); Benth. in Hook., Lond. J. Bot. 3 : 93 (1844); Harv. in F.C. 2 : 285 (1862). Syntypes: Natal, between Umzimkulu [Omsamculo] and Umkomaas, *Drege* (P!); Durban [Port Natal], *Drege* (K!; OXF!; P!).

Inga fastigiata (E. Mey.) Steud., Nom. Bot. ed.2 : 809 (1840). Types as for *Zygia fastigiata*.

Albizia fastigiata (E. Mey.) Oliv. in F.T.A. 2 : 361 (1871); Benth. in Trans. Linn. Soc. Lond. 30 : 570 (1875), pro parte excl. syn. *Inga sassa* et *Mimosa sassa*; Hiern, Cat. Afr. Pl. Welw. 1 : 317 (1896);

Wood & Evans, Natal Plants 1 : 24, t.27 (1898); Sim, For. Fl. Cape Col. 213, t.62 (1907); Sim, For. Fl. P.E. Afr. 59, t.58 (1909); Marloth, Fl. S.Afr. 2, 1 : fig.30 (1925). Types as for *Zygia fastigiata*. *A. gummifera* sensu C.A. Sm. in Kew Bull. 1930 : 218 (1930), pro parte quoad syn. *Mimosa adianthifolia* [*"adiantifolia"*], *Zygia fastigiata*, *Albizia fastigiata*; Burtt Davy, Fl. Transv. 2 : 349 (1932); Henkel, Woody Pl. Natal 236 (1934), non (J.F. Gmel.) C.A. Sm. sensu stricto.

Tree to 20 m high, crown typically flattened and spreading. *Bark* grey to yellowish-brown, smooth or rough; young branchlets densely and rather coarsely and persistently rusty- to fulvous-pubescent, indumentum sometimes becoming grey with age. *Leaves* rusty- to fulvous-pubescent, indumentum sometimes becoming grey with age: petiole 2,4–7 cm long, a large raised gland usually situated immediately above or a short distance above the pulvinus; rhachis (3)5,5–14 cm long, a small rounded gland usually present at or just below the junction of the top pinna pair; pinnae 4–8 pairs; rhachillae (2)3–10 cm long; leaflets (4)6–15 pairs, 7–20 × 4–8(11) mm (in our area), obliquely rhombic-quadrate or -oblong, proximal margin usually ± rounded into the pulvinule basally and not auriculate, midrib diagonal, usually obtuse and mucronate apically, sometimes subacute, upper surface dark green and thinly pubescent, lower surface paler, usually ± appressed-pubescent throughout but especially on midrib and margins. *Stipules* and bracts at base of peduncles 5–12 × 3–8(11) mm, ovate. *Inflorescences* globose; peduncles 2,5–4,5(6) cm long, densely rusty- to fulvous-pubescent. *Flowers* whitish, on pedicels 0,5–1 mm long; bracteoles 5–8 mm long, linear-spathulate to oblanceolate, exceeding the flower-buds, variably persistent and sometimes deciduous before the flowers open. *Calyx* 2,5–5 mm long, fulvo-pubescent. *Corolla* 6–10 mm long, pubescent, tube 5–8 mm long, lobes 2–3 mm long. *Stamens* 2–3,2 cm long, united into a narrow tube for most of their length, tube exserted beyond the corolla for 1,3–2,5 cm. *Ovary* ± 2,5 mm long, glabrous, subsessile. *Pods* pale brown, (7,5)9–19 × 1,9–3,4 cm (in our area), oblong, valves thin textured, densely and persistently pubescent, not glossy, umbonate over the seeds, p ominently venose, margins thickened, longitudinally dehiscent. *Seeds* 7–9,5 × 6,5–8,5 mm, flattened.

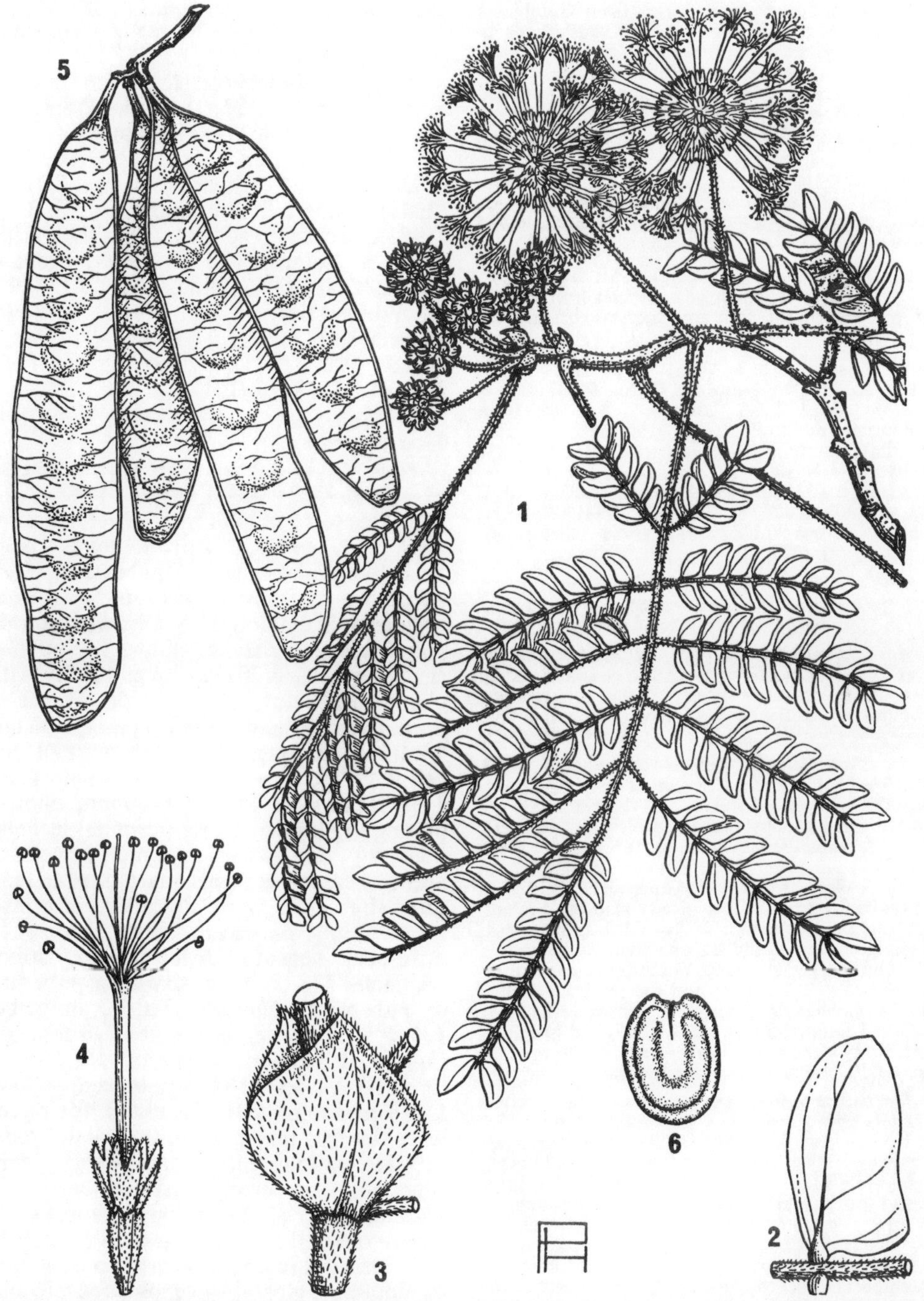

FIG. 2.—**Albizia adianthifolia.** 1, flowering branch, × ⅔, *Ross* 1369; 2, leaflet, undersurface, × 2; 3, stipules × 2, both from *Ross* 420; 4, flower, × 2, *Moll* 1925; 5, pods, × ⅔; 6, seed, × 2, both from *Ross* 2119,

Widespread in tropical Africa from Gambia in the west to Kenya in the east and southwards to Angola, Rhodesia, Mozambique, the Transvaal, Swaziland, Natal and eastern Cape Province. Occurs in forest, wooded ravines, woodland and wooded grassland; often, but by no means always, near water.

TRANSVAAL.—2230 (Messina): Kruger National Park, Shipudza, N.W. of Punda Milia, *Van der Schijff 3784*. 2330 (Tzaneen): Tshakoma, *Obermeyer 975*. Grid ref. unknown: Soutpansberg District, Shewass, *Legat sub PRE 4867*; Pepiti Falls, *Curson & Irvine 97*.

SWAZILAND.—Compton, in J. S.Afr. Bot., Suppl. 6 : 45 (1966), recorded *A. adianthifolia* from Swaziland, but I have not seen any specimens from this territory.

NATAL.—2632 (Bela Vista): 10,4 km N.E. of Maputa on road to Ponta do Ouro, *De Winter & Vahrmeijer 8619*. 2732 (Ubombo): 6,4 km from Hluhluwe on road to False Bay Park, *Ross 2119*. 2831 (Nkandla): forest adjoining Eshowe, *Codd 1864*. 2832 (Mtubatuba): Nyalazi, 27,2 km S. of Hluhluwe on Mtubatuba road, *Ross 1369*. 2930 (Pietermaritzburg): Shongweni, *Ross 420* (K, NH, NU). 2931 (Stanger): Umhlali, *Repton 1838*. 3030 (Port Shepstone): Uvongo, *Strey 6842*.

CAPE.—3129 (Port St. Johns): 4 km N. of Embotyi, *Codd 9740*; Egossa forest, *Sim 2365*; Port St. Johns, *Bull sub Sim 19961*.

A. adianthifolia, with its flattened spreading crown, is a conspicuous tree in the coastal areas of Natal. This characteristically flattened spreading crown has earned the species the common name "flat crown".

A. adianthifolia is very closely related to *A. gummifera* (J. F. Gmel.) C.A. Sm. The specific limits within this complex of species and the synonymy are dealt with in detail by Brenan in Kew Bull. 7 : 507 (1953). *A. gummifera*, which so far has not been recorded in our area, differs from *A. adianthifolia* in that the leaflets are markedly auriculate basally on the proximal side and are typically glabrous beneath apart from the pubescence on the midrib and margins; the young branchlets and leaf-rhachides are finely and shortly brownish-pubescent; the stipules are lanceolate, ± 6–7 × 2–2,5 mm; and the pods are glabrescent. *A. gummifera* is considered in more detail in F.T.E.A. Legum.–Mimos. : 157 (1959) and in F.Z. 3, 1 : 129 (1970). In territories outside our area *A. adianthifolia* and *A. gummifera* hybridize.

The nature of the bark was one of the differential characters between *A. adianthifolia* and *A. gummifera* noted by Brenan l.c. : 510 (1953). In *A. gummifera* the bark is said to be usually smooth and only rarely rough, while in *A. adianthifolia* the bark in tropical east and south tropical Africa is said to be rough and only rarely smooth. Detailed field observations on bark characteristics in our area are necessary to establish whether our plants are predominantly smooth-barked or rough-barked, and whether this character is of any ecological or distributional significance.

Sim, For. Fl. Cape Col. 214 (1907), reported that the wood of *A. adianthifolia* is "straight grained, lighter than yellow-wood, easily worked, soft, with yellowish heart and whiter sapwood." The wood was formerly in great demand for use in wagon-building and other purposes.

12. **Albizia lebbeck** (*L.*) *Benth.* in Hook., Lond. J. Bot. 3 : 87 (1844); in Trans. Linn. Soc. Lond. 30 : 562 (1875); Sim, For. Fl. P.E. Afr. 60 (1909); Bak.f., Leg. Trop. Afr. 3 : 862 (1930); Brenan, Checklist Tang. Terr. 342 (1949); Gilbert & Boutique in F.C.B. 3 : 187 (1952); Torre in C.F.A. 2 : 292 (1956); Codd in Bothalia 7 : 81 (1958); Brenan in F.T.E.A. Legum.–Mimos. : 147 (1959); F. White, For. Fl. N. Rhod. 90 (1962); Brenan in F.Z. 3, 1 : 118 (1970); Ross, Fl. Natal 192 (1973). Type: Egypt, *Herb. Linnaeus* 1228.16 (LINN!).

Mimosa lebbeck L., Sp. Pl. ed.1 : 516 (1753). Type as above.

Acacia lebbeck (L.) Willd., Sp. Pl. 4 : 1066 (1806); DC., Prodr. 2 : 466 (1825). Type as above.

Tree to 15 m high with a rounded crown. *Bark* grey to light brown, rough; young branchlets glabrous or pubescent. *Leaves* subglabrous, puberulous or pubescent: petiole 4–8 cm long, usually with a raised gland a short distance above the pulvinus; rhachis (0)1,3–8(20) cm long, usually with a gland just below the junction of the top pinna pair; pinnae (1)2–4(5) pairs; rhachillae 4–11(16) cm long, often with a small gland below the junction of each leaflet pair; leaflets 3–9(11) pairs, 15–45 × (6)8–24 mm, obliquely oblong or elliptic-oblong (terminal leaflets ± obovate), somewhat asymmetric with the midrib nearer the upper margin, rounded or somewhat emarginate apically, glabrous or rarely thinly pubescent above, glabrous to pubescent beneath. *Inflorescences* globose; peduncles 4–8 cm long, sparingly puberulous to pubescent. *Flowers* whitish, on pedicels 1,5–4,5 mm long; bracteoles 2–3 mm long, rapidly deciduous. *Calyx* (2,5)3,5–5 mm long, grey- to fulvous-pubescent. *Corolla* 5,5–9 mm long, glabrous except for puberulence towards the apices of the lobes. *Stamens* 1,5–2,5 cm long, united basally for ± 5 mm, tube not or scarcely exserted beyond the corolla. *Ovary* ± 2 mm long, glabrous. *Pods* straw-coloured to light brown, (9)12–25(33) × 3–4,5 cm, oblong, valves coriaceous, glabrous or almost so, glossy, ± venose, conspicuously umbonate over the seeds, margins thickened, very tardily dehiscent and often only after falling to the ground. *Seeds* 7–11,5 × 7–9 mm, flattened.

Pantropical, probably a native of tropical Asia and nowhere indigenous in Africa. Introduced into our area and now naturalized along parts of the north coast of Natal.

NATAL.—2831 (Nkandla): Mtunzini, *Ward 2996*; *Lawn 2118*. 2930 (Pietermaritzburg): Oakford Clinic, *Moll 3285*. 2931 (Stanger): 3,2 km S. of Verulam, *Codd 9653*; Gingindhlovu, *Lawn 1863* (NH).

The stiff straw-coloured or light brown pods with prominent bumps over the seeds are very characteristic. When agitated by wind the mature pods, and the seeds inside, are said to produce an incessant rattle that, according to Brenan in F.Z. l.c. : 118, has been likened to women's chatter and the sound of fish being fried.

The specific epithet is said to be derived from the Arabian name "Labach" for the tree. The epithet *"lebbeck"* has often been misspelt *"lebbek"*.

A. lebbeck has in the past been confused in the herbarium with *A. tanganyicensis*. The two species have very different distributional ranges in our area, and the morphological differences are discussed under *A. tanganyicensis*. *A. lebbeck* bears a slight resemblance to *A. suluensis*, but the latter may be readily distinguished by the crisped margins of the leaflets and by the golden indumentum on the calyx and corolla.

13. **Albizia procera** (*Roxb.*) *Benth.* in Hook., Lond. J. Bot. 3 : 89 (1844); in Fl. Austral. 2 : 422 (1864); in Trans. Linn. Soc. Lond. 30 : 564 (1875); Bak. in Hook.f., Fl. Brit. Ind. 2 : 299 (1878); Brenan, Checklist Tang. Terr. 342 (1949); Ross, Fl. Natal 192 (1973). Type: India, *Roxburgh* (K, painting of holotype material, No. 485!).

Mimosa procera Roxb., Pl. Corom. 2 : 12, t.121 (1799); Fl. Ind. 2 : 548 (1832). Type as above.

Acacia procera (Roxb.) Willd., Sp. Pl. 4 : 1063 (1806); DC., Prodr. 2 : 466 (1825); Wight & Arn., Prodr. Fl. Ind. 1 : 275 (1834). Type as above.

Tree to 12 m high; bark smooth, yellowish to grey; young branchlets glabrous or subglabrous. *Leaves* glabrous or subglabrous to sparingly puberulous: petiole 5–8,5 cm long, with a large slightly raised elongated gland up to 11 mm long situated a short distance above the pulvinus; rhachis 3,5–13 cm long; pinnae 2–4 pairs; rhachillae 7–19(24) cm long; leaflets 5–11 pairs, 15–60 × 8–21 mm, obliquely oblong, elliptic-oblong or obovate, apex obtuse or rounded, often somewhat emarginate, upper surface very sparingly appressed-pubescent, lower surface paler, appressed-pubescent. *Inflorescences* globose, in axillary or terminal panicles; peduncles glabrous or sparingly puberulous. *Flowers* whitish, sessile or almost so; bracteo-

les rapidly deciduous. *Calyx* 2–3 mm long, glabrous or almost so. *Corolla* 4–6 mm long, apices of lobes with a conspicuous tuft of hairs. *Stamens* up to 1,5 cm long, united basally, tube not or scarcely exserted beyond the corolla. *Ovary* glabrous. *Pods* light to dark brown, 8–16 × 1,5–2 cm, linear-oblong, acuminate apically, glabrous or almost so, obscurely venose, slightly umbonate over the seeds, longitudinally dehiscent. *Seeds* 6–8 × 5–6,5 mm, flattened.

A native of India. Introduced into our area and now naturalized in parts of Natal.

NATAL.—2931 (Stanger): 15,2 km S.E. of Mapumulo, *Codd 9654*; Thring's Post, *Ward 3186*.

A. procera bears a superficial resemblance to *A. lebbeck*, but differs in having the midrib of the leaflets ± centric, a panicled inflorescence, ± sessile flowers, and much smaller pods.

14. **Albizia lophantha** (*Willd.*) *Benth.* in Hook., Lond. J. Bot. 3 : 86 (1844); in Fl. Austral. 2 : 421 (1864); in Trans. Linn. Soc. Lond. 30 : 559 (1875); Codd in Bothalia 7 : 81 (1958); Ross, Fl. Natal 192 (1973). Type: It is not known if the specimen on which Ventenat, Desc. Pl. Jard. Cels : 20, t.20, based his description and illustration is preserved; if not, t.20 will suffice as the type.

Acacia lophantha Willd., Sp. Pl. 4 : 1070 (1806); DC., Prodr. 2 : 457 (1825). Type as above.

Mimosa distachya Vent., Desc. Pl. Jard. Cels : 20, t.20 (1800 or 1801), non *M. distachya* Cav., Icon. 3(2) : 48, t.295 (1795 or 1796). Type as above.

Albizia distachya (Vent.) MacBride in Contr. Gray Herb. 59 : 3 (1919); Salter in Adamson & Salter, Fl. Cape Penins. 452 (1950). Type as above.

Tree to 7 m high; young branchlets usually ± golden-pubescent, becoming sparingly pubescent to glabrescent with age. *Leaves* ± golden-pubescent when young but often becoming grey-pubescent with age: petiole 4–7,5 cm long, a conspicuous elongated gland usually situated ± midway along the length of the petiole; rhachis 10–20 cm long, usually with a gland just below the junction of the top pinna pair; pinnae 7–12 pairs; rhachillae 5–14 cm long; leaflets 20–35 pairs, 7–13 × 1,75–3 mm, obliquely-oblong or sometimes slightly falcate, the midrib nearer the upper margin, apex obtuse to rounded, mucronate, often turned towards the apex of the pinna, glabrous or sparingly pubescent, especially on the midrib. *Inflorescences* spicate; spikes axillary, solitary or fascicled, 4–8 cm long; peduncles 0,8–1,5 cm long, fulvo-pubescent. *Flowers* creamy to yellowish-white, on pedicels 1–3 mm long. *Calyx* 2–2,5 mm long, pubescent, conspicuously toothed. *Corolla* 5–7 mm long, appressed-pubescent. *Stamens* 1,2–1,6 cm long, shortly united basally, tube not or scarcely exserted beyond the corolla. *Ovary* ± 2,5 mm long, glabrous, shortly stipitate. *Pods* light to dark brown, 5,5–11 × 1,4–1,7 cm, linear-oblong, glabrous or almost so, obscurely venose, umbonate over the seeds, margins thickened. *Seeds* ± 7 × 5 mm, flattened.

A native of the south-western coastal region of Western Australia. Introduced into our area and now naturalized, particularly in the Cape along the coast from Humansdorp to the Cape Peninsula.

NATAL.—Grid ref. unknown: south coast, *Wood 10588* (K, NH).

CAPE.—3318 (Cape Town): Rondebosch, *Gerstner 6135*. 3418 (Simonstown): Simonstown, *Watt & Brandwijk 1682*; Kogel Bay, *Parker 4206* (BOL, K). 3421 (Riversdale): Corente River farm, *Muir sub Galpin 5092*. 3423 (Knysna): Knysna, *District Forest Officer sub PRE 8707*. 3424 (Humansdorp): Humansdorp, *Rogers 3021*.

A. lophantha favours river banks, forest margins and wooded ravines. It is the only species of *Albizia* in our area with a spicate inflorescence.

3446

2. ACACIA

Acacia *Mill.*, Gard. Dict. abridg. ed.4 (1754); Neck., Elem. Bot. 1297 : 458 (1790); Willd., Sp. Pl. 4 : 1049 (1806); Willd., Enum. 1049 (1809); DC., Prodr. 2 : 448 (1825); G. Don, Gen. Syst. 2 : 401 (1832); Harv., Gen. Pl. ed.1 : 90 (1838); Benth. in Hook., Lond. J. Bot. 1 : 318 (1842); Harv. in F.C. 2 : 279 (1862); Benth. & Hook.f., Gen. Pl. 1 : 594 (1865); Harv., Gen. Pl. ed.2 : 92 (1868); Oliv. in F.T.A. 2 : 337 (1871); Benth. in Trans. Linn. Soc. Lond. 30 : 444 (1875); Taub. in Pflanzenfam. 3, 3 : 108 (1891); Sim, For. Fl. Cape Col. 210 (1907); Sim, For. Fl. P.E. Afr. 54 (1909); Harms in Engl., Pflanzenw. Afr. 3, 1 : 344 (1915); Marloth, Fl. S. Afr. 2 : 51 (1925); Bak.f., Leg. Trop. Afr. 3 : 815 (1930); Burtt Davy, Fl. Transv. 2 : 333 (1932); Phill., Gen. 391 (1951); Gilbert & Boutique in F.C.B. 3 : 145 (1952); Torre in C.F.A. 2 : 269 (1956); Keay in F.W.T.A. ed.2, 1 : 496 (1958); Brenan in F.T.E.A. Legum.–Mimos. : 49 (1959); Dale & Greenway, Kenya Trees & Shrubs 279 (1961); F. White, For. Fl. N. Rhod. 76 (1962); Hutch., Gen. Fl. Pl. 1 : 280 (1964); Schreiber in F.S.W.A. 58 : 2 (1967); Brenan in F.Z. 3, 1 : 53 (1970); Ross, Acacia Spp. Natal 5 (1971). Type species: *A. nilotica* (L.) Willd. ex Del.

Mimosa L., Sp. Pl. 1 : 516 (1753) pro parte; Thunb., Fl. Cap. ed. Schult. : 432 (1823).

Phyllodoce Link, Handb. ii : 132 (1831), non Salisb. (1806).

Vachellia Wight & Arn., Prodr. Fl. Ind. 272 (1834); Small in Bull. N.York Bot. Gard. 2 : 94 (1901); Britton & Rose in N. Am. Fl. 23 : 87 (1928).

Farnesia Gasparr., Desc. Nuov. Gen. (1838), non Heist. ex Fabr. (1763).

Faidherbia A. Chev. in Rev. Bot. Appl. Agric. Trop. 14 : 876 (1934); Aubrév., Fl. Forest. Soudano-Guin. 280, t.51/3, t.53/6 (1950); Gilbert & Boutique in F.C.B. 3 : 169 (1952).

Trees or shrubs, sometimes scandent or climbing; the indigenous species in our area almost always armed with stipular spines or recurved prickles, the introduced species usually unarmed. *Leaves* bipinnate or (in some introduced species) often modified to phyllodes (flattened leaf-like organs without pinnae or leaflets); each pinna with one to many pairs of leaflets; gland on the upper side of the petiole usually present; glands also often present at point of attachment of pinnae (at least some) to the rhachis. *Inflorescences* usually axillary, racemose or paniculate; flowers in spikes or spiciform racemes (spicate) or in round heads (capitate), rarely (in *A. redacta*) only 2–4 flowers per head, flowers all hermaphrodite or male and hermaphrodite. *Calyx* (in our species) gamosepalous, usually with 4–5 teeth or lobes or subtruncate. *Corolla* 4–5(7)-lobed. *Stamens* many, fertile, their filaments free or (in *A. albida* and in *A. redacta*) shortly united into a tube at the base only; anthers (at least some) glandular apically at least in bud, or all eglandular (in all indigenous species glandular except in *A. albida*, introduced species mostly eglandular). *Ovary* stipitate to sessile, glabrous to puberulous. *Pods* very variable, usually dehiscent but sometimes indehiscent, flattened, ±compressed, or sometimes almost cylindric, straight, curved, spiral or contorted, margins entire or moniliform, glabrous to pubescent. *Seeds* unwinged, exendospermous, often with a hard smooth testa.

A genus of 850–900 species, mostly tropical or subtropical; ± 620 in Australia, ± 115 in Africa, many in America and fewer in Asia. 44 indigenous species occur in our area and several Australian species have become naturalized.

The generic name *Acacia* is derived from the Greek word *akis*, meaning a sharp point.

In a few of our species the stipular spines are distinctly and characteristically swollen into structures commonly known as ant-galls. There is evidence that these structures may not be galls at all, but natural outgrowths of the plant itself. The presence of these structures is certainly taxonomically important. Following Brenan's decision in F.T.E.A. Legum.–Mimos. : 49 (1959), I have retained the familiar term "ant-gall" but have enclosed it in inverted commas. A new and more appropriate term may have to be devised.

The indigenous species of *Acacia* occurring in our area can be divided into groups and these groups form the basis of the sequence in which the species are arranged here. The primary division of the species is on the nature of the inflorescence, that is, whether the flowers are in spikes or in round heads. For use in a regional flora this is a convenient character to employ for the primary division of the species into two main artificial groups, but, as discussed in Bothalia 11 : 107–113 (1973), when a more natural systematic division is sought it is the nature of the stipules, that is, whether they are spinescent or non-spinescent, that must be employed for separating the two main groups of species.

As *Acacia* is such an important genus in our area, and as many of the species seldom bear flowers and fully-formed pods at the same time, three alternative keys to the identification of the indigenous species are provided. The first key is a comprehensive one for use when flowering and fruiting (young fruits are often adequate) material is available. The second key is for use when only flowering material is available, and the third is for use when only fruiting material is present.

A few of the indigenous species often produce flowers when the plants are quite leafless. As difficulty may be experienced in keying out leafless flowering specimens, it may help to know which species most regularly flower when leafless. If the flowers are in elongate spikes (sometimes in short ellipsoid heads but even then the axis is clearly elongate), and the plants are armed with paired recurved prickles, the specimen is probably *A. mellifera*, *A. nigrescens*, *A. galpinii* or *A. erubescens*. If the flowers are in round heads and the plants are armed with paired straight stipular spines or a mixture of long and straight and short and recurved spines, then the specimen is probably *A. tortilis*, *A. reficiens*, *A. hebeclada* or *A. stuhlmannii*.

Leaves bipinnate; plants armed with stipular spines or with recurved prickles........Indigenous species (below)

Leaves modified to phyllodes (entire, leaf-like flattened organs) or bipinnate but then plants unarmed; stipular spines and prickles absent (except in *A. armata* but then leaves phyllodic).............................
....Naturalized and cultivated species (p. 39).

For convenience, *A. farnesiana* is keyed out under the indigenous species as it is more closely allied to some of the indigenous species than it is to any of the naturalized Australian species.

Keys to the indigenous species

Key to the groups

Flowers in spikes or spiciform racemes, rarely the inflorescences short and ellipsoid but even then the axis clearly elongate:

Stipules spinescent, straight or almost so...A (sp.1)

Stipules not spinescent; plants armed with prickles:

Prickles irregularly scattered along the internodes...B (sp.2)

Prickles at or near the nodes:

Prickles in threes or solitary...C (sp.3)

Prickles in pairs:

Flowers distinctly pedicellate...D (sp.4)

Flowers not distinctly pedicellate..E (spp. 5–15)

Flowers in round heads, very rarely inflorescence reduced, apparently capitate and with only 2–4 flowers per head:

Stipules not spinescent, plants armed with irregularly scattered recurved prickles..........F (spp. 16–18)

Stipules spinescent:
  Inflorescences with ± 35–200 flowers per head:
    Flowers bright or golden yellow:
      Pods dehiscent...G (spp. 19–26)
      Pods indehiscent:
        Valves of pod markedly thickened, woody or pulpy in texture..................H (spp. 27–30)
        Valves of pod thin...I (sp. 31)
    Flowers white or pale yellowish-white, rarely pink:
      Pods ±transversely jointed, each joint bearing a wart-like projection....................J (sp. 32)
      Pods not transversely jointed, without wart-like projections:
        Pods spirally coiled...K (sp. 33)
        Pods not spirally coiled:
          Pods dehiscent...L (spp. 34–40)
          Pods tardily dehiscent or indehiscent....................................M (spp. 41–43)
  Inflorescences greatly reduced, 2–4 flowers per head....................................N (sp. 44)

Comprehensive key

a Flowers in spikes or spiciform racemes; rarely (in *A. mellifera* subsp. *detinens*) the inflorescences short and ellipsoid but even then the axis clearly elongated:

Stipules spinescent, straight or only slightly curved; petiolar gland absent; stamen filaments united basally for ±1 mm; anthers eglandular even in bud; pods orange or chestnut- to reddish-brown, falcate or curled into a circular coil or variously twisted.............................1. *A. albida*

Stipules not spinescent, plants armed with recurved prickles; petiolar gland usually present; stamen filaments free to the base; anthers glandular, at least in bud; pods not as above:

Prickles scattered irregularly along the internodes; stipules obliquely ovate, up to 12 × 7 mm; petiolar gland stalked; ovary densely pilose, on a long stipe that elevates it above the corolla; pods reddish-brown to purplish, usually distinctly acuminate at both ends, brittle; areole of seed a small central depression..2. *A. ataxacantha*

Prickles confined to near the nodes, very rarely *and in addition* a few casually and irregularly scattered along the internodes; stipules±linear, smaller than above; petiolar gland not stalked; ovary and pods not as above; areole horse-shoe shaped:

Prickles in threes, the two laterals curved upwards and the central one down, or else the prickles solitary, the two laterals being absent......................................3. *A. senegal*

Prickles in pairs near the nodes:

  b Well developed leaves of mature shoots with 1–7 pinnae pairs:

  Flowers shortly but distinctly pedicellate (pedicels (0,5) 0,75–1,5 mm long); inflorescence ellipsoid to distinctly spicate; calyx cupular, 0,6–1 mm long, yellow, lobes often with a purplish tinge; leaflets 1–2(3) pairs per pinna; prickles of each pair often lying almost parallel to each other...4. *A. mellifera*

  Flowers sessile or subsessile (pedicel 0–0,3 mm long); inflorescence distinctly spicate; calyx 1,5–3,5 mm long (except in *A. galpinii* where it is shorter but red or purplish); leaflets 3 or more pairs per pinna (except in *A. nigrescens*); prickles spreading:

    c Calyx glabrous, rarely with occasional scattered hairs:

    Leaflets 1–2 (rarely 3) pairs per pinna, obovate or obliquely-obovate to broadly-elliptic or orbicular, (7)12–50 mm wide; leaves with 2–3(4) pinnae pairs; trunk and larger branches usually with scattered irregularly shaped knobs...............8. *A. nigrescens*

    Leaflets 3–35(45) pairs per pinna, 1–5 mm wide; leaves with 1–7 pinnae pairs:

      Calyx cupular, 0,75–1,25 mm long, red or purplish; leaflets 12–35(45) pairs per pinna; pods 2,3–3,5 cm wide..10. *A. galpinii*

      Calyx 1,3–3,5 mm long, not red or purplish; leaflets 3–13 pairs per pinna; pods 1–2,4 cm wide:

        Leaflets 3–8 pairs per pinna, elliptic, broadly elliptic or somewhat ovate, usually broadest at or below the middle, 2,5–5 mm wide; young branchlets dark grey to brownish-black; prickles strongly recurved; occurs in the eastern Transvaal
        ....7. *A. welwitschii*

Leaflets (4)6–13 pairs per pinna, linear to linear-oblong or obovate, sometimes slightly falcate, 0,9–3,5(5) mm wide; young branchlets not as above; prickles often spreading laterally and almost at right angles to the stem; confined to S.W.A.:

Leaves with 1–2 pinnae pairs; pods 3,5–6,8 cm long, with a fine closely reticulate venation; shrub with several erect slender branches which tend to droop apically or a slender tree with whip-like branches; young branchlets white to reddish-brown or purple, often as though whitewashed over a purplish background . . . . .5. *A. robynsianai*

Leaves with (2)3–6 pinnae pairs; pods 7,6–18 cm long, inconspicuously venose or venation coarse; slender tree branching from near the base, "broom-like"; young branchlets olive- to reddish-brown, bark often papery . . . .6. *A. montis-ust*

cc Calyx sparingly to densely puberulous:

Calyx cupular, 0,75–1,25 mm long, red or purplish; pods 2,3–3,5 cm wide . .10. *A. galpinii*

Calyx 1,3–4,5 mm long, greenish-yellow to yellowish-white (sometimes with a distinct pinkish tinge in *A. burkei*); pods 1–2,4 cm wide:

Leaflets (1)3–12 (rarely to 15) pairs per pinna, (1)2–13 mm wide, obovate, obovate-oblong or oblanceolate to orbicular, seldom linear-oblong; bracts in very young inflorescences projecting beyond the buds; calyx 1,3–2,5 mm long, often tinged with pink; stamen filaments 4–6 mm long; pods inconspicuously venose or venation coarse . . . . .9. *A. burkei*

Leaflets (10)12–30 pairs per pinna, 0,75–1,6(2,5) mm wide, linear to linear-oblong, straight or ±falcate; bracts not projecting beyond the buds; calyx 2–4,5 mm long, greenish-yellow; stamen filaments 6–10 mm long; pods with a fine close reticulate venation:

Petiole (0,8)1,2–2,5(4) cm long; petiolar gland 0,3–0,7(1) mm long, usually slightly raised or stalked; leaflets often±falcate, mostly 5–10 mm long; leaf-rhachis usually with a gland at the junction of some of the pinnae pairs or the top pinna pair only, occasionally eglandular; flowers sometimes produced before the leaves . . . . .15. *A. erubescens*

Petiole 0,3–1,3 cm long; petiolar gland 0,8–2,2 mm long, flattened, elliptic or oval; leaflets straight, mostly <5 mm long; leaf-rhachis eglandular; flowers produced after or with the leaves . . . . .14. *A. fleckii*

bb Well developed leaves of mature shoots with 8 or more pinnae pairs:

Calyx cupular, 0,75–1,25 mm long, red or purplish; pods 2,3–3,5 cm wide . . . . .10. *A. galpinii*

Calyx 1,3–4,5 mm long, greenish-yellow to yellowish-white (sometimes with a distinct pinkish tinge in *A. burkei*); pods 1–2,4 cm wide:

Petiolar gland large, 1,5–4 × 1,5–3 mm, usually slightly flattened, discoid or oblong; petiole 0,5–4 cm long; rhachis glandular between the top 3–16 pinnae pairs . . . . . .11. *A. polyacantha*

Petiolar gland small to medium, 0,3–1,5 × 0,1–0,7 mm, rarely up to 2,2 mm long but then rhachis eglandular and petiole 0,3–1,3 cm long:

Leaf-rhachis eglandular; petiolar gland 0,8–2,2 mm long; petiole 0,3–1,3 cm long . . . . .14. *A. fleckii*

Leaf-rhachis glandular, with a gland at the junction of the top pinna pair only or at the junction of all or some of the pinna pairs; petiolar gland 0,3–1,5 mm long; petiole 0,4–4 cm long:

Leaflets 4–15 (rarely to 19) pairs per pinna, linear-oblong or obovate to obovate-oblong, 0,8–7 mm wide; pinnae 8–13 pairs per leaf; pods glabrous except sometimes near margins and stipe, eglandular . . . . .9. *A. burkei*

Leaflets 16–64 pairs per pinna, linear to linear-oblong, 0,5–2,3(3,8) mm wide; pinnae 8–38 pairs per leaf; pods usually puberulous or pubescent, glandular:

Leaf-rhachis (2,7)5–15(23) cm long; pinnae >4 mm apart (rarely less), giving the leaf an "open" appearance; leaflets 0,7–2,3(3,8) mm wide; pods 0,7–1,5 (rarely up to 2,7) cm wide, margins usually entire . . . . .12. *A. caffra*

Leaf-rhachis 1,7–6,5(10) cm long, pinnae <4 mm apart, giving the leaf a compact appearance; leaflets 0,5–1,1 mm wide; pods 1,2–2,3 cm wide, margins often irregularly constricted between the seeds . . . . .13. *A. hereroensis*

aa Flowers in round heads; very rarely inflorescences greatly reduced, apparently capitate with only 2–4
flowers per head:

  d Stipules not spinescent, plant armed with irregularly scattered recurved prickles:

   Leaves with 1–6 pinnae pairs; leaflets 6–17 per pinna, the lowest pair very much reduced and bract-
   like, leaflets large, 5–15(23) mm long, (2)3–6(8) mm wide.......................16. *A. kraussiana*

   Leaves with 8–30 pinnae pairs; leaflets 20–60 pairs per pinna, the lowest pair not reduced and bract-
   like, leaflets much smaller, <5,5 mm long and <2 mm wide:

    Leaflets 0,6–1,2 mm wide, lower surface densely appressed-pubescent throughout or sometimes
    on portion only, rarely leaflets entirely glabrous; petiole 0,5–3,5 cm long; petiolar gland not
    distinctly humped; young branchlets greyish-brown.........................17. *A. brevispica*

    Leaflets 0,8–2 mm wide, glabrous apart from the appressed marginal cilia; petiole 2,6–5,5 cm
    long; petiolar gland humped, usually situated immediately above the pulvinus; young branch-
    lets olive-green to olive-brown.......................................18. *A. schweinfurthii*

  dd Stipules spinescent, in pairs at or near the nodes, spines either long and straight or short and strongly
   recurved or hooked, "ant-galls" sometimes present:

   Inflorescences greatly reduced, with 2–4 flowers per head; dwarf shrub; leaves with only 1 pinna pair
   and 2–4 pairs of leaflets per pinna; pods 2,6–3,2 cm long, densely appressed grey-puberulous
   ....44. *A. redacta*

   Inflorescences with ± 35–200 flowers per head:

    e Flowers bright- or golden-yellow:

     Leaves bipinnate but the leaflets so tightly compressed laterally that the leaves appear simply
     pinnate, each pinna resembling a single linear, crenulate, densely grey-puberulous leaflet;
     leaflets 0,25–0,8 mm long, up to 0,5 mm wide; involucel apical; pods densely grey-velutinous
     ....28. *A. haematoxylon*

     Leaves bipinnate but leaflets larger than above and quite distinct from one another, not laterally
     compressed as above:

      f Involucel apical, rarely slightly below the capitulum; leaflets usually with the lateral nerves
       visible and somewhat raised beneath, rounded apically, (7)9–25 pairs per pinna:

      Midrib and lateral nerves of leaflets visible and somewhat raised beneath; foliage light or
       dark green; leaflets 3–11,5 × 0,75–2,5 mm, glabrous throughout or with marginal cilia;
       leaf-rhachides, rhachillae and peduncles glabrous or sometimes sparingly pubescent:

       Petiole without a gland on the upper side, but leaf-rhachis with a gland at the junction of
        each pinna pair; leaves and peduncles glabrous, seldom sparingly pubescent; spines
        stout, usually fused basally and often swollen into enlarged "ant-galls" 1,5–2 ×
        2–2,5 cm, tapering apically; anthers glandular in bud; pods densely grey-velutinous
        ....27. *A. erioloba*

       Petiole with a small gland on the upper side; leaf-rhachis eglandular except for a gland
        at the junction of the top pinna pair; leaf-rhachides, rhachillae and peduncles sparingly
        pubescent; spines slender, never forming "ant-galls"; anthers eglandular even in
        bud; pods glabrous, subterete.......................................48. *A. farnesiana*

      Midrib and lateral nerves of leaflets indistinct beneath; foliage greyish; leaflets 1–4 ×
       0,75–1,75 mm, sparingly to densely grey-puberulous; leaf-rhachides, rhachillae and
       peduncles densely grey-puberulous; pods densely grey-velutinous...................
       ....29. *A. erioloba* × *A. haematoxylon*

    ff Involucel basal to over halfway up the peduncle, seldom up to ⅔ths way up the peduncle
     but then leaflets spinulose-mucronate apically; leaflets without raised lateral nerves beneath
     (except in *A. swazica* which has (3)4–6(7) pairs of leaflets):

     Bark on trunk and branches greenish-yellow or lemon, becoming powdery and flaking
      minutely; bark of twigs soon becoming greenish-yellow; pods thin-valved, straight or
      slightly curved, transversely segmented, segments mostly longer than wide...........
      ....31. A. *xanthophloea*

     Bark not greenish-yellow or lemon:

      Leaflets glandular-punctate with conspicuous glands on the lower surface and on the
       margins, the margins appearing crenulate-glandular; pods slightly to strongly falcate
       or curled into an almost complete circle, glandular, viscid...............25. *A. borleae*

      Leaflets eglandular on the surface and margins, or at most with a few inconspicuous
       glands on the margin near the apex:

Well developed leaves with 12–27 pinnae pairs (reduced leaves with as few as 8 pinnae pairs sometimes also present); leaflets 20–44 pairs per pinna, up to 1 mm wide; bark on trunk and branches usually corky; pods straight or slightly curved, 0,5–0,8 cm wide, eglandular...26. *A. davyi*

All leaves with less than 13 pinnae pairs; leaflets fewer than 20(25) pairs per pinna, up to 5,5 mm wide; bark not corky; pods not as above:

  g Leaflets (at least mostly) spinulose-mucronate apically; pinnae 1–7 pairs per leaf; leaflets 3–10 pairs per pinna; pods yellowish to reddish-brown, up to 7,5(9) × 1,4 cm, valves thin, usually glandular, viscid; small shrubs or slender trees:

Young branchlets densely tomentose with spreading whitish hairs 0,75–2 mm long ....24. *A. permixta*

Young branchlets glabrous or occasionally very sparingly pubescent but hairs not more than 0,5 mm long:

Involucel at or near the base of the peduncle; leaves usually with only 1 pinna pair, rarely 2–3 pairs.....................................23. *A. nebrownii*

Involucel at or above the middle of the peduncle; leaves with (1)2–6 pinnae pairs:

Leaflets with lateral nerves conspicuous and somewhat raised on the lower surface...22. *A. swazica*

Leaflets with lateral nerves inconspicuous on the lower surface:

Rhizomatous shrub 0,5–1,2(2,4) m high; leaflets 0,9–1,5 mm wide, 3–9 pairs per pinna; spines slender, up to 1,5 mm in diameter basally; pods ± densely glandular................................20. *A. tenuispina*

Small tree or slender shrub, not rhizomatous; leaflets 1,5–4,5 mm wide, 3–6 pairs per pinna; spines frequently enlarged and swollen basally; bark peeling off in strips; pods eglandular or with few scattered glands ....21. *A. exuvialis*

  gg Leaflets not spinulose-mucronate apically; pinnae 2–13 pairs per leaf; leaflets 6–36 pairs per pinna; pods rarely glandular and viscid; robust shrubs or trees:

Young branchlets sparingly to densely pubescent, rarely subglabrous; leaflets 12–27(36) pairs per pinna, 0,5–1,5 mm wide, linear-oblong; peduncle sparingly to densely pubescent; spines often deflexed; pods straight or almost so, distinctly moniliform or jointed, each segment marked with a distinct raised bump which corresponds to the seed inside, indehiscent, valves markedly thickened, ± fleshy when young..........................30. *A. nilotica*

Young branchlets glabrous, rarely sparingly to densely pubescent; leaflets 6–15(22) pairs per pinna, 1–5 mm wide, linear- to obovate-oblong; peduncle glabrous, rarely densely pubescent; pods slightly to strongly falcate, sometimes ± straight, constricted between the seeds to ± moniliform but not jointed as above, dehiscent, valves ± coriaceous...................................19. *A. karroo*

ee Flowers yellowish-white or cream, very rarely pale pink:

 h Spines short and strongly recurved or hooked, often intermixed with some long straight spines or some enlarged swollen "ant-galls":

Pinnae 1–3(4) pairs per leaf; rhachides and rhachillae subglabrous or appressed-puberulous; leaflets 5–13 pairs per pinna, margins entirely glabrous or occasionally cilia present but then appressed; "ant-galls" and usually long straight spines absent; pods straight or almost so, valves coriaceous...35. *A. reficiens*

Pinnae (3)5–13 pairs per leaf; rhachides and rhachillae with spreading hairs; leaflets 6–26 pairs per pinna, margins glabrous or with spreading cilia; "ant-galls" or long straight spines sometimes present; pods contorted or spirally twisted or ± straight:

Corolla tube 1,2–2 mm long; stamens up to 4,5 mm long; leaflets 1–4 × 0,6–1 mm; spines slender, usually <1,5 mm in diameter basally; pods contorted or spirally twisted ....33. *A. tortilis*

Corolla tube 1,8–3,2 mm long; stamens up to 6,5 mm long; leaflets (1,5)2–5(7) × 0,5–1,5(2) mm; spines usually 2–3 mm in diameter basally; "ant-galls" sometimes present; pods straight or almost so:

Greatly enlarged swollen ashen or whitish (often purplish when young) "ant-galls" or long straight spines often present; hairs on peduncle shorter than its diameter, often ± appressed; valves of pods coriaceous, not markedly thickened, subglabrous to finely puberulous; pods pendulous; shrub or large tree...................34. *A. luederitzii*

"Ant-galls" and usually long straight spines absent; hairs on peduncle equal to or exceeding
its diameter; valves of pods markedly thickened, woody, densely grey-tomentellous;
pods usually held erect, sometimes pendulous; low spreading shrub branching near
the ground or sometimes a tree....................................42. *A. hebeclada*

hh Spines straight or only slightly curved, never strongly recurved or hooked, long or short:

i Leaves with 1–14 pinnae pairs:

Involucel apical or in upper half of the peduncle; young branchlets, leaves and peduncles
usually densely golden or greyish pubescent; bark yellowish-brown, usually papery and
peeling off; pod valves woody...41. *A. sieberana*

Involucel basal or in lower third of the peduncle, very seldom more than halfway up the
peduncle but then peduncle and leaves not densely pubescent; pod valves not woody:

j Young branchlets glabrous or occasionally very sparingly puberulous:

Leaflets 0,5–1,25 mm wide; bark yellowish- or greyish-brown, flaking irregularly,
papery, often revealing a greenish inner layer; peduncles usually with conspicuous
sessile glands throughout; corolla usually pinkish-red; pods transversely segmented, each segment with a wart-like projection in the centre......32. *A. kirkii*

Leaflets mostly 1,5–3,5(8,5) mm wide; bark not as above; peduncles eglandular or
glands present but inconspicuous; corolla not pinkish-red; pods not transversely
segmented and without wart-like projections:

Leaf-rhachides, rhachillae and peduncles sparingly to densely pubescent..37. *A. robusta*

Leaf-rhachides, rhachillae and peduncles glabrous:

Peduncle 1,2–5,4 cm long; calyx 1,6–3 mm long; corolla 3,2–4,2 mm long;
leaves usually with 3–5(7) pinnae pairs; young branchlets thick and robust,
with well developed "cushions" at the nodes; spines typically not swollen;
pods straight to falcate, 1,2–2,4(3) cm wide..............37. *A. robusta*

Peduncle 1,2–2(2,5) cm long; calyx up to 1,5 mm long; corolla up to 2,5 mm
long; leaves usually with 2–3(5) pinnae pairs; young branchlets slender,
"cushions" poorly developed; spines typically slightly swollen and fused
basally, pods usually falcate 0,6–1,1 cm wide........38. *A. grandicornuta*

jj Young branchlets sparingly to densely pubescent:

Young branchlets with dense spreading villous hairs 1,5–3 mm long, hairs golden when
young but becoming greyish-white with age; peduncle 0,6–1,2 cm long; pods with
dense spreading hairs 2–4 mm long; obconical shrub branching from the base;
inflorescences sometimes produced before the leaves...............43. *A. stuhlmannii*

Hairs on young branchlets neither villous nor markedly golden, <1,5 mm long;
peduncle (0,5)1,2–5,2 cm long; pods glabrous to densely grey-puberulous or
-tomentellous, hairs <1,5 mm long:

Bark yellowish- or greyish-brown, flaking irregularly, papery, often revealing a
greenish inner layer; leaflets 0,5–1,25 mm wide; peduncles usually with conspicuous sessile glands throughout; corolla usually pinkish-red; pods transversely
segmented, each segment with a wart-like projection in the centre....32. *A. kirkii*

Bark not as above; leaflets (0,75)0,9–3 mm wide; peduncles eglandular or glands
present but inconspicuous; corolla not pinkish-red; pods not as above:

Leaves with (2)4–14 pinnae pairs; rhachillae usually <2,5 cm long; leaflets mostly
<1,5 mm wide; pods sparingly to densely grey-puberulous or -tomentellous,
pendulous or erect, straight or falcate:

Epidermis of young branchlets often splitting and peeling away to reveal a rusty
red inner layer; leaflets 12–26 pairs per pinna; pods falcate, pendulous, valves
rather thin; tree with an irregularly flattened crown, less frequently a shrub;
occurs in the Transvaal, Swaziland and Natal..............36. *A. gerrardii*

Epidermis of young branchlets sometimes splitting but inner layer usually not
rusty-red; leaflets 7–18 pairs per pinna; pods straight or almost so, erect or
sometimes pendulous, valves thick, hard; low spreading shrub branching
near the ground or sometimes a tree; occurs in S.W.A., Transvaal, O.F.S.
and Cape..42. *A. hebeclada*

Leaves with 3–5 (rarely to 7) pinnae pairs; rhachillae usually 2,5–6 cm long;
leaflets (1,25)1,5–3,5 mm wide; pods glabrous, straight or falcate..37. *A. robusta*

Well developed leaves with 15–44 pinnae pairs (reduced leaves with fewer pairs of pinnae
often also present):

Involucel apical or in upper half of the peduncle:

Calyx short, up to 1,2 mm long, shorter than the projecting part of the corolla; flowers white or pale pink; pods arcuate or falcate, 0,5–0,8 cm wide, valves thin; young branchlets sparingly to densely puberulous..........................39. *A. arenaria*

Calyx >1,5 mm long, longer than the projecting part of the corolla; flowers pale yellowish-white; pods straight or almost so, 1,3–3,3 cm wide, valves thick, ± woody; young branchlets usually densely golden-pubescent........................41. *A. sieberana*

Involucel basal or in lower half of the peduncle:

Leaflets 24–48 pairs per pinna, 1,2–2,8 × 0,4–0,9 mm; pinnae 15–43 pairs per leaf; pods glabrous to sparingly pubescent.................................40. *A. rehmanniana*

Leaflets 6–15 pairs per pinna, 2,5–5,5 × 0,6–1,5 mm; pinnae 15–19 pairs per leaf; pods with dense spreading hairs 2–4 mm long..........................43. *A. stuhlmannii*

Key based on vegetative and floral characters

Flowers in spikes or spiciform racemes; rarely (in *A. mellifera* subsp. *detinens*) the inflorescences short and ellipsoid but even then the axis clearly elongate:

Stipules spinescent, straight or only slightly curved; petiolar gland absent; rhachis with a gland at the junction of each pinna pair; stamen filaments united basally for ± 1 mm; anthers eglandular even in bud...1. *A. albida*

Stipules not spinescent, plants armed with recurved prickles; petiolar gland usually present; stamen filaments free to the base; anthers glandular, at least in bud:

Prickles scattered irregularly along the internodes; stipules obliquely ovate, up to 12 × 7 mm; petiolar gland stalked; ovary densely pilose, on a long stipe that elevates it above the corolla..2. *A. ataxacantha*

Prickles confined to near the nodes, very rarely *and in addition* a few casually and irregularly scattered along the internodes; stipules ± linear, smaller than above; petiolar gland not stalked; ovary not as above:

Prickles in threes, the two laterals curved upwards and the central one down, or else the prickles solitary, the two laterals being absent...3. *A. senegal*

Prickles in pairs near the nodes:

Flowers shortly but distinctly pedicellate (pedicels (0,5)0,75–1,5 mm long); inflorescence ellipsoid to distinctly spicate; calyx cupular, 0,6–1 mm long, yellow, lobes often with a purplish tinge; leaflets 1–2(3) pairs per pinna; prickles of each pair often lying almost parallel to each other and not spreading..4. *A. mellifera*

Flowers sessile or subsessile (pedicels 0–0,3 mm long); inflorescence distinctly spicate; calyx 1,5–3,5 mm long (except in *A. galpinii* where it is shorter but red or purplish); leaflets 3 or more pairs per pinna (except in *A. nigrescens*); prickles spreading:

b Well developed leaves of mature shoots with 1–7 pinnae pairs:

c Calyx glabrous, rarely with occasional scattered hairs:

Leaflets 1–2 (rarely 3) pairs per pinna, obovate or obliquely-obovate to broadly-elliptic or orbicular, (7)12–50 mm wide; leaves with 2–3(4) pinnae pairs; trunk and larger branches usually with scattered irregularly shaped knobs.............8. *A. nigrescens*

Leaflets 3–13 pairs per pinna, 1–5 mm wide; leaves with 1–6 pinnae pairs:

Leaflets 3–8 pairs per pinna, elliptic, broadly elliptic or somewhat ovate, usually broadest at or below the middle, 2,5–5 mm wide; young branchlets dark grey- to brownish-black; prickles strongly recurved; occurs in the eastern Transvaal....7. *A. welwitschii*

Leaflets (4)6–13 pairs per pinna, linear to linear-oblong or obovate, sometimes slightly falcate, 0,9–3,5(5) mm wide; young branchlets not as above, prickles often spreading laterally and almost at right angles to the stem; confined to S.W.A.:

Leaves with 1 or 2 pinnae pairs; shrub with several erect slender branches which tend to droop apically or a slender tree with whip-like branches; young branchlets white to reddish-brown or purple, often as though whitewashed over a purplish background...5. *A. robynsiana*

Leaves with (2)3–6 pinnae pairs; slender tree branching from near the base, "broom-like"; young branchlets olive- to reddish-brown; bark often papery:.6. *A. montis-usti*

cc Calyx sparingly to densely puberulous:

Calyx cupular, 0,75–1,25 mm long, red or purplish.....................10. *A. galpinii*

Calyx 1,3–4,5 mm long, greenish-yellow to yellowish-white (sometimes with a distinct pinkish tinge in *A. burkei*):

Leaflets (1)3–12 (rarely to 15) pairs per pinna, (1)2–13 mm wide, obovate, obovate-oblong or oblanceolate to orbicular, seldom linear-oblong; bracts in very young inflorescences projecting beyond the buds; calyx 1,3–2,5 mm long, often tinged with pink; stamen filaments 4–6 mm long...9. *A. burkei*

Leaflets (10)12–30 pairs per pinna, 0,75–1,6(2,5) mm wide, linear to linear-oblong, often ± falcate; bracts not projecting beyond the buds; calyx 2–4,5 mm long, greenish-yellow; stamen filaments 6–10 mm long:

Petiole (0,8)1,2–2,5(4) cm long; petiolar gland 0,3–0,7(1) mm long, usually slightly raised or stalked; leaflets often ± falcate, mostly 5–10 mm long; leaf-rhachis usually glandular, with a gland at the junction of some of the pinnae pairs or the top pinna pair only, occasionally eglandular; flowers sometimes produced before the leaves...15. *A. erubescens*

Petiole 0,3–1,3 cm long; petiolar gland 0,8–2,2 mm long, flattened, elliptic or oval; leaflets straight, mostly <5 mm long; leaf-rhachis eglandular; flowers produced after or with the leaves......................................14. *A. fleckii*

bb Well developed leaves of mature shoots with 8 or more pinnae pairs:

Calyx cupular, 0,75–1,25 mm long, red or purplish.........................10. *A. galpinii*

Calyx 1,3–4,5 mm long, greenish-yellow to yellowish-white (sometimes with a distinct pinkish tinge in *A. burkei*):

Petiolar gland large, 1,5–4 × 1,5–3 mm, usually slightly flattened, discoid or oblong; petiole 0,5–4 cm long; rhachis glandular between the top 3–16 pinnae pairs..........

....11. *A. polyacantha*

Petiolar gland small to medium, 0,3–1,5 × 0,1–0,7 mm, rarely up to 2,2 mm long but then rhachis eglandular and petiole 0,3–1,3 cm long:

Leaf-rhachis eglandular; petiolar gland 0,8–2,2 mm long; petiole 0,3–1,3 cm long

....14. *A. fleckii*

Leaf-rhachis glandular, with a gland at the junction of the top pinna pair only or at the junction of all or some of the pinna pairs; petiolar gland 0,3–1,5 mm long; petiole 0,4–4 cm long:

Leaflets 4–15 (rarely to 19) pairs per pinna, linear-oblong or obovate to obovate-oblong, 0,8–7 mm wide; pinnae 8–13 pairs per leaf..................9. *A. burkei*

Leaflets 16–64 pairs per pinna, linear to linear-oblong, 0,5–2,3(3,8) mm wide; pinnae 8–38 pairs per leaf:

Leaf-rhachis (2,7)5–15(23) cm long; pinnae >4 mm apart (rarely less), giving the leaf an "open" appearance; leaflets 0,7–2,3(3,8) mm wide........12. *A. caffra*

Leaf-rhachis 1,7–6,5(10) cm long; pinnae <4 mm apart, giving the leaf a compact appearance; leaflets 0,5–1,1 mm wide......................13. *A. hereroensis*

aa Flowers in round heads or very rarely inflorescences greatly reduced, apparently capitate with only 2–4 flowers per head:

d Stipules not spinescent, plant armed with irregularly scattered recurved prickles:

Leaves with 1–6 pinnae pairs; leaflets 6–17 pairs per pinna, the lowest pair very much reduced and bract-like, leaflets large, 5–15(23) mm long, (2)3–6(8) mm wide...................16. *A. kraussiana*

Leaves with 8–30 pinnae pairs; leaflets 20–60 pairs per pinna, the lowest pair not reduced and bract-like, leaflets much smaller, <5,5 mm long and <2 mm wide:

Leaflets 0,6–1,2 mm wide, lower surface densely appressed-pubescent throughout or sometimes on portion only, rarely leaflets entirely glabrous; petiole 0,5–3,5 cm long; petiolar gland not distinctly humped; young branchlets greyish-brown........................17. *A. brevispica*

Leaflets 0,8–2 mm wide, glabrous apart from the appressed marginal cilia; petiole 2,6–5,5 cm long; petiolar gland humped, usually situated immediately above the pulvinus; young branchlets olive-green to olive-brown.......................................18. *A. schweinfurthii*

dd Stipules spinescent, in pairs at or near the nodes, spines either long and straight or short and strongly recurved or hooked, "ant-galls" sometimes present:

Inflorescences greatly reduced, with 2–4 flowers per head; dwarf shrub; leaves with only 1 pinna pair and 2–4 pairs of leaflets per pinna..44. *A. redacta*

Inflorescences with $\pm$ 35–200 flowers per head:

 e Flowers bright- or golden-yellow:

  Leaves bipinnate but the leaflets so tightly compressed laterally that the leaves appear simply pinnate, each pinna resembling a single linear, crenulate, densely grey-puberulous leaflet; leaflets 0,25–0,8 mm long, up to 0,5 mm wide; involucel apical . . . . 28. *A. haematoxylon*

  Leaves bipinnate but the leaflets larger than above and quite distinct from one another, not laterally compressed as above:

  f Involucel apical, rarely slightly below the capitulum; leaflets usually with the lateral nerves visible and somewhat raised beneath, rounded apically, (7)9–25 pairs per pinna:

   Midrib and lateral nerves of leaflets visible and somewhat raised beneath; foliage light or dark green; leaflets 3–11,5 × 0,75–2,5 mm, glabrous throughout or with marginal cilia; leaf-rhachides, rhachillae and peduncles glabrous or sometimes sparingly pubescent:

    Petiole without a gland on the upper side, but leaf-rhachis with a gland at the junction of each pinna pair; leaves and peduncles glabrous, seldom sparingly pubescent; leaflets 4–11,5 × 0,7–2,4 mm; spines stout, usually fused basally and often swollen into enlarged "ant-galls" 1,5–2 × 2–2,5 cm, tapering apically; anthers glandular in bud . . . . 27. *A. erioloba*

    Petiole with a small gland on the upper side; leaf-rhachis eglandular except for a gland at the junction of the upper pinna pair; leaf-rhachides, rhachillae and peduncles sparingly pubescent; leaflets 3–7 × 0,75–1,75 mm; spines slender, never forming "ant-galls" anthers eglandular even in bud . . . . 48. *A. farnesiana*

   Midrib and lateral nerves of leaflets indistinct beneath; foliage greyish; leaflets 1–4 × 0,4–1,1 mm, sparingly to densely grey-puberulous; leaf-rhachides, rhachillae and peduncles densely grey-puberulous . . . . 29. *A. erioloba* × *A. haematoxylon*

  ff Involucel basal to over halfway up the peduncle, seldom up to 4/5ths way up the peduncle but then the leaflets spinulose-mucronate apically; leaflets without raised lateral nerves beneath (except in *A. swazica* which has (3)4–6(7) pairs of leaflets):

  Bark on trunk and branches greenish-yellow or lemon, becoming powdery and flaking minutely . . . . 31. *A. xanthophloea*

  Bark not greenish-yellow or lemon:

   Leaflets glandular-punctate with conspicuous glands on the lower surface and on the margins, the margins appearing crenulate-glandular . . . . 25. *A. borleae*

   Leaflets eglandular on the surface and margins, or at most with a few inconspicuous glands on the margin near the apex:

    Well developed leaves with 12–27 pinnae pairs (reduced leaves with as few as 8 pinnae pairs sometimes also present); leaflets 20–44 pairs per pinna, linear, up to 1 mm wide; bark on trunk and branches usually corky . . . . 26. *A. davyi*

    All leaves with less than 13 pinnae pairs; leaflets fewer than 25 pairs per pinna, up to 5,5 mm wide; bark not corky:

    g Leaflets (at least mostly) spinulose-mucronate apically; pinnae 1–7 pairs per leaf; leaflets 3–10 pairs per pinna; young branchlets often glutinous; small shrubs or slender trees:

     Young branchlets densely tomentose with spreading whitish hairs 0,75–2 mm long . . . . 24. *A. permixta*

     Young branchlets glabrous or occasionally very sparingly pubescent but hairs not more than 0,5 mm long:

      Involucel at or near the base of the peduncle; leaves usually with only 1 pinna pair, rarely 2–3 pairs . . . . 23. *A. nebrownii*

      Involucel at or above the middle of the peduncle; leaves with (1)2–6 pinnae pairs:

       Leaflets with lateral nerves conspicuous and somewhat raised on the lower surface . . . . 22. *A. swazica*

       Leaflets with lateral nerves inconspicuous on the lower surface:

        Rhizomatous shrub 0,5–1,2(2,4) m high; leaflets 0,9–1,5 mm wide, 3–9 pairs per pinna; spines slender, up to 1,5 mm in diameter basally . . . . 20. *A. tenuispina*

        Small tree or slender shrub, not rhizomatous; leaflets 1,5–4,5 mm wide, 3–6 pairs per pinna; spines frequently enlarged and swollen basally; bark peeling off in strips . . . . 21. *A. exuvialis*

gg Leaflets not spinulose-mucronate apically; pinnae 2–13 pairs per leaf; leaflets 6–36
   pairs per pinna; young branchlets not glutinous; robust shrubs or trees:

Young branchlets sparingly to densely pubescent, rarely subglabrous; leaflets
   12–27(36) pairs per pinna, 0,5–1,5 mm wide, linear-oblong; inflorescences
   solitary or fascicled, usually terminal; peduncle sparingly to densely pubescent;
   spines often deflexed; tree, usually with a somewhat flattened crown 30. *A. nilotica*

Young branchlets glabrous, rarely sparingly to densely pubescent; leaflets 6–15(22)
   pairs per pinna, 1–5 mm wide, linear- to obovate-oblong; inflorescences fascicled, sometimes forming a terminal raceme or panicle; peduncle glabrous,
   rarely densely pubescent; shrub or tree, variable in habit..........19. *A. karroo*

ee Flowers yellowish-white or cream, very rarely pale pink:

h Spines short and strongly recurved or hooked, often intermixed with some long straight spines
   or some enlarged swollen "ant-galls":

Pinnae 1–3(4) pairs per leaf; rhachides and rhachillae subglabrous or appressed-puberulous;
   leaflets 5–13 pairs per pinna, margins entirely glabrous or occasionally cilia present but
   then appressed; "ant-galls" and usually long straight spines absent..........35. *A. reficiens*

Pinnae (3)5–13 pairs per leaf; rhachides and rhachillae with spreading hairs; leaflets 6–26 pairs
   per pinna, margins glabrous or with spreading cilia; "ant-galls" or long straight spines
   sometimes present:

Corolla tube 1,2–2 mm long; stamens up to 4,5 mm long; leaf-rhachides 0,2–1,8 cm long;
   rhachillae 0,3–1,4 cm long; leaflets 1–4 × 0,6–1 mm; spines slender, usually <1,5 mm
   in diameter basally; "ant-galls" absent...............................33. *A. tortilis*

Corolla tube 1,8–3,2 cm long; stamens up to 6,5 cm long; leaf-rhachides 1–4,2 cm long;
   rhachillae 0,8–2,8 cm long; leaflets (1,5)2–5(7) × 0,5–1,5(2) mm; spines usually 2–3 mm
   in diameter basally; "ant-galls" sometimes present:

Greatly enlarged, swollen, ashen or whitish (often purplish when young) "ant-galls" or
   long straight spines often present; hairs on the peduncle shorter than its diameter,
   often ± appressed; shrub or large tree.......................34. *A. luederitzii*

"Ant-galls" and usually long straight spines absent; hairs on the peduncle equal to or
   exceeding its diameter; low spreading shrub branching near the ground or sometimes
   a tree..42. *A. hebeclada*

hh Spines straight or only slightly curved, never strongly recurved or hooked, long or short:

i Leaves with 1–14 pinnae pairs:

Involucel apical or in upper half of the peduncle; young branchlets, leaves and peduncles
   usually densely golden or greyish pubescent; bark yellowish-brown, usually papery
   and peeling off...41. *A. sieberana*

Involucel basal or in lower third of the peduncle, very seldom more than halfway up the
   peduncle but then peduncle and leaves not densely pubescent:

j Young branchlets glabrous or occasionally very sparingly puberulous:

Leaflets 0,5–1,25 mm wide; bark yellowish- or greyish-brown, flaking irregularly,
   papery, often revealing a greenish inner layer; peduncles usually with conspicuous
   sessile glands throughout; corolla usually pinkish-red.................32. *A. kirkii*

Leaflets mostly 1,5–3,5(8,5) mm wide; bark not as above; peduncles eglandular or
   glands present but inconspicuous; corolla not pinkish-red:

Leaf-rhachides, rhachillae and peduncles sparingly to densely pubescent..37. *A. robusta*
Leaf-rhachides, rhachillae and peduncles glabrous:

Peduncle 1,2–5,4 cm long; calyx 1,6–3 mm long; corolla 3,2–4,2 mm long;
   leaves usually with 3–5(7) pinnae pairs; young branchlets thick and robust,
   with well developed "cushions" at the nodes; spines typically not swollen
   ....37. *A. robusta*

Peduncle 1,2–2(2,5) cm long; calyx up to 1,5 mm long; corolla up to 2,5 mm long;
   leaves usually with 2–3(5) pinnae pairs; young branchlets slender, "cushions"
   poorly developed; spines typically slightly swollen and fused basally
   ....38. *A. grandicornuta*

jj Young branchlets sparingly to densely pubescent:

Young branchlets with dense spreading villous hairs 1,5–3 mm long; hairs golden when
   young but becoming greyish-white with age; peduncle 0,6–1,2 cm long; obconical
   shrub branching from the base; leaflets 9–14 pairs per pinna; inflorescences sometimes produced before the leaves..............................43. *A. stuhlmannii*

Hairs on young branchlets neither villous nor markedly golden, <1,5 mm long; peduncle (0,5)1,2–5,2 cm long:

Bark yellowish-brown or greyish-brown, flaking irregularly, papery, often revealing a greenish inner layer; leaflets 0,5–1,25 mm wide; peduncles usually with conspicuous sessile glands throughout; corolla usually pinkish-red......32. *A. kirki*

Bark not as above; leaflets (0,75)1–3,5 mm wide; peduncles eglandular or glands present but inconspicuous; corolla not pinkish-red:

Leaves with (2)4–14 pinnae pairs; rhachillae usually <2,5 cm long; leaflets mostly <1,5 mm wide; tree up to 10 m high or a shrub:

Epidermis of young branchlets often splitting and peeling away to reveal a rusty-red inner layer; leaflets 12–26 pairs per pinna; tree with an irregularly flattened crown, less frequently a shrub; occurs in the Transvaal, Swaziland and Natal..36. *A. gerrardii*

Epidermis of young branchlets sometimes splitting but inner layer usually not rusty-red; leaflets 7–18 pairs per pinna; low spreading shrub branching near the ground or sometimes a tree; spines often slightly recurved apically; occurs in S.W.A., Transvaal, O.F.S. and Cape..............42. *A. hebeclada*

Leaves with 3–5 (rarely to 7) pinnae pairs; rhachillae usually 2–5,6 cm long; leaflets (1,25)1,5–3,5 mm wide; tree to 20 m high with ascending branches. .37. *A. robusta*

ii Well developed leaves with 15–44 pinnae pairs (reduced leaves with fewer pairs of pinnae often also present):

Involucel apical or in upper half of the peduncle:

Calyx short, up to 1,2 mm long, shorter than the projecting part of the corolla; flowers white or pale pink; young branchlets sparingly to densely puberulous....39. *A. arenaria*

Calyx >1,5 mm long, longer than the projecting part of the corolla; flowers pale yellowish-white; young branchlets usually densely golden-pubescent............41. *A. sieberana*

Involucel basal or in lower half of the peduncle:

Leaflets 24–48 pairs per pinna, 1,2–2,8 × 0,4–0,9 mm; pinnae 15–43 pairs per leaf; tree or shrub...40. *A. rehmanniana*

Leaflets 6–15 pairs per pinna, 2,5–5,5 × 0,6–1,5 mm; pinnae 15–19 pairs per leaf; obconical shrub...43. *A. stuhlmannii*

Key based on vegetative and pod characters

a Stipules not spinescent, plants armed with recurved prickles:

b Prickles irregularly scattered along the internodes:

Leaves with 1–6 pinnae pairs; leaflets 6–17 pairs per pinna, the lowest pair very much reduced and bract-like, leaflets large, 5–15(23) mm long, (2)3–6(8) mm wide; pods tardily dehiscent or indehiscent...16. *A. kraussiana*

Leaves with 8–30 pinnae pairs; leaflets 20–60 pairs per pinna, the lowest pair not reduced and bract-like, leaflets much smaller, <5,5 mm long and <2 mm wide:

Pods reddish-brown to purplish, usually distinctly acuminate at both ends, brittle; areole of seed a small central depression; stipules obliquely ovate, up to 12 mm long and 7 mm wide; petiolar gland stalked...2. *A. ataxacantha*

Pods not as above; areole large, elliptic, conforming to the outline of the seed; stipules linear-oblong, up to 5 × 2 mm; petiolar gland not stalked:

Leaflets 0,6–1,2 mm wide, lower surface densely appressed-pubescent throughout or sometimes on portion only, rarely leaflets entirely glabrous; petiole 0,5–3,5 cm long; petiolar gland not distinctly humped; young branchlets greyish-brown; pods usually densely puberulous and with numerous minute reddish glands, longitudinally dehiscent..............17. *A. brevispica*

Leaflets 0,8–2 mm wide, glabrous apart from the appressed marginal cilia; petiole 2,6–5,5 cm long; petiolar gland humped, usually situated immediately above the pulvinus; young branchlets olive-green to olive-brown; pods glabrous, with minute glands, tardily dehiscent or indehiscent...18. *A. schweinfurthii*

bb Prickles confined to near the nodes, very rarely *and in addition* a few casually and irregularly scattered along the internodes:

Prickles in threes, the two laterals curved upwards and the central one down, or else the prickles solitary, the two laterals being absent...3. *A. senegal*

Prickles in pairs near the nodes:

c Well developed leaves with 1–7 pinnae pairs:

 d Leaflets 1–2 (rarely 3) pairs per pinna:

  Pods olive- to yellowish-brown, 2,5–8(9,3) cm long, valves somewhat papery, with a conspicuous closely reticulate venation; prickles of each pair often lying almost parallel to each other and not spreading; obconical shrub or small tree................4. *A. mellifera*

  Pods dark brown or reddish-brown to blackish, (4)6–18 cm long, valves not papery, without a closely reticulate venation; prickles spreading; usually a fairly large tree:

   Leaflets > 5,5 mm wide, asymmetric basally, obovate, obovate-orbicular to broadly obovate-elliptic, usually sparingly to densely appressed-pubescent above and/or below:

    Leaves with 2–3(4) pinnae pairs; leaflets (7)12–50 mm wide; remnant of calyx glabrous; trunk and larger branches usually with scattered irregularly shaped knobs 8. *A. nigrescens*

    Leaves with 2–7 pinnae pairs; leaflets 5,5–13 mm wide; remnant of calyx sparingly to densely pubescent; trunk and large branches usually without knobs......9. *A. burkei*

   Leaflets 2,5–5 mm wide, almost symmetric basally except for the terminal ones, elliptic or broadly elliptic, sometimes somewhat ovate, usually glabrous throughout..7. *A. welwitschii*

 dd Leaflets 4–45 pairs per pinna:

  Pods with a fine closely reticulate venation, yellowish-brown or brown, up to 1,8 cm wide (rarely wider) and up to 13,5 cm long:

   Leaves with 1 or 2 pinnae pairs; leaflets 1,2–2,8(3,5) mm wide; low shrub with several slender erect branches which tend to droop apically or a slender tree; young branchlets white to reddish-brown or purple, often as though whitewashed over a purplish background..5. *A. robynsiana*

   Leaves with 3–7 pinnae pairs; leaflets 0,75–2(2,8) mm wide; much branched shrub or tree:

    Petiole (0,8)1,2–2,5(4) cm long; petiolar gland 0,3–0,7(1) mm long, usually slightly raised or stalked; leaflets often ± falcate, mostly 5–10 mm long; leaf-rhachis usually with a gland at the junction of some of the pinnae pairs or the top pinna pair only, occasionally eglandular...15. *A. erubescens*

    Petiole 0,3–1,3 cm long; petiolar gland 0,8–2,2 mm long, flattened, elliptic or oval; leaflets straight, mostly < 5 mm long; leaf-rhachis eglandular..........14. *A. fleckii*

  Pods without a fine close reticulate venation, inconspicuously venose or venation coarse, olive- or reddish- to purplish-brown or blackish, 1–3,5 cm wide and up to 28 cm long:

   Pods 2,3–3,5 cm wide, 11,5–28 cm long, reddish to purplish-brown when mature, valves thinly woody, brittle; leaflets 13–45 pairs per pinna......................10. *A. galpinii*

   Pods 1–2,4 cm wide, up to 11(16,9) cm long, brown or reddish-brown to blackish; leaflets 4–13(15) pairs per pinna:

    Leaf-rhachides and rhachillae with spreading hairs; leaflets usually with marginal cilia and often appressed-pubescent beneath, typically with a basal tuft of hairs to one side of the midrib; leaflets 2–13 mm wide.................................9. *A. burkei*

    Leaf-rhachides, rhachillae and leaflets glabrous throughout; leaflets 0,9–5 mm wide:

     Leaflets 2,5–5 mm wide; elliptic or broadly elliptic, sometimes somewhat ovate; prickles strongly recurved; young branchlets dark grey or reddish-brown to blackish; usually a fairly large tree with a rounded or spreading crown; confined (in our area) to the eastern Transvaal.................................7. *A. welwitschii*

     Leaflets 0,9–3(5) mm wide, linear-oblong to obovate, sometimes slightly falcate; prickles ± straight or very slightly curved, spreading laterally and almost at right angles to the stem; young branchlets pale olive- to reddish-brown; slender tree branching from near the base, "broom-like"; confined to S.W.A..............6. *A. montis-usti*

cc Well developed leaves with 8 or more pinnae pairs:

 Pods 2,3–3,5 cm wide, reddish- to purplish-brown when mature, valves thinly woody, brittle, glabrous..10. *A. galpinii*

 Pods 0,7–2,2 cm wide (very rarely to 2,7 cm wide), yellowish-brown, brown or reddish-brown to blackish:

  Petiolar gland large, 1,5–4 × 1,5–3 mm, usually slightly flattened, discoid or oblong; petiole 0,5–4 cm long; rhachis with a gland at the junction of the top 3–16 pinnae pairs........ ....11. *A. polyacantha*

  Petiolar gland small to medium, 0,3–1,5 × 0,1–0,7 mm, rarely up to 2,2 mm long but then rhachis eglandular and petiole 0,3–1,3 cm long:

Leaf-rhachis eglandular; petiolar gland 0,8–2,2 mm long; petiole 0,3–1,3 cm long; pods yellowish-brown or brown, with a fine closely reticulate venation............14. *A. fleckii*

Leaf-rhachis glandular, with a gland at the junction of each pinna pair, at the junction of some or the top pinna pair only; petiolar gland 0,3–1,5 mm long; petiole 0,4–4 cm long:

Leaflets 4–15 (rarely up to 19) pairs per pinna; pinnae 8–13 pairs per leaf; leaflets linear-oblong to obovate or obovate-oblong, 0,8–7 mm wide; pods dark brown or reddish-brown to blackish, glabrous or almost so.............................9. *A. burkei*

Leaflets 18–64 pairs per pinna; pinnae 8–38 pairs per leaf; leaflets linear to linear-oblong, 0,5–2,3(3,8) mm wide; pods brown or olive-brown, usually puberulous, glandular:

Leaf-rhachis (2,7)5–15(23) cm long; pinnae >4 mm apart (rarely less), giving the leaf an "open" appearance; leaflets 0,7–2,3(3,8) mm wide; pods 0,7–1,5 (rarely up to 2,7) cm wide, margins usually entire...................................12. *A. caffra*

Leaf-rhachis 1,7–6,5(10) cm long; pinnae <4 mm apart, giving the leaf a compact appearance; leaflets 0,5–1,1 mm wide; pods 1,2–2,3 cm wide, margins often irregularly constricted between the seeds........................13. *A. hereroensis*

aa Stipules spinescent, spines in pairs at or near the nodes:

e Spines short and strongly recurved or hooked, usually intermixed with some long straight spines or some enlarged swollen "ant-galls":

Pods contorted or spirally twisted...33. *A. tortilis*

Pods straight or almost so, not contorted or spirally twisted:

Valves of pods coriaceous, not markedly thickened, brown or reddish-brown, subglabrous to finely puberulous, with fine longitudinal venation, dehiscent; pods pendulous:

Pods 0,9–1,9 cm wide; pinnae (3)5–13 pairs per leaf; rhachides and rhachillae with spreading hairs; leaflets 7–26 pairs per pinna, margins usually with spreading cilia; "ant-galls" sometimes present...34. *A. luederitzii*

Pods 0,6–1,1 cm wide; pinnae 1–3(4) pairs per leaf; rhachides and rhachillae subglabrous to appressed-puberulous; leaflets 5–13 pairs per pinna, margins glabrous, seldom cilia present but then appressed; "ant-galls" absent................................35. *A. reficiens*

Valves of pods markedly thickened, woody, yellowish- or greyish-brown, densely grey-tomentellous, tardily dehiscent; pods usually held erect, sometimes pendulous.................42. *A. hebeclada*

ee Spines straight or only slightly curved, never strongly recurved or hooked, long or short:

Pods indehiscent and thin-valved (except sometimes for tubercles in the centre of the segments), jointed and breaking up transversely, ±transversely or net-veined, glabrous except usually for sessile glands; bark on trunk usually yellow to green, sometimes grey or brown, powdery or papery and peeling:

Segments of pods not tubercled, mostly longer than wide; bark on trunk and branches lemon-coloured or greenish-yellow, becoming powdery and flaking minutely; bark of twigs soon becoming greenish-yellow...31. *A. xanthophloea*

Segments of pods with a wart-like projection up to 4,5 mm high in the centre, mostly as wide as or wider than long; bark on trunk grey or yellowish-brown, sometimes flaking to reveal a greenish inner layer; bark of twigs grey-brown to plum-coloured.......................32. *A. kirkii*

Pods dehiscent or indehiscent, if indehiscent then the valves markedly thickened, woody or pulpy in texture, not venose and glandular as above:

f Valves of pods markedly thickened, woody or pulpy in texture, pods indehiscent or slowly dehiscent:

Pods orange or chestnut- to reddish-brown, falcate or curled into a circular coil or variously twisted, 1,4–4,5 cm wide; petiolar gland absent; rhachis with a conspicuous gland at the point of attachment of each pinna pair.....................................1. *A. albida*

Pods not as above; petiolar gland usually present:

Pods distinctly moniliform or jointed, often irregularly so, each segment marked with a distinct raised bump which corresponds to the seed inside, green when young, drying and becoming black with age, straight or almost so...................................30. *A. nilotica*

Pods not as above:

Pods glabrous or very sparingly pubescent:

Pods glabrous or very sparingly pubescent, flattened to oval in cross-section, 1,5–3,5 cm wide, yellowish- or reddish-brown, not longitudinally striate; young branchlets usually densely pubescent...41. *A. sieberana*

Pods glabrous, subterete and turgid, 0,9–1,5 cm in diameter, dark brown to blackish, finely longitudinally striate; young branchlets glabrous or nearly so......48. *A. farnesiana*

Pods ± densely puberulous, pubescent, tomentellous or villous:

  g Indumentum on pods short, the hairs <1 mm long:

    h Pods densely and continuously grey-tomentellous, often with numerous minute reddish-brown glands scattered in amongst the hairs:

      Pods falcate, semi-lunate, suborbicular or curled into an almost complete circle, rarely straightish, usually persistently greyish, indehiscent, pendulous:

        Leaves bipinnate but the leaflets so tightly compressed laterally that the leaves appear simply pinnate, each pinna resembling a single linear, crenulate, densely grey-puberulous leaflet; leaflets 0,25–0,8 mm long, up to 0,5 mm wide; pods 0,6–1,4 cm wide...............................28. *A. haematoxylon*

        Leaves bipinnate but the leaflets larger than above and quite distinct from one another, not laterally compressed as above:

          Pods (1,8)2,5–5 cm wide; foliage dark green; pinnae 1–6 pairs per leaf; leaflets 4–11,5 mm long, glabrous or subglabrous, midrib and lateral nerves visible and somewhat raised beneath; spines stout, usually fused basally and often swollen into an enlarged "ant-gall" 1,5–2 × 2–2,5 cm, tapering apically....................................27. *A. erioloba*

          Pods 1,2–2,3 cm wide; foliage greyish; pinnae 3–12 pairs per leaf; leaflets 1–4 mm long, sparingly to densely puberulous, midrib and lateral nerves indistinct beneath; spines slender, "ant-galls" absent...................
          ....29. *A. erioloba* × *A. haematoxylon*

      Pods straight or almost so, yellowish-brown or brown, ultimately dehiscent, erect or occasionally pendulous......................................42. *A. hebeclada*

    hh Pods not densely and continuously grey-tomentellous:

      Pods pendulous, without longitudinal nervation; indumentum on young branchlets, leaves and pods usually golden; pinnae (4)8–30 pairs per leaf; leaflets 13–45 pairs per pinna; tree to 20 m high; bark usually yellowish-brown, papery and peeling...41. *A. sieberana*

      Pods erect or occasionally pendulous, often conspicuously longitudinally nerved; indumentum not golden; pinnae (1)4–9(12) pairs per leaf; leaflets 7–18 pairs per pinna; low spreading shrub branching near the ground and often forming thickets, or a small tree; bark not as above.....................42. *A. hebeclada*

  gg Indumentum on pods long, the hairs spreading, 2–4 mm long; indumentum on branchlets spreading, golden, becoming greyish with age; obconical shrub to 3 m...
  ....43. *A. stuhlmannii*

ff Valves of pods membranous to subcoriaceous or coriaceous, not markedly thickened, dehiscent:

  i Well developed leaves of mature shoots with 13–45 pinnae pairs:

    Leaflets glandular-punctate, with conspicuous glands on the lower surface and on the margins, the margins appearing crenulate-glandular; pods ± falcate or curled into an almost complete circle, ± moniliform and constricted between the seeds, glandular, viscid..25. *A. borleae*

    Leaflets not glandular-punctate; pods not as above:

      Pods 0,5–0,8 cm wide, straightish to falcate; young branchlets glabrous to densely puberulous:

        Pods yellowish-brown or sometimes reddish-brown, straightish to slightly falcate; bark yellowish-brown, soft, corky; occurs in Transvaal, Swaziland and Natal....26. *A. davyi*

        Pods deep reddish-brown, arcuate or falcate; bark dark grey- or reddish-brown to black; occurs in S.W.A...39. *A. arenaria*

      Pods 1,2–2,3 cm wide, ±straight, margins flattened; young branchlets clothed with spreading villous hairs 1–2,5 mm long, hairs golden when young but becoming greyish-white with age...40. *A. rehmanniana*

  ii Well developed leaves of mature shoots with 1–12 (rarely 13) pinnae pairs:

    Leaflets glandular-punctate, with conspicuous glands on the lower surface and on the margins, the margins appearing crenulate-glandular; pods ± falcate or curled into an almost complete circle, ± moniliform and constricted between the seeds, glandular, viscid....25. *A. borleae*

    Leaflets not glandular-punctate, or at most with a few inconspicuous glands on the margin near the apex:

      j Leaflets (at least mostly) spinulose-mucronate apically; pinnae 1–7 pairs per leaf; leaflets 3–10 pairs per pinna; pods yellowish- to reddish-brown, up to 7,5(9) × 1,4 cm, valves thin, usually glandular, viscid; small shrubs or slender trees:

        Young branchlets densely tomentose with spreading whitish hairs 0,75–2 mm long
        ....24. *A. permixta*

Young branchlets glabrous or occasionally very sparingly pubescent, but hairs not more than 0,5 mm long:

Leaflets with lateral nerves conspicuous and somewhat raised on the lower surface ....22. *A. swazica*

Leaflets with inconspicuous lateral nerves on the lower surface:

Rhizomatous shrub 0,5–1,2(2,4) m high; leaflets 0,9–1,5 mm wide, 3–9 pairs per pinna; spines slender, up to 1,5 mm in diameter basally..........20. *A. tenuispina*

Small tree or slender shrub, not rhizomatous; leaflets (0,8)1,5–4,5 mm wide, 3–6 pairs per pinna; spines sometimes swollen basally:

Leaves usually with only 1 pinna pair, rarely 2–3 pairs; pods (0,6)0,9–1,3 cm wide, with numerous dark glands scattered over the surface; spines not swollen basally ....23. *A. nebrownii*

Leaves with (1)2–4(6) pinna pairs; pods 0,4–0,9 cm wide, eglandular or almost so; spines frequently enlarged and swollen basally; bark peeling off in strips ....21. *A. exuvialis*

jj Leaflets not spinulose-mucronate apically; pinnae 1–12(13) pairs per leaf; leaflets 2–24 pairs per pinna; pods rarely viscid:

Pods glabrous or subglabrous:

Margins of pods usually entire, occasionally slightly constricted between some of the seeds; valves thinly woody; seeds 6–11 × 5–8,5 mm; leaves with 2–5(6) pinnae pairs:

Pods 0,7–1,1 cm wide, falcate; leaves usually with 2–3 pinnae pairs, leaf-rhachides glabrous; spines typically slightly swollen and fused basally; young branchlets slender, "cushions" poorly developed at the nodes..............38. *A. grandicornuta*

Pods 1,2–2,4(3) cm wide, ± straight to falcate; leaves usually with 3–6 pinnae pairs; leaf-rhachides glabrous to densely pubescent; spines not swollen; young branchlets robust, usually with well developed "cushions" at the nodes....37. *A. robusta*

Margins of pods invariably distinctly constricted between the seeds, regularly or irregularly so, often ± moniliform, valves not thinly woody; seeds 3,5–6,5(9) × 2–5(7) mm; leaves with 2–13 pinnae pairs...........................19. *A. karroo*

Pods sparingly to densely grey-puberulous or -tomentellous:

Pods straight, 2,6–3,2 cm long; dwarf shrub up to 0,6 m high; leaves with 1 pinna pair; 2–4 pairs of leaflets per pinna.................................44. *A. redacta*

Pods curved or falcate, longer than above; robust shrub or large tree; leaves with 3–12 pinnae pairs; > 4 pairs of leaflets per pinna:

Pods falcate; seeds ± quadrate, 9–12 × 5–7 mm; young branchlets with well developed "cushions" at the nodes; usually a tree with an irregularly flattened crown ....36. *A. gerrardii*

Pods slightly curved to ±falcate; seeds oblong-elliptic, up to 8 × 6 mm; "cushions" on young branchlets poorly developed or absent; shrub or tree, crown often irregularly rounded...19. *A. karroo*

Naturalized and cultivated exotic species

Leaves bipinnate:

Plant armed with paired stipular spines..48. *A. farnesiana*

Plant unarmed:

Leaflets large, mostly 2–6 cm long, 0,4–1 cm wide, lanceolate to linear-lanceolate, often somewhat falcate; leaves large, 30–40 cm long.....................................60. *A. elata*

Leaflets 1,5–15 mm long, < 2 mm wide; leaves smaller than above:

Leaves with 1–4 pinnae pairs:

Petiole and rhachis together < 2,5 cm long; pinnae crowded, glaucous; midrib ± central in the leaflet or inconspicuous; inflorescence an axillary raceme or panicle, longer than the leaves ....52. *A. baileyana*

Petiole and rhachis together > 3 cm long; pinnae neither crowded nor glaucous; midrib almost marginal throughout the length of the leaflet, pubescent; inflorescences axillary, solitary, paired or fascicled...61. *A. visite*

Leaves with (5)8–26 pinnae pairs:

Leaflets 1,5–5,5 mm long:

Leaf-rhachis with a gland at the junction of each pinna pair and usually also with additional glands between the pinnae pairs; pods ± moniliform....................49. *A. mearnsii*

Leaf-rhachis with a gland at the junction of all or most pairs of pinnae but not between the pinnae pairs; pods not or only slightly moniliform........................50. *A. dealbata*

Leaflets 6–15 mm long:

Young branchlets prominently angled, sometimes with wing-like ridges 1–2 mm high; pinnae up to 15 pairs per leaf; leaflets 6–15 × 0,3–0,75 mm, midrib ± central, glabrous; inflorescence an axillary raceme or panicle...51. *A. decurrens*

Young branchlets not prominently angled; pinnae up to 7 pairs per leaf; leaflets 6–9 × 0,8–1,25(2) mm, midrib almost marginal throughout its length and pubescent; inflorescences axillary, solitary, paired or fascicled...............................61. *A. visite*

Leaves apparently simple, modified to phyllodes by dilation of the petiole and rhachis:

Plant armed with stipular spines..53. *A. armata*

Plant unarmed:

Phyllodes each with one main longitudinal nerve:

Phyllodes < 4cm long, 0,5–2 cm wide, seldom up to 4,5 cm long but then < 0,5 mm wide and linear-oblong:

Phyllodes 2–5 mm wide, linear-oblong, the margins typically densely ciliate........64. *A. fimbriata*

Phyllodes 6–20 mm wide, not linear-oblong:

Young branchlets and phyllodes densely grey-pubescent, especially when young; phyllodes ovate to elliptic or elliptic-oblong, mostly 10–20 mm wide; pods 1,5–2 cm wide ....54. *A. podalyriifolia*

Young branchlets and phyllodes glabrous; phyllodes obliquely obovate-lanceolate to ovate-triangular, 6–11 mm wide; pods 5–7 mm wide........................62. *A. cultriformis*

Phyllodes 5 –22 cm long:

Phyllodes 1,5–3 mm wide..65. *A. adunca*

Phyllodes > 5 mm wide:

Phyllodes ± straight or sometimes slightly falcate, linear-lanceolate to linear-oblong or oblanceolate, mostly 0,6–1,4 cm wide:

Flower-heads > 6,5 mm in diameter; peduncles 6–22 mm long; naturalized species. .55. *A. saligna*

Flower-heads < 6 mm in diameter; peduncles 3–6 mm long; cultivated only....63. *A. retinodes*

Phyllodes distinctly falcate, obovate-lanceolate, mostly 1,3–3 cm wide..........56. *A. pycnantha*

Phyllodes each with 2–7 or more longitudinal nerves:

Flowers in spikes:

Pods straight or slightly curved, glabrous; inflorescence axis glabrous or almost so; naturalized species..57. *A. longifolia*

Pods variously coiled or twisted, pubescent; inflorescence axis pubescent; cultivated only........ ....66. *A. maidenii*

Flowers in round heads:

Phyllodes < 4 mm wide:

Phyllodes green; inflorescences axillary, solitary or paired, rarely fascicled; pods 3–3,5 mm wide, sparingly to densely pubescent, margins not winged...................67. *A. viscidula*

Phyllodes greyish or glaucous; inflorescence a short axillary raceme; pods 8–18 mm wide; the margins distinctly winged...68. *A. pendula*

Phyllodes > 5 mm wide:

Phyllodes 0,6–2,5 cm wide, green, longitudinal nerves conspicuous; pods falcate or variously coiled or spirally twisted; margins not winged, naturalized species:

Flowers bright yellow; petals free; phyllodes usually±straight, sometimes slightly falcate, with anastomosing longitudinal venation between the main longitudinal nerves 58. *A. cyclops*

Flowers pale yellowish-white; petals united to above the middle; phyllodes usually ± falcate, sometimes±straight, with conspicuous reticulate venation between the main longitudinal nerves...59. *A. melanoxylon*

Phyllodes 0,5–0,9 cm wide, greyish or glaucous, longitudinal nerves not prominent; pods straight or slightly curved, the margins distinctly winged; cultivated only......68. *A. pendula*

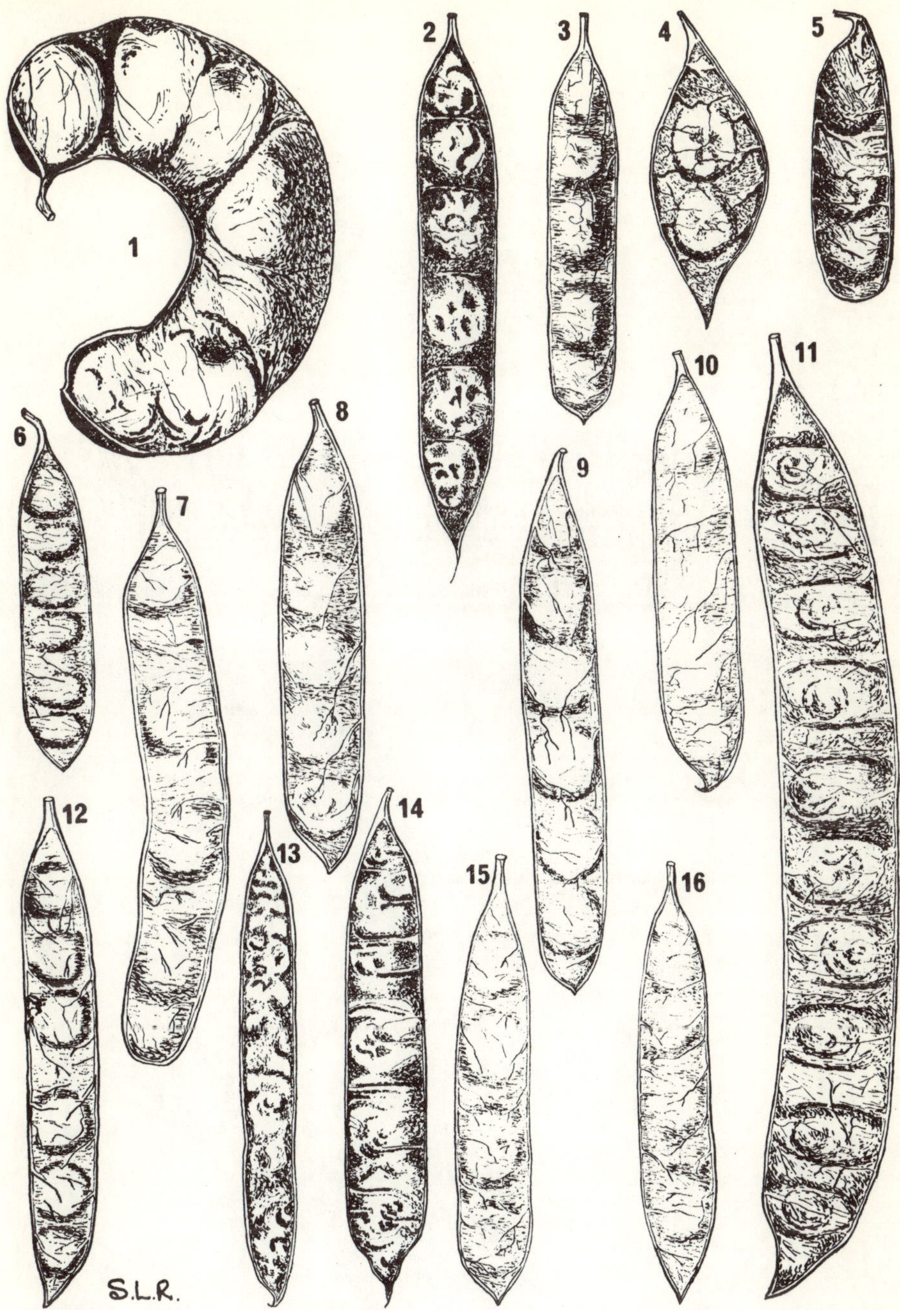

FIG. 3.—Pods of spicate-flowered **Acacia** species, × ⅔. 1, **A. albida** (*Van der Schijff* 1425); 2, **A. ataxacantha** (*Ross* 972); 3, **A. senegal** var. **leiorhachis** (*Verdoorn* 2326); 4, **A. senegal** var. **rostrata** (*Ross* 1632); 5, **A. mellifera** subsp. **detinens** (*Ross* 1491); 6, **A. robynsiana** (*De Winter & Hardy* 8161); 7, **A. montis-usti** (*De Winter & Leistner* 5841); 8, **A. welwitschii** subsp. **delagoensis** (*Codd & Verdoorn* 5480); 9, **A. nigrescens** (*Ross* 1156); 10, **A. burkei** (*Codd* 4021); 11, **A. galpinii** (*Galpin* 14009); 12, **A. polyacantha** subsp. **campylacantha** (*Codd* 3024); 13, **A. caffra** (*Edwards* 2834); 14, **A. hereroensis** (*De Winter* 2344A); 15, **A. fleckii** (*De Winter* 3029); 16, **A. erubescens** (*Ross* 1504).

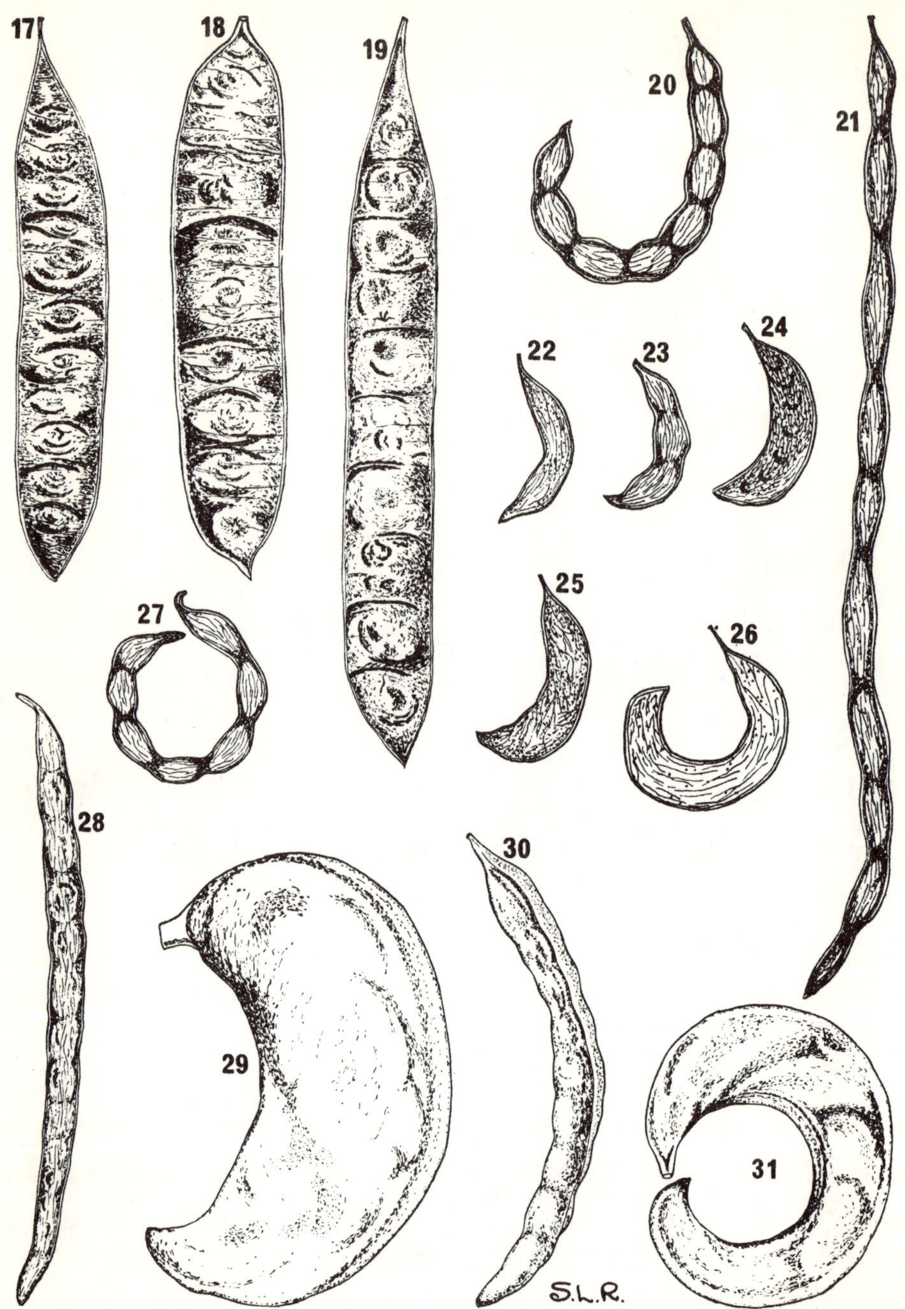

FIG. 4.—Pods of capitate-flowered **Acacia** species, × ⅔. 17, **A. kraussiana** (*Strey* 6559); 18, **A. brevispica** subsp. **dregeana** (*Ross* 766); 19, **A. schweinfurthii** (*Ross* 1022); 20, **A. karroo** (*Ross* 1617); 21, **A. karroo** (*Ross* 1673); 22, **A. tenuispina** (*Codd* 6370); 23, **A. exuvialis** (*Codd & Verdoorn* 5467); 24, **A. swazica** (*Codd & De Winter* 4970); 25, **A. nebrownii** (*Codd* 5824); 26, **A. permixta** (*Codd* 5916); 27, **A. borleae** (*Codd* 4312); 28, **A. davyi** (*Ross* 264); 29, **A. erioloba** (*Meeuse* 10143); 30, **A. haematoxylon** (*De Winter* 3433); 31, **A. erioloba** × **A. haematoxylon** (*Acocks* 12689).

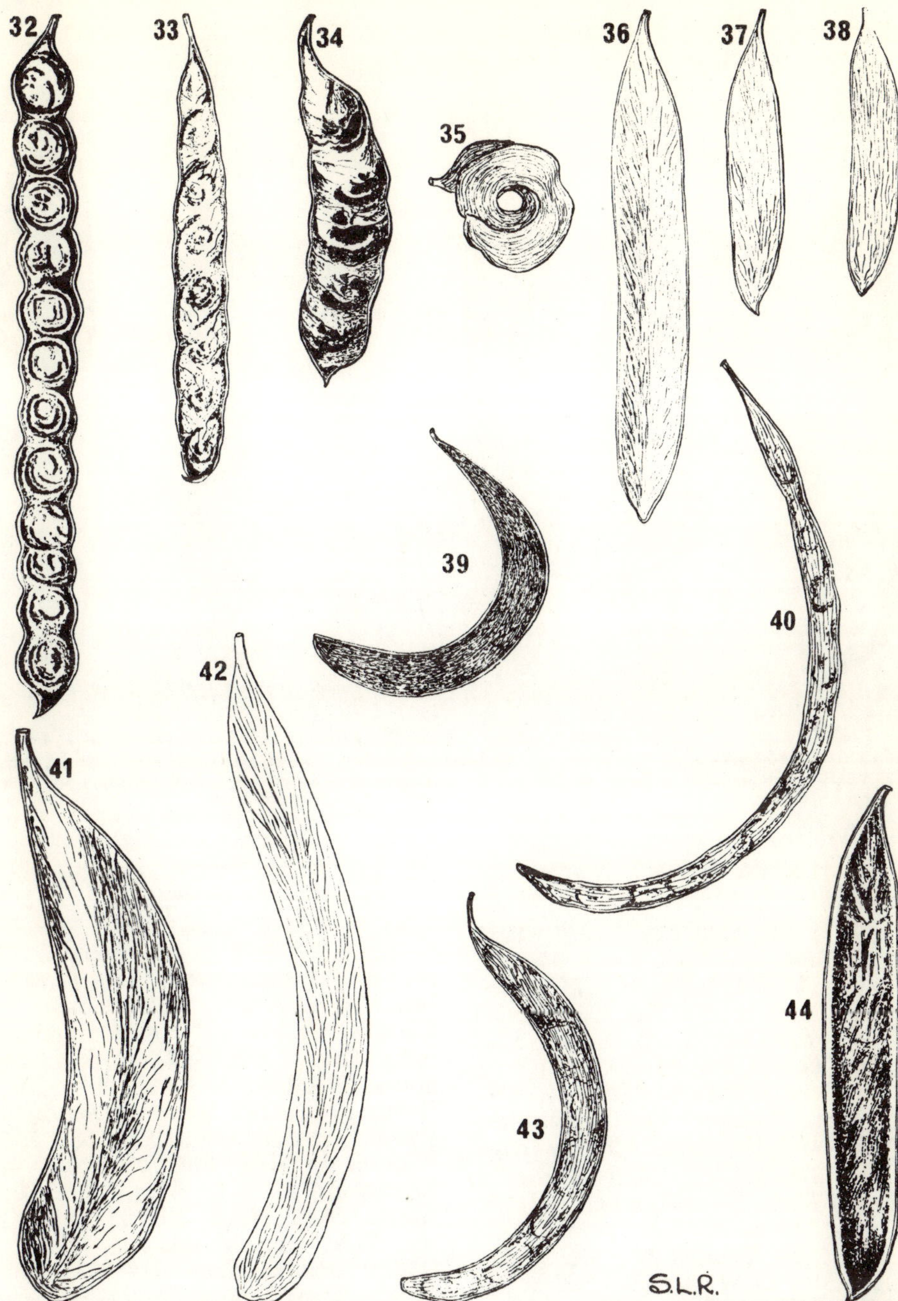

FIG. 5.—Pods of capitate-flowered **Acacia** species, × ⅔. 32, **A. nilotica** (*Ross* 927); 33, **A. xanthophloea** (*Lang* sub TRV 32250); 34, **A. kirkii** subsp. **kirkii** (*De Winter & Leistner* 5869); 35, **A. tortilis** subsp. **heteracantha** (*Ross* 819); 36, **A. luederitzii** var. **luederitzii** (*Acocks* 18767); 37, **A. luederitzii** var. **retinens** (*Edwards* 3260); 38, **A. reficiens** subsp. **reficiens** (*De Winter & Leistner* 5911); 39, **A. gerrardii** var. **gerrardii** (*Mogg* 17309); 40, **A. arenaria** (*De Winter & Leistner* 5609); 41, **A. robusta** subsp. **robusta** (*Marais* 903); 42, **A. robusta** subsp. **clavigera** (*Ross* 1021); 43, **A. grandicornuta** (*Codd & Verdoorn* 5493); 44, **A. rehmanniana** (*Galpin* 14008).

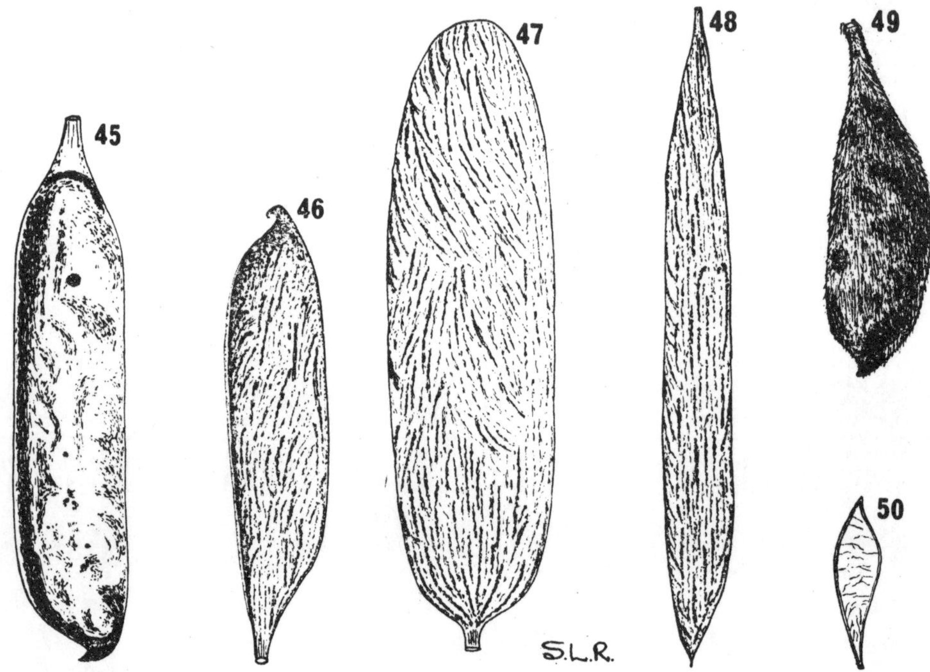

FIG. 6.—Pods of capitate-flowered **Acacia** species, × ⅔. 45, **A. sieberana** var. **woodii** (*Ross* 1187); 46, **A. hebeclada** subsp. **hebeclada** (*Morris* 1044); 47, **A. hebeclada** subsp. **chobiensis** (*Codd* 7091); 48, **A. hebeclada** subsp. **tristis** (*De Winter & Leistner* 5305); 49. **A. stuhlmannii** (*Van der Schijff* 5331A); 50, **A. redacta** (*Werger* 1518).

1. **Acacia albida** *Del.*, Fl. Égypte Expl. Planches : 286, t.52, fig. 3 (1813); DC., Prodr. 2 : 459 (1825); Oliv. in F.T.A. 2 : 339 (1871); Benth. in Trans. Linn. Soc. Lond. 30 : 515 (1875); Engl. in Bot. Jahrb. 10 : 20 (1888); Marloth in Trans. S. Afr. Phil. Soc. 5 : 271 (1893); Schinz in Mém. Herb. Boiss. 1 : 104 (1900); Sim, For. Fl. P.E. Afr. 54, t.34 (1909); Dinter, Fl. Deutsch-Sudwest-Afrika 74(1909); Dinter, Veg. Veldkost Deutsch-Südwest-Afrikas 32 (1912); Harms in Engl., Pflanzenw. Afr. 3, 1 : 376, fig. 220 (1915); Glover in Ann. Bolus Herb. 1 : 146, t.18/12 (1915); Bak.f., Leg. Trop. Afr. 3 : 825 (1930); Burtt Davy, Fl. Transv. 2 : 335 (1932); Pönninghaus in J.S.W. Afr. Sci. Soc. 6 : 12 (1933); Boss, Pflanzenleben Südwestafrika's 30 (1934); Henkel, Woody Pl. Natal 232 (1934); Stapleton, Common Transv. Trees 5 (1937); Hutch., Botanist in S. Afr. 391, 392 (1946); Codd, Trees & Shrubs Kruger Nat. Park 38, fig. 32 (1951); O. B. Miller in J.S. Afr. Bot. 18 : 18 (1952); Young in Candollea 15 : 89 (1955); Torre in C.F.A. 2 : 272 (1956); Keay in F.W.T.A. ed. 2, 1 : 499 (1958); Brenan in F.T.E.A. Legum.–Mimos. : 78, fig. 14/1 (1959); Palmer & Pitman, Trees S. Afr. 148, t. v, 31 (1961); F. White, For. Fl. N. Rhod. 82, fig. 17B, C(1962); Von Breitenbach, Indig. Trees S. Afr. 2 : 272 (1965); Gomes & Sousa, Dendrol. Mocamb. 1 : 232, t.36 (1966); Ross in Bol. Soc. Brot., Sér. 2, 40 : 187 (1966); Schreiber in F.S.W.A. 58 : 7 (1967); Wickens in Kew Bull. 23 : 181 (1969); Brenan in F.Z. 3, 1 : 63, t.17 (1970); Ross, Acacia Spp. Natal 20, fig. 1/1 (1971); Van Wyk, Trees Kruger Nat. Park 1 : 126 (1972); Ross, Fl. Natal 192 (1973); Palmer & Pitman, Trees S. Afr. 2 : 741 (1973); Schreiber in Mitt. Bot. Staatssamml. Munchen 11 : 115 (1973). Type: Egypt, above Philae, *Nectoux* (MPU, holo.).

A. mossambicensis Bolle in Peters, Reise Mossamb. Bot. 1 : 5 (1861). Type: Mozambique, Rios de Sena and R. Chimazo, W. of Tete, *Peters* (B, ? syn. †).

Prosopis ? *kirkii* Oliv. in F.T.A. 2 : 332 (1871).
Type: Malawi, Shire River, *Kirk* (K, holo.!).

Faidherbia albida (Del.) A. Chev. in Rev. Bot. Appl.
Agric. Trop. 14 : 876 (1934); Aubrév., Fl. For. Soud.
Guin. 280, t.51/3, t.53/6 (1950); Gilbert & Boutique in
F.C.B. 3 : 169 (1952). Type as for *Acacia albida* Del.

Tree to 30 m high; trunk to 2 m diam.;
crown rounded, branches spreading and
frequently drooping to the ground in mature
trees, young plants often irregularly branched
and spindle-like. *Bark* dark brown to
greenish-grey or ashen, rough; young branchlets greenish-white or ashen, subglabrous to
pubescent. *Stipules* spinescent, in pairs, up to
3,2 cm long, straight or slightly curved,
greenish-white to light grey-brown, tips
often reddish-brown when young; no prickles
below the stipules. *Leaves*: petiole 0,5–3,7 cm
long, adaxial gland absent; rhachis (1,3)
3–6(7,5) cm long, subglabrous or puberulous,
with a single conspicuous gland at the
junction of each of the 2(3)–10 pairs of
pinnae; rhachillae (1,5)2,5–5,5(8,9) cm
long; leaflets grey-green, 6–23 pairs per pinna,
(2,5)4,5–9(13) × 0,75–3(5) mm, linear or
linear-oblong to slightly obovate-oblong,
apex rounded to subacute or mucronate,
margin with or without white ciliolate hairs,
glabrous or sparingly to densely appressed-
pubescent ab- and/or adaxially. *Inflorescences*
spicate, usually produced singly in the axil of
a leaf, collectively forming a terminal panicle
or raceme. *Flowers* yellowish-white to pale
cream, sessile or to 0,5(2) mm pedicellate;
spikes 3,5–15,7 cm long; peduncles (0,8)
2–4(6,3) cm long, subglabrous to pubescent.
Calyx campanulate, glabrous to pubescent,
tube 0,5–1,8 mm long, lobes 0,3–0,7 mm
long. *Corolla* often a delicate pink inside
basally, tube 0,8–2,5 mm long, lobes
divided almost to the base, up to 3 mm long,
glabrous to pubescent. *Stamen-filaments* 4–6
mm long, shortly connate basally for ± 1
mm; anthers 0,2–0,4 mm across, eglandular
even in bud. *Ovary* 0,7–1,4 mm long,
shortly stipitate, pilose; style glabrous or
subglabrous. *Pods* bright orange to reddish-
brown, falcate or curled into a circular coil
or variously twisted, indehiscent, thick,
6–35 × (1,4)2–3,5(4,5) cm, glabrous or very
rarely puberulous. *Seeds* light to dark brown,
9–12 × 4–8 mm, elliptic-lenticular; areole
7–9 × 4–6 mm, elliptic-lenticular.

Widespread in tropical and subtropical Africa
from Senegal, Gambia and Egypt southwards to
South West Africa, Botswana, the Transvaal and

Natal (Tongaland). Found usually on alluvial
floodplains, in riverine fringing vegetation, on the
margins of pans or swamps or, in more arid localities,
along dry watercourses or where a fairly high water
table exists.

S.W.A.—1712 (Posto Velho): bank of Kunene
River, *Story 5803*. 1713 (Swartbooisdrift): Ombazu,
51,2 km N. of Ohopoho on road to Swartbooisdrift,
De Winter & Leistner 5895. 1724 (Katima Mulilo):
Lisikili, 25 km E. of Katima Mulilo, *Codd 7100*. 1913
(Sesfontein): near Sesfontein, *Hall 497*. 2014 (Wel-
witschia): bed of Ugab River at Sorris-Sorris, near
Brandberg, *De Winter 3151*. 2115 (Karibib): Usakos,
Pole Evans sub PRE 19303. 2216 (Otjimbingwe):
Kuiseb, *Strey 2635*. Grid ref. unknown: bed of the
Khan River, *Pillans 5919* (BOL); Damaraland,
Marloth 1194 (BOL, GRA)—the labels on these two
specimens bear different localities in Damaraland.

TRANSVAAL.—2230 (Messina): Messina, *Pole
Evans sub PRE 13118*. 2231 (Pafuri): Kruger National
Park, Makuleka, *Lang sub TRV 32262*. 2329 (Pietersburg): 32 km N. of Pietersburg on Kalkbank road,
Story 1557. 2330 (Tzaneen): 80 km N. of Gravelotte
station, on bank of Great Letaba River, *Galpin
13533*—the precise locality of this specimen is
uncertain because the Great Letaba River is only
32 km north of Gravelotte by road; the Small Letaba
River, however, is ± 80 km north of Gravelotte.
2428 (Nylstroom): 14,4 km N.W. of Potgietersrust on
road to Zaaiplaats, *Codd 4185*. 2531 (Komatipoort):
Lebombo flats near Swaziland border, *Keet 1495*.
Grid ref. unknown: northern Soutpansberg, Sand
River, *Legat 49*; N.E. Soutpansberg, Nwanedzi River,
Pole Evans 191; junction of Crocodile and Magalak-
win rivers, *Pole Evans 3968*.

NATAL.—2632 (Bela Vista): Ndumu Game
Reserve, Usutu floodplain, *Ross 699*. 2732 (Ubombo):
9,6 km N. of Otoboteni, P.O. at Mfongozi, *Codd
2085*.

A. albida displays a number of unusual characters, some of which are peculiar to this species alone
amongst the African acacias. *A. albida* differs in
having eglandular petioles but a gland on the rhachis
at the junction of each pair of pinnae, stamen-
filaments which are shortly connate basally, large
anthers which are eglandular even when in bud, and
typically falcate or spirally coiled indehiscent pods.
However, each of these characters may be found in
other species of *Acacia* although in no other species
are all of these characters associated together.
Chevalier in Rev. Bot. Appl. Agric. Trop. 14 : 876
(1934) considered the species to be sufficiently distinct
from all others to transfer it to the monotypic genus
Faidherbia.

A. albida is not closely related to any of the other
African species and there are the above differences,
coupled with differences in pollen morphology and
seedling development, to suggest that the species
should be excluded from *Acacia*. However, although
differing from the other African acacias, *A. albida*
does nevertheless share many characters in common
with them. In deciding whether or not the species
should be excluded from *Acacia*, it depends upon
whether the emphasis is placed on the differences or on
the similarities. It may ultimately prove better to
transfer the species to *Faidherbia*.

Another unusual feature of *A. albida* is the tendency for plants to sometimes shed their leaves at the commencement of the rainy season and remain leafless throughout the summer, finally producing new leaves towards the beginning of the dry season.

A. albida, popularly known as the "ana tree" or "anaboom", is one of our largest acacias. A fine group of trees north-west of Potgietersrust on the road to Zaaiplaats in the Transvaal is protected under the National and Historical Monuments Act. These trees have been referred to as "Livingstone's Trees" but it seems very unlikely that Livingstone ever saw them.

The pods of *A. albida* are regarded as a valuable fodder and are relished by game and domestic stock.

In our area *A. albida* has the leaflets ± pubescent on the surface and ± pubescent young branchlets, inflorescence-axes, calyces and (often) corollas, thus corresponding to Race B as defined by Brenan in F.T.E.A. Legum.–Mimos. : 79 (1959). Race A is not known in our area.

2. **Acacia ataxacantha** *DC.*, Prodr. 2 : 459 (1825); Oliv. in F.T.A. 2 : 343 (1871); Benth. in Trans. Linn. Soc. Lond. 30 : 520 (1875); Burtt Davy in Kew Bull. 1908 : 156 (1908); Glover in Ann. Bolus Herb. 1 : 147, t.18/3 (1915); Bews, Fl. Natal 114 (1921); Bak.f., Leg. Trop. Afr. 3 : 834 (1930); Henkel, Woody Pl. Natal 234 (1934); Codd, Trees & Shrubs Kruger Nat. Park 40, fig. 33a (1951); Gilbert & Boutique in F.C.B. 3 : 153 (1952); Torre in C.F.A. 2 : 278 (1956); Keay in F.W.T.A., ed. 2, 1 : 499 (1958); Brenan in F.T.E.A. Legum.–Mimos. : 82, fig. 14/5 (1959); Von Breitenbach, Indig. Trees S. Afr. 2 : 270 (1965); Ross in Webbia 21 : 629 (1966); Schreiber in F.S.W.A. 58 : 7 (1967); Brenan in F.Z. 3, 1 : 65, t.15/1 (1970); Ross, Acacia Spp. Natal 21, fig. 1/2 (1971); Flow. Pl. Afr. 42 : t.1652 (1972); Ross, Fl. Natal 192 (1973); Palmer & Pitman, Trees S. Afr. 2 : 747 (1973); Schreiber in Mitt. Bot. Staatssamml. Munchen 11 : 116 (1973). Syntypes: Senegal, *Bacle* (G—DC) & *Perrottet* (G—DC).

A. eriadenia Benth. in Hook., Lond. J. Bot. 5 : 98 (1846); Harv. in F.C. 2 : 283 (1862); Benth. in Trans. Linn. Soc. Lond. 30 : 520 (1875); Schinz in Mém. Herb. Boiss. 1 : 108 (1900) (sphalm *ariadenia*); Burtt Davy in Kew Bull. 1908 : 157 (1908); Glover in Ann. Bolus Herb. 1 : 147 (1915); Dinter in Feddes Repert. 15 : 79 (1917); Bews, Fl. Natal 114 (1921); Hutch., Botanist in S. Afr. 308 (1946). Type: Transvaal, Crocodile River, Magaliesberg *Burke* 130 (K, holo.!; BM!; PRE!; Z!). *A. lugardiae* N.E. Br. in Kew Bull. 1909 : 107 (1909); Bak.f., Leg. Trop. Afr. 3 : 834 (1930); O. B. Miller in J. S. Afr. Bot. 18 : 23 (1952). Type: Botswana, Kwebe Hills, *Mrs E. J. Lugard* 195 (K, holo.!; GRA!; Z!). ? *A. caffra* var. *rupestris* Sim, For. Fl. P.E.Afr. 56, t.39B (1909). Type: Mozambique,

"on and below the Lebombo's, in Maputa and Marracuene, less common nearer Lourenco Marques", *Sim* 6235 (whereabouts unknown). Sim's description and plate indicate that the prickles were scattered along the stem as in the case of *A. ataxacantha*; in *A. caffra* the prickles are in pairs near the nodes. Sim's description of the pubescence, pod, growth form and distribution suggest that he was recognizing in var. *rupestris* the tree growth form of *A. ataxacantha*. *A. ataxacantha* var. *australis* Burtt Davy in Kew Bull. 1922 : 324 (1922); Bak.f., Leg. Trop. Afr. 3 : 834 (1930); Burtt Davy, Fl. Transv. 2 : 335 (1932); O. B. Miller in J. S. Afr. Bot. 18 : 19 (1952); Young in Candollea 15 : 84 (1955); Brenan in F.T.E.A. Legum.–Mimos. : 83 (1959); F. White, For. Fl. N.Rhod. 82, fig. 17A (1962); Von Breitenbach, Indig. Trees S.Afr. 2 : 270 (1965). Type: Transvaal, Letaba Distr., Magoebas Kloof, Houtboschberg, *Burtt Davy* 5231 (K, holo.!; BOL!).

Scandent shrub up to 15 m high, often with many stems arising from a common base, a non-climbing shrub or occasionally a tree up to 10 m high, crown often slightly rounded in arborescent forms; trunk to 0,5 m diam. *Bark* pale to dark yellowish- or grey-brown, rough, slightly fissured, sometimes flaking; young branchlets pale yellowish- or grey- to reddish-brown, sparingly to densely pubescent, indumentum frequently slightly golden. *Stipules* not spinescent, in pairs above the nodes, obliquely ovate (rarely almost linear), up to 12 × 7 mm, soon deciduous. *Prickles* scattered along the internodes, usually strongly recurved, reddish-brown to purplish, often broad-based, up to 15 mm long. *Leaves*: petiole (0,4)1–2(3,4) cm long, adaxial gland usually present (sometimes two), variable in position, usually stalked, up to 2 mm high; rhachis (2,3)6–12(16,5) cm long, sparingly to densely pubescent, rarely subglabrous, with or without recurved prickles abaxially, a gland often present at the junction of the top pair of pinna only or between the top 1–5 and occasionally the lowest 1–3 pairs; pinnae (6)8–20(29) pairs; rhachillae (0,9)2,6–4,2(7,3) cm long; leaflets 19–62 pairs per pinna, 2–6,9 × 0,5–1,3 mm, linear to linear-oblong, often slightly falcate, apex obtuse to acute, margins with or without ciliate hairs, glabrous abaxially or appressed-pubescent with a tendency for an apical and/or basal tuft. *Inflorescences* spicate, fascicled or crowded into an irregular terminal raceme, occasionally solitary. *Flowers* yellowish-white, pedicellate or appearing sessile; spikes 2,2–11,5 cm long; peduncles 0,3–2,5 cm long, sparingly to densely pubescent. *Calyx* cupular, glabrous

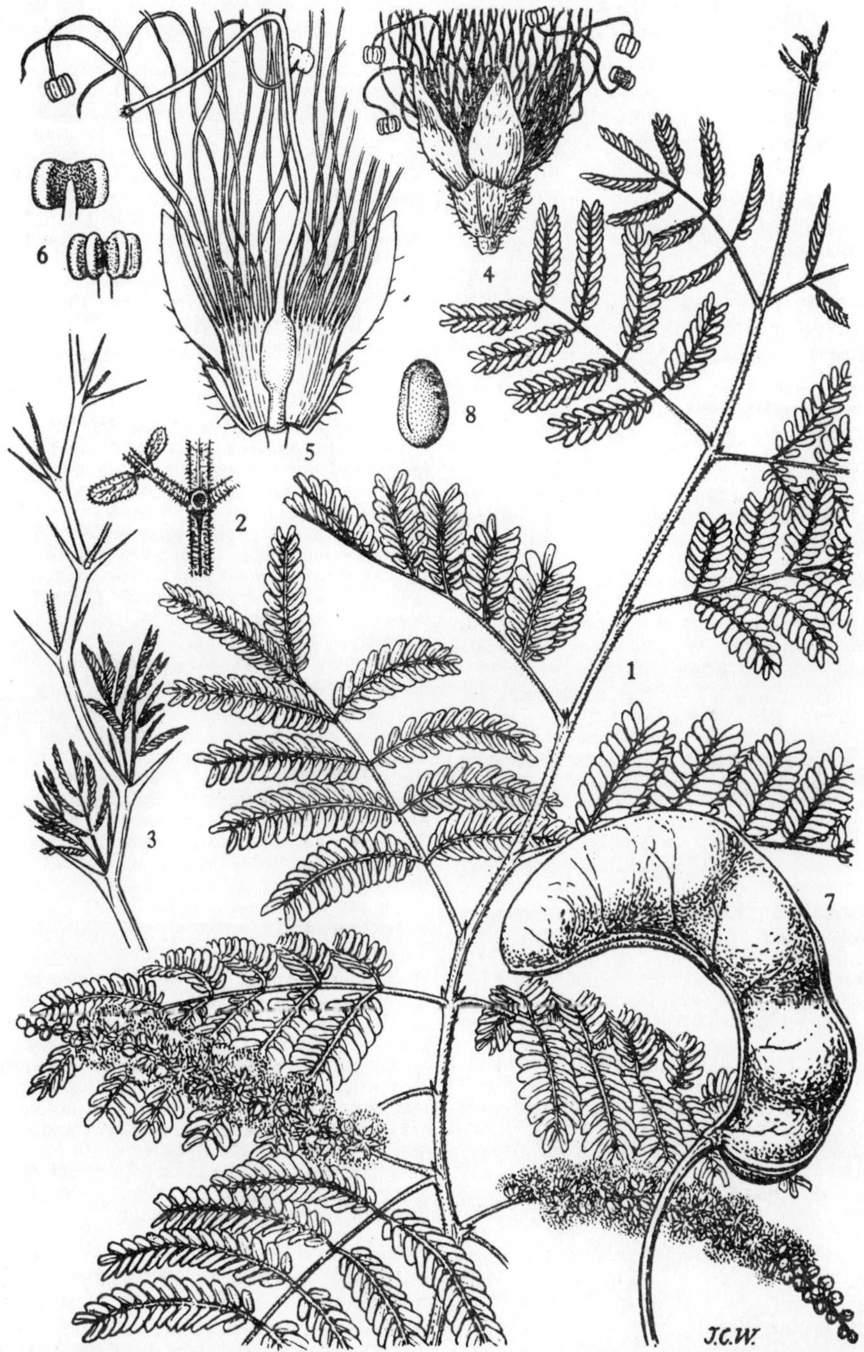

FIG. 7.—**Acacia albida.** 1, flowering branch, × ⅔; 2, part of leaf-rhachis showing gland, × 4; both from *Robinson* 269; 3, juvenile shoot, × ⅔, *Lovemore* 68; 4, flower, × 6; 5, flower opened out to show ovary, × 8; 6, anthers, × 20, all from *Robinson* 269; 7, pod, × ⅔, *Meikle* s.n.; 8, seed, × 1, *Kirk* s.n. Reproduced by permission of the Editorial Board of Flora Zambesiaca.

48

to sparingly pubescent, usually $\frac{1}{3}$–$\frac{1}{2}$ as long as corolla, tube 0,6–1,7 mm long, lobes 0,2–0,6 mm long. *Corolla* campanulate, glabrous to sparingly pubescent, tube 1,4–2,2 mm long, lobes up to 0,8 mm long. *Stamen-filaments* free, up to 6 mm long; anthers 0,15 mm across, with a deciduous apical gland. *Ovary* densely pubescent, 0,6–1,4 mm long, on a stipe longer than itself. *Pods* reddish- or purplish-brown, 5,3–20,4 × 0,9–2,4 cm, linear-oblong, straight, very acuminate at both ends or merely acute, longitudinally dehiscent, brittle, umbonate over the seeds, glabrous or subglabrous. *Seeds* olive-brown, subcircular-lenticular, 6–9 mm in diam., compressed; areole central, small, 2,5–3 mm in diam., obscure.

Widespread in tropical and subtropical Africa from Senegal in the west to the Sudan Republic in the north-east, southwards to South West Africa, Botswana, the Transvaal, Swaziland, Natal and the Eastern Cape Province. Occurs in riverine fringing vegetation, on forest margins, in forest clearings, as a component of mesophytic scrub in shaded kloofs and ravines, in dry river valley scrub and thornveld, on boulder-strewn slopes or, occasionally, in open grassland. Often forms dense impenetrable thickets.

S.W.A.—1716 (Enana): 7 km S.E. of Oshandi, *De Winter & Giess 7025.* 1719 (Runtu): 8 km W. of Runtu on road to Kapako, *De Winter 3737.* 1819 (Karakuwisa): Cigarette, E. of Karakuwisa, *Maguire 2315* (K, NBG). 1820 (Tarikora): 16 km E. of Nyangana mission station, *De Winter 4199.* 1917 (Tsumeb): Farm Heidelberg, near Tsumeb, *Walter 521* (M). 1918 (Grootfontein): 8 km from Narugas, *Merxmuller & Giess 1777* (K, M). 1920 (Tsumkwe): western foot of Aha mountains, *Story 6374.*

TRANSVAAL.—2229 (Waterpoort): northern entrance to Wyllie's Poort, *Hutchinson 2062.* 2230 (Messina): Sibasa, *Van Warmelo sub PRE 53312/2.* 2329 (Pietersburg): Veekraal, near Woodbush, 57,6 km E. of Pietersburg, *Mogg 14740.* 2330 (Tzaneen): Tzaneen, *Rogers 12407* (BOL, GRA). 2427 (Thabazimbi): Thabazimbi, *Rogers 255* (GRA). 2428 (Nylstroom): near Zandriverspoort, near Vaalwater, *Smuts 376.* 2429 (Zebediela): Chuniespoort, *Van Vuuren 1604.* 2430 (Pilgrim's Rest): Vaalhoek, *Rogers 25073.* 2527 (Rustenburg): pass near Meerhof, Hartebeespoort Dam, *De Winter 7658.* 2528 (Pretoria): Pretoria, Grosvenor Square, *Repton 2119* (K). 2529 (Witbank): Roossenekal, *Morris 1179.* 2530 (Lydenburg): near Lydenburg, *Wilms 447* (BM,K). 2531 (Komatipoort): Kruger National Park, Shabin Kop, *Van der Schijff 1240.* 2627 (Potchefstroom): Farm Gladysvale, 14,4 km W. of Krugersdorp, *Rodin 3869.* 2629 (Bethal): N. of Amersfoort, *Strey 7880.* 2730 (Vryheid): 3,2 km from Brauschweig, *Devenish 989.*

SWAZILAND.—2531 (Komatipoort): Piggs Peak, *Compton 30605.* 2631 (Mbabane): Black Umbeluzi valley, *Compton 31867.* 2731 (Louwsburg): Hluti, *Ross 1528* (NU).

NATAL.—2730 (Vryheid): 19,2 km from Natal Spa on Paulpietersburg road, *Ross 1245* (NH, NU). 2731 (Louwsburg): Nongoma, *Ross 1076* (K, NH, NU). 2732 (Ubombo): 0,5 km from Ubombo on Mkuze–Ubombo road, *Ross 262* (K, NU). 2829 (Harrismith): Ladysmith, *Ross 559* (K, NH, NU). 2830 (Dundee): 11,2 km from Muden on Weenen road, *Ross 644* (K, NH, NU). 2831 (Nkandla): Eshowe forest, *Kotze 156.* 2832 (Mtubatuba): Hluhluwe Game Reserve, *Ward 2877.* 2930 (Pietermaritzburg): Bisley, near Pietermaritzburg, *Ross 1603* (NH, NU). 2931 (Stanger): below Bulwer farm, opposite confluence of Nembe and Tugela Rivers, *Ross 206* (NU). 3030 (Port Shepstone): Shelley Beach, *Strey 7072.* 3130 (Port Edward): 4,8 km from Port Edward on cliffs above Umtamvuna River, *Hilliard 1663* (NU).

CAPE.—Grid ref. unknown: between Umtata and Port St. Johns, *Bolus 8900* (BOL); between Umtata and Umzimvubu, *Drege 5552* (P).

Variety *australis* was held by Burtt Davy to differ from typical *A. ataxacantha* in having larger leaves with more numerous pinnae and leaflet pairs and leaflets "thinly pilose with scattered, appressed hairs, especially on the margins." However, the abundant material collected in recent years contains so many intermediates between var. *australis* and typical *A. ataxacantha* that it is no longer considered desirable to uphold var. *australis.* Despite this, there is an overall tendency for an increase in the degree of pubescence of the young branchlets, leaf-rhachides, leaflets and inflorescence-axes in the southern part of the species range in Africa.

Mention must be made of some atypical specimens from the Kaokoveld in South West Africa, for example, *De Winter & Leistner 5494, Merxmuller & Giess 1409* (K, M) and *Walter 1/260* (M). These specimens differ from more typical *A. ataxacantha* in our area in lacking stalked petiolar glands, in having few pinnae pairs, glabrous or very sparingly pubescent leaf-rhachides, and short inflorescences. It is thought that these differences may be the response to drier conditions or to some edaphic factor. For the present it is not intended to accord these specimens any formal taxonomic recognition.

A. ataxacantha has in the past often been confused with *A. caffra* (Thunb.) Willd. *A. caffra* is usually easily separated from *A. ataxacantha* in having prickles in pairs near the nodes, linear stipules, sessile petiolar glands, the calyx much longer in proportion to the corolla, a glabrous shortly stipitate ovary and brownish pods. Very occasionally *A. caffra* may have a few scattered prickles in addition to the normal paired prickles and then confusion is perhaps possible. However, the other characters enumerated above should enable the two species to be distinguished without much difficulty.

3. **Acacia senegal** (*L.*) *Willd.*, Sp. Pl. 4 : 1077 (1806) pro parte excl. syn. fere omnibus; DC., Prodr. 2 : 459 (1825); Benth. in Trans. Linn. Soc. Lond. 30 : 516 (1875); Hutch. & Dalz. in F.W.T.A., ed. 1, 1 : 361 (1928) pro parte; Bak.f., Leg. Trop. Afr. 3 : 827 (1930) pro majore parte; Burtt Davy, Fl. Transv. 2 : 337 (1932); Codd, Trees

& Shrubs Kruger Nat. Park 50 (1951); Gilbert & Boutique in F.C.B. 3 : 149 (1952); Young in Candollea 15 : 93 (1955); Torre in C.F.A. 2 : 273 (1956); Keay in F.W.T.A., ed. 2, 1 : 498, fig. 159 (1958); Brenan in F.T.E.A. Legum.–Mimos. : 92, fig. 14/17 (1959); Von Breitenbach, Indig. Trees S. Afr 2 : 270 (1965); Ross in Bol. Soc. Brot., Sér. 2, 42 : 207 (1968); Brenan in F.Z. 3, 1 : 79 (1970); Palmer & Pitman, Trees S. Afr. 2 : 745 (1973); Ross in Bothalia 11(4) : 449 (1975). Type: Senegal, *Herb. Adanson* No. 16899 (P, neo.).

Mimosa senegal L., Sp. Pl. 1 : 521 (1753) excl. syn. fere omnibus. Type uncertain, presumably a specimen collected by Adanson in Senegal.

Shrub or tree up to 8 m high (in our area) with a slightly rounded or flattened and somewhat spreading crown, or a slender spindly tree with irregular virgate branches. *Bark* yellowish- or greyish-brown to purplish-black, rough, often corky and flaking off or smooth, papery and peeling off in strips; young branchlets yellowish- or greyish-brown to purplish-black, sometimes as though whitewashed over a purplish background, smooth or rough and flaking minutely or exfoliating to reveal a yellow inner layer, glabrous to densely pubescent. *Stipules* not spinescent, in pairs, linear, up to 5 mm long, soon deciduous. *Prickles* just below the nodes, up to 9 mm long, typically in threes, the central one hooked downwards and the two laterals $\pm$ curved upwards, or else solitary, the laterals being absent. *Leaves*: petiole 0,2–2 cm long, sparingly to densely pubescent, rarely subglabrous, adaxial gland usually present, sometimes two, variable in position, 0,5–0,75 mm in diam.; rhachis 0,7–6,9 cm long, sparingly to densely pubescent, lower surface with or without recurved prickles, with a gland at the junction of the top 1–5 pinnae pairs, between each pair or absent from some; pinnae (2)3–8(12) pairs; rhachillae 0,5–2,5 cm long, sparingly to densely pubescent; leaflets 7–25 pairs per pinna, 1–7 × 0,5–1,75 mm, linear to linear- or elliptic-oblong, apex obtuse to subacute, margins with or without cilia, sparingly appressed-pubescent above and beneath or entirely glabrous, lateral nerves invisible beneath or sometimes $\pm$ prominent. *Inflorescences* spicate, solitary or fascicled. *Flowers* yellowish-white, sessile; spikes 1,5–10 cm long, axis glabrous to densely pubescent; peduncle 0,1–2 cm long. *Calyx* sometimes tinged with pink, glabrous to somewhat pubescent, tube 0,9–2,6 mm long, lobes 0,2–0,8 mm long. *Corolla* glabrous or subglabrous, tube 2–3,2 mm long, lobes up to 0,9 mm long. *Stamen-filaments* free, up to 7 mm long; anthers 0,2–0,25 mm across, with a deciduous apical gland. *Ovary* glabrous, very shortly stipitate. *Pods* yellowish- or greyish-brown to brown, 1,8–9,5 × 1,2–3,4 cm, oblong, straight or almost so, longitudinally dehiscent, venose, apex rounded to acuminate or distinctly rostrate, sparingly to densely appressed-pubescent or puberulous. *Seeds* olive-brown to brown, 8–12 mm in diam., $\pm$subcircular-lenticular, compressed; central areole 2,5–6 × 2,5–5 mm, impressed, horse-shoe shaped.

Widespread in tropical Africa from Senegal in the west to Ethiopia in the north-east, southwards to the Transvaal, Swaziland and Natal. Four varieties are at present recognized within *A. senegal*, two of which occur in our area.

Inflorescence axis sparingly to densely pubescent throughout; pods up to 3,4 cm wide, usually rostrate or acuminate apically, sparingly to densely pubescent or puberulous, yellowish- or greyish-brown to brown; a shrub branching from or near the base or a tree with a slightly rounded or flattened and somewhat spreading crown; fairly widespread in South West Africa, the Transvaal, Swaziland and Natal..........(a) var. *rostrata*

Inflorescence axis glabrous or sometimes with a few basal hairs; pods up to 1,9 cm wide, rounded to acute apically, sparingly puberulous, yellowish to olive-brown; a slender spindly tree with irregular virgate branches; confined to the northern and eastern Transvaal.................(b) var. *leiorhachis*

(a) var. **rostrata** *Brenan* in Kew Bull. 8 : 99 (1953); Young in Candollea 15 : 96 (1955); Von Breitenbach, Indig. Trees S. Afr. 2 : 271 (1965); Schreiber in F.S.W.A. 58 : 12 (1967); Ross in Bol. Soc. Brot. Sér. 2, 42 : 233 (1968); Brenan in F.Z. 3, 1 : 79 (1970); Ross, Acacia Spp. Natal 39, fig. 1/6 (1971); Van Wyk, Trees Kruger Nat. Park 1 : 160 (1972); Ross, Fl. Natal 193 (1973). Type: Transvaal, Soutpansberg Distr., Dongola Reserve, *Verdoorn* 2264 (K, holo.!; PRE!).

A. spinosa Marloth & Engl. in Bot. Jahrb. 10 : 20 (1888), non *A. spinosa* E. Mey, Comm. 1 : 170 (1836); Dinter in Feddes Repert. 15 : 81 (1917); O. B. Miller in J. S. Afr. Bot. 18 : 24 (1952). Type: South West Africa, Karibib Distr., near Usakos, *Marloth* 1257 (K!, PRE!, iso.). *A. trispinosa* Marloth & Engl.

[Marloth in Trans. S. Afr. Phil. Soc. 5 : 269 (1893) nom. nud.] ex Schinz in Mém. Herb. Boiss. 1 : 115 (1900) sphalm, non *A. trispinosa* Stokes in Bot. Mat. Med. 3 : 168 (1812); Dinter, Deutsch-Südwest-Afrika 73 (1909); Ponnighaus in J. S.W. Afr. Sci. Soc. 6 : 16 (1933). Type: South West Africa, near Usakos, *Marloth 1257* (K!, PRE!, iso.). *A. rostrata* Sim, For. Fl. P.E.Afr. 55, t.37a (1909), non *A. rostrata* Humb. & Bonpl. ex Willd., in L., Sp. Pl. ed. 4, 4 : 1060 (errore typogr. 1054) (1806); Bak.f., Leg. Trop. Afr. 3 : 827 (1930). Type: Mozambique, Lourenco Marques and Maputo, *Sim 6263* (whereabouts unknown, perhaps destroyed). *A. senegal* sensu Burtt Davy, Fl. Transv. 2 : 337 (1932) pro parte; Henkel, Woody Pl. Natal 231 (1934); Hutch., Botanist in S. Afr. 664 (1946). *A. senegal* (L.) Willd. subsp. *trispinosa* (Stokes) Roberty in Candollea 11 : 155 (1948) pro parte quoad specim. *Dinter 222. A. senegal* var., Codd, Trees & Shrubs Kruger Nat. Park 51 (1951). *A. senegal* sensu Wild, S. Rhod. Bot. Dict. 49 (1953). *A. volkii* Suesseng. in Mitt. Bot. Staatssamml. München 2 : 40 (1954); Walter & Volk, Grundlagen Weidewirtschaft Südwestafrika 211, t. 68a (1954). Type: South West Africa, near Usakos, *Marloth 1257* (K!, PRE!. iso.).

Found in Angola, South West Africa, Botswana, Rhodesia, Mozambique, the Transvaal, Swaziland and Natal. It is quite probable that this variety has a wider range than indicated here as several specimens from Kenya and the Somali Republic, previously referred to the rather heterogeneous var. *kerensis* Schweinf., have pods similar to those of var. *rostrata*. The possibility exists that *rostrata* is not the earliest epithet available at varietal rank for this taxon (see Bothalia 11 : 301,302, 1974). If this proves to be the case then the name of this taxon will have to be altered. The situation in north-eastern tropical Africa needs clarifying. Occurs in woodland, bushveld, thornveld and river valley scrub; often found on alluvium.

S.W.A.—1712 (Posto Velho): Etanga, *De Winter & Leistner 5445.* 1715 (Ondangua): 40 km W. of Ndola Store on road to Ombalantu, *De Winter 3635.* 2015 (Otjihorongo): 32 km W. of Outjo, *Esterhuyse 449.* 2016 (Otjiwarongo): Omatjenne, *Keet 1681.* 2115 (Karibib): Karibib, *Dinter 6926.* 2116 (Okahandja): Okahandja, *Dinter 222* (BM, E, GRA, K). 2117 (Otjosondu): Otjosondu, *Seydel 4454* (K). 2215 (Trekkopje): farm Nudis, *Seydel 437* (K). 2216 (Otjimbingwe): farm Dusternbrook, *Wiss & Kinges 807.* 2316 (Nauchas): Marienhof to Rehoboth, *Walter 181.* 2416 (Maltahöhe): Bullsport, *Strey 2314.* 2417 (Mariental): 49,6 km W. of Mariental on Maltahöhe road, *Hardy 1952.* 2418 (Stampriet): 21 km E.N.E. of Mariental on road to Witbooisvlei, *De Winter 3542.* 2517 (Gibeon): Gibeon, *Pearson 9207* (BOL, K).

TRANSVAAL.—2229 (Waterpoort): 19,3 km from Dongola Camp on Weipe Flats, *Verdoorn 2262.* 2327 (Ellisras): 6,4 km S. of Ellisras, *Codd 8489.* 2429 (Zebediela): 41 km from Zebediela on road to Olifants River via Gompies, *Story 1582.* 2431 (Acornhoek): Kruger National Park, 8,8 km E. of Skukuza on Lower Sabie road, *Codd & De Winter 5030.* 2530 (Lydenburg): Schoemanskloof, *Pole Evans sub PRE 30008.* 2531 (Komatipoort): along main road

9,6 km W. of Malelane flat area, *Buitendag 785.* 2731 (Louwsburg): 1,6 km N. of Pongola River on road to Gollel, *Ross 1702* (K, NH, NU).

SWAZILAND.—2631 (Mbabane): Umtintegwa (Sipofaneni road), *Compton 27345.* 2731 (Louwsburg): Hluti-Gollel, *Pole Evans 3393* (2).

NATAL.—2632 (Bela Vista): Ndumu Game Reserve, *Ward 2202.* 2731 (Louwsburg): 19,2 km N. of Nongoma, *Acocks 11681.* 2732 (Ubombo): Mkuzi Game Reserve, *Ross 1658* (NH, NU). 2831 (Nkandla): Umfolozi Game Reserve, western area, *Ross 910* (K, NH, NU). 2832 (Mtubatuba): Hluhluwe Game Reserve, *Skead & Ward 17.* 2931 (Stanger): 4,8 km S. of Mandini on old main road, *Ross 1630* (NH, NU).

(b) var. **leiorhachis** *Brenan* in Kew Bull. 8 : 98 (1953); Young in Candollea 15 : 95 (1955); Boughey in J. S. Afr. Bot. 30 : 158 (1964) excl. syn.; Von Breitenbach, Indig. Trees S. Afr. 2 : 271 (1965); Ross & Brenan in Kew Bull. 21 : 69 (1967); Ross in Bol. Soc. Brot., Sér. 2, 42 : 231 (1968); Brenan in F.Z. 3, 1 : 80, t.15/5 (1970); Van Wyk, Trees Kruger Nat. Park 1 : 158 (1972). Type: Tanzania, Tanga Province, Pare District, Same, *Greenway 2192* (K, holo.!, EA).

A. circummarginata Chiov. in Ann. Bot., Roma 13 : 394 (1915); Bak.f., Leg. Trop. Afr. 3 : 834 (1930); Brenan in F.T.E.A. Legum.–Mimos. : 94, fig. 14/18 (1959); Dale & Greenway, Kenya Trees & Shrubs 286, fig. 58h (1961). Syntypes: Ethiopia, Ogaden, *Paoli 794, 913 bis, 920, 1010* (FI!). *A. senegal* sensu Codd, Trees & Shrubs Kruger Nat. Park 50 (1951). *A. senegal* var. *senegal* sensu Brenan in F.T.E.A. Legum.–Mimos. : 93 (1959) pro parte quoad syn. *A. senegal* var. *leiorhachis. Acacia* sp. 1, White, For. Fl. N. Rhod. 82 (1962).

Found in Ethiopia, Kenya, Tanzania, Zambia, Rhodesia, Mozambique and the northern and eastern Transvaal. Occurs in thornveld, scrub and woodland, sometimes with *Colophospermum mopane.*

TRANSVAAL.—2229 (Waterpoort): Dongola Reserve, Saker's homestead, Hartjiesveld, *Verdoorn 2326.* 2230 (Messina): Messina, *Legat 53* (K). 2329 (Pietersburg): 5,6 km S. of P.O. Vivo, *Codd 4320.* 2331 (Phalaborwa): Kruger National Park, near the Gorge Camp, *Codd 6190.* 2429 (Zebediela): 15,2 km E. of Chuniespoort on road to Burgersfort, *Codd 1701.* 2430 (Pilgrim's Rest): 4,8 km S. of Burgersfort, *Codd 6679.* 2531 (Komatipoort): Malelane, *Pole Evans sub PRE 15768.*

The specimen, *Hutchinson 2141,* cited as *Acacia seyal* Del. in Hutch., Botanist in S. Afr. 327 (1946), is in fact a specimen of *A. senegal.* Unfortunately the specimen is sterile and cannot be referred to either of the above varieties with certainty.

4. **Acacia mellifera** (*Vahl*) *Benth.* in Hook., Lond. J. Bot. 1 : 507 (1842); Oliv. in F.T.A. 2 : 340 (1871); Benth. in Trans. Linn. Soc. Lond. 30 : 517 (1875); Harms in Notizbl. Bot. Gart. Berl. 4 : 208, fig. 6

(1906); Harms in Engl., Pflanzenw. Afr. 3, 1 : 382, fig. 222 (1915); Bak.f., Leg. Trop. Afr. 3 : 828 (1930); Brenan, Checklist Tang. Terr. 329 (1949); Torre in C.F.A. 2 : 273 (1956); Brenan in F.T.E.A. Legum.–Mimos. : 84 (1959); in F.Z. 3, 1 : 67 (1970); Palmer & Pitman, Trees S. Afr. 2 : 751 (1973). Type: Arabia, Surdad and elsewhere, *Forskal* (C, holo.).

Much branched, dense, ±obconical shrub or a tree to 6 m high. *Bark* grey-brown to purplish-black, smooth or rough and fissured; young branchlets olive- or reddish-brown to greyish-brown or purplish-black, with numerous somewhat transversely elongated pale lenticels, glabrous or pubescent. *Stipules* not spinescent, in pairs, linear, $1-2,5 \times 0,2-0,8$ mm, soon deciduous. *Prickles* in pairs just below each node, strongly recurved, often not spreading but lying almost parallel to each other, up to 6 mm long, grey-brown to blackish. *Leaves:* petiole $0,2-1,2(2)$ cm long, glabrous or pubescent, adaxial gland usually present, variable in position, $0,2-0,7 \times 0,2-0,6$ mm; rhachis $(0,2)0,5-2(4)$ cm long, glabrous to pubescent, frequently with a gland at the junction of the top 1–2 pinnae pairs; pinnae 2–3, rarely 4 pairs; rhachillae $0,1-1,6$ cm long; leaflets 1–2, rarely 3, pairs per pinna, $3,5-15(20) \times 2-12$ mm, very variable in shape, obliquely obovate to obovate-elliptic or -oblong, apex rounded to emarginate or subacute and often apiculate, venose, glabrous or appressed-pubescent below, margins with or without cilia. *Inflorescences* subglobose to ellipsoid but the axes clearly elongate, or in elongated spikes, fascicled or solitary. *Flowers* yellowish-white or cream, buds often with a purplish tinge, on pedicels $(0,5)0,75-1,5$ mm long; axes $0,15-3,5$ cm long; peduncles $0,3-1,4$ cm long, glabrous or pubescent. *Calyx* cupular, often tinged with purple apically, glabrous or sometimes slightly pubescent near the apices of the lobes, tube $0,4-1$ mm long, lobes very small, up to $0,2$ mm long. *Corolla* glabrous or sometimes slightly pubescent near the apices of the lobes, tube up to $2,5$ mm long, lobes to $1,4$ mm long, often tinged with purple. *Stamen-filaments* free, $4-6,5$ mm long; anthers $0,15-0,25$ mm across, with a deciduous apical gland. *Ovary* up to $1,1$ mm long, glabrous, shortly stipitate. *Pods* greenish-yellow to straw-coloured or brown, $2,5-8$

$(9,3) \times 1,3-2,5(3,2)$ cm, oblong, straight, rounded to acute, acuminate or sometimes rostrate apically, longitudinally dehiscent, chartaceous, venose, glabrous. *Seeds* $7-10 \times 6-8$ mm, subcircular-lenticular, compressed; central areole small, $2-3 \times 2-3$ mm, horseshoe shaped.

Found in Arabia, in north-east Africa from Egypt, the Sudan, Ethiopia southwards to South West Africa, Botswana, the Transvaal, Orange Free State and the Cape Province. Two subspecies are recognized within *A. mellifera*, both of which occur in our area.

Leaves with normally 2 pinnae pairs; inflorescences ± elongate, the $0,4-1,8$ cm long peduncles usually shorter than the $0,5-3$ cm long axes; leaf-rhachides, rhachillae and leaflets glabrous; bark on young branchlets chestnut- or reddish-brown. . (a) subsp. *mellifera*

Leaves with normally 3 (rarely 4) pinnae pairs; inflorescences very short, ± subglobose to ellipsoid, the $0,4-1,1$ cm long peduncles usually longer than the very short $1,5-6,5$ mm long axes; leaf-rhachides and rhachillae usually pubescent and leaflets usually with cilia or sometimes appressed-pubescent beneath; bark on young branchlets olive- or greyish-brown to purplish-black. . . . .
. . . .(b) subsp. *detinens*

(a) subsp. **mellifera**.

Brenan in Kew Bull. 11 : 191 (1956); in F.T.E.A. Legum.–Mimos. : 84 (1959); Schreiber in F.S.W.A. 58 : 10 (1967).

Mimosa mellifera Vahl, Symb. Bot. 2 : 103 (1791). Type as above.

Inga mellifera (Vahl) Willd., Sp. Pl. 4 : 1006 (1806). Type as above.

Acacia senegal (L.) Willd. subsp. *mellifera* (Vahl) Roberty in Candollea 11 : 153 (1948). Type as above.

Subsp. *mellifera* is found in Arabia, in north-east Africa from Egypt, the Sudan, Ethiopia southwards to Tanzania, and in Angola and South West Africa. Within our area subsp. *mellifera* is found only in the Kaokoveld in South West Africa. Occurs in dry localities, often in broken mountainous country.

S.W.A.—1712 (Posto Velho); 11 km S. of the Kunene River at Otjinungua, *De Winter & Leistner 5755*; on the Kunene at Otjinungua, Hartmann mountains, *De Winter & Leistner 5769*; Kunene Gorge, *Davies, Thompson & Miller 103B*.

Subsp. *mellifera* not infrequently produces larger leaflets than any seen in subsp. *detinens*; it also tends to be glabrous and the bark tends to be chestnut- or reddish-brown and smooth. In South West Africa subsp. *mellifera* sometimes has rather slender elongate branchlets which are often unarmed.

Over most of their ranges subsp. *mellifera* and subsp. *detinens* occur exclusively and present no difficulty in their recognition. However, in northern South West Africa and in Angola (and again in north-central Tanzania) they meet and in Tanzania intermediates showing various combinations of characters occur so that difficulty is sometimes experienced in referring specimens to either subspecies with certainty. Within our area, however, no such difficulty is usually experienced.

(b) subsp. **detinens** (*Burch.*) *Brenan* in Kew Bull. 11 : 191 (1956); in F.T.E.A. Legum.–Mimos. : 85 (1959); Palmer & Pitman, Trees S. Afr. 159, t.38, 39 (1961); F. White, For. Fl. N. Rhod. 82, fig. 17D (1962); Von Breitenbach, Indig. Trees S. Afr. 2 : 274 (1965); Leistner, Mem. Bot. Surv. S. Afr. 38 : 123, t.16, 18, 21, 28, 33 (1967); Schreiber in F.S.W.A. 58 : 10 (1967); Brenan in F.Z. 3, 1 : 67 (1970). Type: Cape Province, Zand valley, *Burchell* 1628 (K, holo!, PRE, fragm.!).

A. detinens Burch., Trav. 1 : 310 (1822); DC., Prodr. 2 : 456 (1825): Benth. in Hook., Lond. J. Bot. 1 : 507 (1842); Harv. in F.C. 2 : 282 (1862); Benth. in Trans. Linn. Soc. Lond. 30 : 517 (1875); Engl. in Bot. Jahrb. 10 : 21 (1888); Marloth in Trans. S. Afr. Phil. Soc. 5 : 268 (1889); Schinz in Mém. Herb. Boiss. 1 : 105 (1900); Harms in Warb., Kunene-Samb. Exped. : 243 (1903); Sim, For. Fl. Cape Col. 210, t.59 (1907); Burtt Davy in Kew Bull. 1908 : 157 (1908); Dinter, Veg. Veldkost Deutsch-Sudwest-Afrikas 36 (1912); Glover in Ann. Bolus Herb. 1 : 147, t.18/2, 18/7 (1915); Harms in Engl., Pflanzenw. Afr. 3, 1 : 381 (1915); Dinter in Feddes Repert. 15 : 78 (1917); Pole Evans in S. Afr. J. Sci. 17 : fig. 22 (1920); Marloth, Fl. S. Afr. 2 : 54, t.18E (1925); Bak.f., Leg. Trop. Afr. 3 : 828 (1930); Burtt Davy, Fl. Transv. 2 : 345 (1932); Hutch., Botanist in S. Afr. 175, 179, 543, 631 (1946); O. B. Miller in J. S. Afr. Bot. 18 : 20 (1952); Torre in C.F.A. 2 : 273, t.52B (1956); Story, Mem. Bot. Surv. S. Afr. 30 : 22 (1958). Type as above. *A. ferox* Benth. in Hook., Lond. J. Bot. 5 : 97 (1846) pro parte quoad fr. specim., nom. illegit., non *A. ferox* Mart. & Gal. (1843); Harv. in F.C. 2 : 282 (1862) pro parte; Benth. in Trans. Linn. Soc. Lond. 30 : 517 (1875) pro parte; Burtt Davy in Kew Bull. 1908 : 157 (1908) pro parte; N.E. Br. in Kew Bull. 1909 : 107 (1909) pro parte; Glover in Ann. Bolus Herb. 1 : 151 (1915). Type: Transvaal, Magaliesberg, *Burke* (K, holo.!). The holotype of *A. ferox* is a mixed gathering consisting of a vegetative twig of *A. burkei* and a fruiting specimen of *A. mellifera* subsp. *detinens*. *A. tenax* Marloth in Bot. Jahrb. 8 : 254 (1887). Type: South West Africa, Otjimbingwe, *Marloth 1258* (BOL!, PRE!, iso.).

Subsp. *detinens* is found in Tanzania and southwards to Angola, South West Africa, Botswana, the Transvaal, Orange Free State and northern Cape Province. Occurs in dry thornveld, bushveld and wooded grassland; frequently found on the Kalahari sands.

S.W.A.—1715 (Ondangua): Oshikango, *Loeb 319.* 1824 (Kachikau): near Linyanti, *Killick & Leistner 3165.* 1920 (Tsumkwe): Nama Pan, *Story 5205.* 2014 (Welwitschia): Franzfontein, *Rodin 2744.* 2116 (Okahandja): Okahandja, *Dinter 268.* 2117 (Otjosondu): Quickborn, *Bradfield 23.* 2216 (Otjimbingwe): farm Otjiseva, *Wiss & Kinges 904.* 2217 (Windhoek): 7 km N. of Windhoek, *Codd 5774.* 2316 (Nauchas): between Gurumanas and Choaberib, *Pearson 9599*

(BOL, K). 2317 (Rehoboth): Uhlenhorst, farm Sib, *De Hoogh s.n.* (K). 2416 (Maltahöhe): farm Bullsport, *Strey 2185.* 2616 (Aus): top of ravine leading down to Kuibis water station, *Pearson 8001* (BOL, GRA). 2617 (Bethanie): Seeheim, *Pillans 5853* (BOL). 2718 (Grunau): Great Karasberg, N.E. of Narudas Sud, *Pearson 8145* (BOL, K). 2818 (Warmbad): 25 km N. of Warmbad, *Pearson 4299* (BOL, K).

TRANSVAAL.—2229 (Waterpoort): Dongola Reserve, *Verdoorn 2274.* 2327 (Ellisras): Moorddrift, *Leendertz 7328.* 2329 (Pietersburg): 4 km S.E. of P.O. Vivo, *Codd 4322.* 2426 (Mochudi): Vleeschfontein, *Gerstner 3130.* 2427 (Thabazimbi): Northam, *Van Nouhuys 30.* 2428 (Nylstroom): Mosdene estate, Naboomspruit, *Galpin M112.* 2526 (Zeerust): Swartruggens, *Sutton 1198.* 2527 (Rustenburg): farm Welgevonden, *Mogg 14628.* 2528 (Pretoria): Rus de Winter, *Gerstner 5536.* 2725 (Bloemhof): from Christiana to Fourteen Streams, *Burtt Davy 1568*b. 2726 (Odendaalsrus): Greylingsdrif, *Morris 1322.*

O.F.S.—2825 (Boshof): Smits Kraal, *Burtt Davy 9935.* 2924 (Hopetown): 22,4 km S.E. of Jacobsdal, farm Waterval-Oost, *Badenhorst 1.* 2925 (Jagersfontein): Fauresmith, *Henrici 4606.*

CAPE.—2520 (Mata Mata): Kalahari Gemsbok National Park, Bayip pan, *Barnard 712.* 2624 (Vryburg): Taungs, *Pole Evans sub PRE 15832.* 2722 (Olifantshoek): 56 km N.W. of Olifantshoek, *Fock sub KMG 8213.* 2723 (Kuruman): Takoon, *Burchell 2266* (K). 2820 (Kakamas): near Kakamas, *Hutchinson 948* (BOL, K). 2821 (Upington): Upington, *Pillans 5879* (BOL). 2823 (Griekwastad): 14,4 km N.W. of Schmidtsdrif, *Leistner & Joynt 2685.* 2824 (Kimberley): Schmidtsdrif, *Acocks 2404.* 2923 (Douglas): Maselsfontein, *E. Anderson 592* (GRA).

A. mellifera subsp. *detinens*, commonly known as "Swart Haak", often forms dense impenetrable thickets.

5. **Acacia robynsiana** *Merxm. & Schreiber* in Bull. Jard. Bot. Brux. 27 : 268, t.7 (1957); Von Breitenbach, Indig. Trees S. Afr. 2 : 276 (1965); Schreiber in F.S.W.A. 58 : 11 (1967); Palmer & Pitman, Trees S. Afr. 2 : 758 (1973). Type: South West Africa, Outjo Distr., Grootberg–Hang, *Walter 2/197* (M, holo.!).

Shrub with several erect slender branches which tend to droop apically or a slender tree to 8 m high with whip-like branches. *Bark* yellowish to reddish-brown or purplish; young branchlets whitish or grey- to reddish-brown or purplish, sometimes as though whitewashed over a purplish background, glabrous or subglabrous, with numerous somewhat transversely elongated lenticels. *Stipules* not spinescent, in pairs, linear, 1,8–3 mm long, 0,2–0,6 mm wide, soon deciduous. *Prickles* in pairs near each node or often absent, spreading laterally, straightish or slightly recurved, up to 4,5 mm long, reddish-brown to purplish-black. *Leaves*: petiole (0,1)0,5–1(1,2) cm long, glabrous or subglabrous;

rhachis 0–1,2 cm long, glabrous, a gland at the junction of each pinna pair; pinnae 1 or 2 pairs; rhachillae 1,2–3,5(4,4) cm long, glabrous; leaflets 6–13 pairs per pinna, 4–13,5 × 1,2–3,5 mm, oblong or obovate-oblong, often slightly to strongly falcate, apex rounded, veins usually fairly conspicuous beneath, completely glabrous. *Inflorescences* spicate, solitary or fascicled. *Flowers* yellowish-white, sessile or very shortly pedicellate; spikes (0,6)3–5 cm long; peduncles up to 0,5 cm long, glabrous. *Calyx* glabrous, tube 1,1–3 mm long, lobes up to 1,8 mm long. *Corolla* glabrous, tube 2,8–4 mm long, lobes up to 2 mm long. *Stamen-filaments* free or sometimes partially united basally, up to 10,5 mm long; anthers with a deciduous apical gland. *Ovary* 1,4–1,8 mm long, glabrous, on a stipe up to 2 mm long. *Pods* pale to dark yellowish-brown, 3,5–6,8 × 1,4–2 cm, oblong, ± straight, longitudinally dehiscent, apex rounded to acute or mucronate, venose, glabrous, sometimes slightly umbonate over the seeds. *Seeds* olive-brown, 8–10 × 8–10 mm, subcircular, compressed; central areole 1,5–2,5 × 2–3,5 mm, horse-shoe shaped.

Confined to South West Africa; found in a fairly restricted area within the Outjo district and in the Kaokoveld. Occurs in rocky ravines, kloofs and on rocky ridges.

S.W.A.—1813 (Ohopoho): 33,2 km from Warmbad on road to Ombombo, *De Winter & Leistner 5827*. 1914 (Kamanjab): Grootberg, *Walter 2/197* (M). 2014 (Welwitschia): 145,6 km W. of Welwitschia on road to Torra Bay, *Giess & Van Vuuren 1176*; 155 km W. of Welwitschia on road to Torra Bay, *De Winter & Hardy 8161*; Atsap, *Wiss 1476a* (M); Twyfelfontein, *Wiss 1476b* (M).

6. Acacia montis-usti *Merxm. & Schreiber* in Bull. Jard. Bot. Brux. 27 : 270, t.8 (1957); Von Breitenbach, Indig. Trees S. Afr. 2 : 278 (1965); Schreiber in F.S.W.A. 58 : 10 (1967); Palmer & Pitman, Trees S. Afr. 2 : 757 (1973). Type: South West Africa, Brandberg, Welwitsch-Tal, *Von Wettstein 95* (M, holo.!).

Tree to 9 m high branching from near the base, "broom-like", crown flattened and spreading or rounded; trunk to 0,5 m in diam. *Bark* olive- or yellowish- to reddish-brown or purplish-black and smooth when young, grey-brown to black, rough and fissured in mature trees; young branchlets olive- or reddish-brown, with numerous somewhat transversely elongated cream or reddish-brown lenticels, glabrous or subglabrous.

Stipules not spinescent, in pairs, linear, up to 3 mm long, 0,2–0,8 mm wide, soon deciduous. *Prickles* in pairs near the nodes or frequently absent (rarely a third prickle a short distance below and between the two lateral ones), spreading laterally, straightish or slightly recurved, up to 8 mm long. *Leaves*: petiole 0,6–2,8 cm long, glabrous or subglabrous, adaxial gland usually present, often just below the lowest pinna pair, slightly raised, 0,3–1,5 × 0,3–1,5 mm; rhachis 0,8–4(6,8) cm long, glabrous or subglabrous, lower surface usually without recurved prickles, a small gland usually at the junction of the top 1–2 pinnae pairs only, occasionally between each pinna pair or absent altogether; pinnae 2–6 pairs; rhachillae 1,4–4(6) cm long, glabrous or subglabrous; leaflets 4–13 pairs per pinna, 5–14,6 × 0,9–3(5) mm, linear-oblong to obovate-oblong, often slightly falcate, apex obtuse to acute, lateral nerves often fairly conspicuous beneath, margins without cilia, surfaces usually glabrous. *Inflorescences* spicate, fascicled or solitary. *Flowers* yellowish-white or pale cream, sessile; spikes 3–8,5 cm long; peduncles 0,2–1,4 cm long, glabrous or subglabrous. *Calyx* glabrous or subglabrous, tube 1,1–1,8 mm long, lobes 0,2–0,8 mm long. *Corolla* glabrous or occasionally minutely pubescent near the apices of the lobes, tube 1,4–2,3 mm long, lobes 0,6–1,2 mm long. *Stamen-filaments* free, up to 8 mm long; anthers with a deciduous apical gland. *Ovary* up to 1,4 mm long, shortly stipitate, glabrous. *Pods* olive- or reddish-brown to blackish, 7,6–18 × 1,6–2,4 cm, oblong, ±straight or slightly curved, apex obtuse to subacute, longitudinally dehiscent, slightly woody but brittle, glabrous or subglabrous, margins conspicuously thickened. *Seeds* olive- to reddish-brown, 10–13 × 8–12 mm, subcircular, compressed; central areole 4–6 × 3–5 mm, horse-shoe shaped.

Confined to South West Africa; found in a fairly restricted area within the Omaruru and Outjo districts and in the Kaokoveld. Occurs in rocky ravines, kloofs and on rocky ridges.

S.W.A.—1812 (Sanitatas): ravine near Nawantes, *Merxmuller & Giess 1436* (M). 1913 (Sesfontein): 19,2 km from Warmbad on road to Ombombo, *De Winter & Leistner 5841*; 30,4 km from Warmbad on road to Ombombo, *De Winter & Leistner 5834*. 1914 (Kamanjab): Kaientes, *Walter 1/256* (M). 2114 (Uis): Brandberg, Tsisab Valley, below the

White Lady, *Nordenstam 2528* (M); Brandberg, Tsisab valley, *Merxmuller & Giess 1662* (M); *Urschler s.n.* (M); *Wiss 1474* (M).

7. **Acacia welwitschii** *Oliv.* in F.T.A. 2 : 341 (1871): Benth. in Trans. Linn. Soc. Lond. 30 : 517 (1875); Bak.f., Leg. Trop. Afr. 3 : 829 (1930); Torre in C.F.A. 2 : 274 (1956); Brenan in F.Z. 3, 1 : 78 (1970). Type: Angola, Luanda District, Barra de Bengo, entre Mutolo e Cacuaco, prox. de Quicuxe, *Welwitsch* 1806 (LISU holo.; BM!, K!, P!).

subsp. **delagoensis** (*Harms*) *Ross & Brenan* in Kew Bull. 21: 67 (1967); Brenan in F.Z. 3, 1: 78 (1970); Van Wyk, Trees Kruger Nat. Park 1: 133 (1972); Palmer & Pitman, Trees S. Afr. 2: 755 (1973). Type: Mozambique, Umbeluzi, *Schlechter* 11718 (B, holo.†; BM!, K!, Z!).

A. welwitschii Oliv. in F.T.A. 2 : 341 (1871) pro parte quoad specim. *Kirk*; Sim, For. Fl. P.E. Afr. 55, t.37B (1909); Bak.f., Leg. Trop. Afr. 3 : 829 (1930) pro parte quoad specim. Rhodesia et Mozambique; Wild, S. Rhod. Bot. Dict. 49 (1953). *A. delagoensis* [Burtt Davy in Kew Bull. 1908 : 157 (1908) nom. nud.; Sim, For. Fl. P.E. Afr. 56 (1909) nom. nud.] Harms in Bot. Jahrb. 51 : 367 (1914); Burtt Davy, Fl. Transv. 2 : 337 (1932); Codd, Trees & Shrubs Kruger Nat. Park 42, fig. 37c, d (1951); Young in Candollea 15 : 117 (1955); Von Breitenbach, Indig. Trees S. Afr. 2 : 278 (1965); Gomes e Sousa, Dendrol. Moçamb. 1 : 234, t.38 (1966). Type as above.

Tree to 13 m high with a rounded, flattened and somewhat spreading or irregularly open crown. *Bark* yellowish- or grey- to reddish-brown or blackish, rough, irregularly fissured; young branchlets grey- to reddish-brown or purplish, glabrous or occasionally sparingly pubescent. *Stipules* not spinescent, in pairs, ±linear, 1–3,5 × 0,3–0,6 mm, glabrous, soon deciduous. *Prickles* in pairs below the nodes, strongly recurved, often broad-based, greyish-brown to blackish, up to 9 mm long. *Leaves*: petiole 0,6–1,9(2,4) cm long, adaxial gland usually present, variable in position, 0,3–0,8 × 0,2–0,5 mm; rhachis (0,9)1,2–4,1(5,8) cm long, glabrous or occasionally sparingly pubescent, eglandular or with a gland at the junction of the top pinna pair; pinnae 3–5 pairs; rhachillae 0,7–2,3(3,4) cm long, glabrous or occasionally sparingly pubescent, usually with a small gland at the junction of the top few leaflet pairs or sometimes each leaflet pair; leaflets 3–9 pairs per pinna, 5–9,5 × 2,5–5 mm, elliptic or broadly elliptic, sometimes somewhat ovate, apex rounded and often slightly emarginate, nearly

symmetrical basally, glabrous, veins rather prominent beneath. *Inflorescences* spicate, fascicled or solitary. *Flowers* yellowish-white, sessile or almost so; spikes 2,6–5,6 cm long; peduncles 0,5–1,2 cm long, glabrous or occasionally sparingly pubescent. *Calyx* sometimes tinged with pink, glabrous, tube 0,9–1,4 mm long, lobes up to 0,9 mm long. *Corolla* glabrous or occasionally with few glandular hairs, tube 1,6–2,3 mm long, lobes up to 1,1 mm long. *Stamen-filaments* free, up to 5,5 mm long; anthers ± 0,15 mm across, with a deciduous apical gland. *Ovary* to 1,3 mm long, glabrous, stipitate. *Pods* grey- to reddish-brown, 5,2–12,2(16,6) × 1,4–2 cm, straight, linear-oblong, rounded to ± acuminate apically, longitudinally dehiscent, coriaceous, venose, glabrous. *Seeds* olive- to reddish-brown, 7–13 × 7–13,5 mm, irregularly subcircular, compressed; central areole 4–8 × 4–9 mm, horse-shoe shaped.

Found in Malawi, eastern Rhodesia, Mozambique and the eastern Transvaal. Occurs in thornveld and woodland.

TRANSVAAL.—2431 (Acornhoek): Kruger National Park, 12,8 km from Satara on Rabelais road (—BC), *Van der Schijff 3293*; Kruger National Park, 32 km N.E. of Skukuza near Tshokwane (—DD), *Codd & Verdoorn 5480*.

Subsp. *welwitschii* is known only from Angola. The geographical ranges of subsp. *welwitschii* and subsp. *delagoensis* are widely separated, and the specimens of each have a markedly different and distinctive "look", yet the only satisfactory morphological difference between the two is in the length of the inflorescence.

A. welwitschii subsp. *delagoensis* differs from *A. burkei* in having a glabrous calyx and in being glabrous in most of its parts. It differs from *A. nigrescens*, which also has a glabrous calyx, in the more numerous pairs of smaller leaflets.

8. **Acacia nigrescens** *Oliv.* in F.T.A. 2 : 340 (1871); Benth. in Trans. Linn. Soc. Lond. 30 : 517 (1875); Sim, For. Fl. P.E. Afr. 54, t.33B (1909); Harms in Engl., Pflanzenw. Afr. 3,1 : 384 (1915); Marloth, Fl. S. Afr. 2 : fig. 31 (1925); Bak.f., Leg. Trop. Afr. 3 : 829 (1930); Milne-Redhead in Kew Bull. 1937 : 417 (1937); Hutch., Botanist in S. Afr. 370, 375, 538, 549 (1946); Codd, Trees & Shrubs Kruger Nat. Park 47, figs. 40, 41, 42, 43c, d, e (1951); O. B. Miller in J.S. Afr. Bot. 18 : 23 (1952); Young in Candollea 15 : 119 (1955); Palgrave, Trees Cent. Afr. 250 (1956); Torre in C.F.A. 2 : 274 (1956): Schreiber in Mitt. Bot. Staatssamml. Munchen 2 : 284 (1957); Brenan in F.T.E.A. Legum.-Mimos. : 85, fig. 14/9 (1959); Palmer &

Pitman, Trees S. Afr. 161, t.viii, 40, 41, 42 (1961); Letty, Wild Flow. Transv. 153, t.77 fig. 1 & 1a (1962); F. White, For. Fl. N. Rhod. 82, fig. 17E (1962); Von Breitenbach, Indig. Trees S. Afr. 2 : 274 (1965) pro parte excl. syn. *A. goetzei* Harms pro parte; De Winter et al, 66 Transv. Trees 52 (1966); Schreiber in F.S.W.A. 58 : 11 (1967); Ross in Bol. Soc. Brot., Sér. 2, 42 : 181 (1968); Brenan in F.Z. 3,1 : 69, t.15/3 (1970); Ross, Acacia Spp. Natal 35, fig. 1/5 (1971); Van Wyk, Trees Kruger Nat. Park 1 : 147 (1972); Ross, Fl. Natal 193 (1973); Palmer & Pitman, Trees S. Afr. 2 : 748 (1973). Type: Malawi, near Mitonda, Shire River, *Kirk s.n.* (K, holo.!).

A. caffra sensu Oliv. in F.T.A. 2 : 345 (1871) pro parte, tantum quoad specim. *McCabe* pro parte; Bak.f., Leg. Trop. Afr. 3 : 833 (1930) etiam pro parte ut praec. *A. nigrescens* var. *pallens* Benth. in Trans. Linn. Soc. Lond. 30 : 517 (1875); Young in Candollea 15 : 119, t.19/14 (1955). Type: Mozambique, near Sena, *Kirk 201* (K, holo.!). *A. passargei* Harms in Passarge, Kalahari : 789 (1904); in Engl., Pflanzenw. Afr. 3,1 : 384 (1915). Type: Presumably from Botswana, *Passarge 22* (B, holo. †; BM drawing!). *A. pallens* (Benth.) Rolfe in Kew Bull. 1907 : 361 (1907); Burtt Davy in Kew Bull. 1908 : 159 (1908); Glover in Ann. Bolus Herb. 1 : 145 (1915); Sim, Native Timbers S. Afr. 35, t.35 & 36 (1921); Bak.f., Leg. Trop. Afr. 3 : 829 (1930); Burtt Davy, Fl. Transv. 2 : 339, fig.57 (1932); Stapleton, Common Transvaal Trees 6 (1937). Type as for *A. nigrescens* var. *pallens. A. mellifera* sensu Henkel, Woody Pl. Natal 232 (1934), non (Vahl) Benth. *A. nigrescens* var. *nigrescens*—Young in Candollea 15 : 119 (1955).

Albizia lugardii N.E. Br. in Kew Bull. 1909 : 109 (1909). Type: Botswana, Ngamiland, Okavango valley, *Lugard 246* (K, holo.!).

Tree to 30 m high; crown rounded or branches ascending and spreading slightly, often cylindrical in young plants; trunk to 0,75 m in diam., typically beset with persistent prickles arising from swollen knobs up to 6,3 cm long. *Bark* yellowish-, grey- or reddish-brown, sometimes almost black, rough, fissured; young branchlets yellowish-, grey- or reddish-brown to blackish, flaking minutely, glabrous to pubescent. *Stipules* not spinescent, in pairs, ± linear, 1–3 × 0,2–0,6 mm, soon deciduous. *Prickles* in pairs below the nodes, strongly recurved, often broad-based, grey- to reddish-brown or black, up to 7 mm long. *Leaves*: petiole 0,5–4,3 cm long, glabrous to pubescent, adaxial gland often absent, variable in position, 0,3–0,7 × 0,2–0,5 mm; rhachis 0,8–10,2 cm long, glabrous to pubescent, usually without a gland at the junction of each pinna pair; pinnae 2–4 pairs; rhachillae

0,3–3,7 cm long; leaflets 1–2(4) pairs per pinna, (6,5)10–30(50) × (5,3)7–30(49,8) mm, very variable in shape, obliquely obovate-orbicular to broadly obovate-elliptic, apex rounded and often emarginate, subcoriaceous, venose, glabrous above and below or sparingly to densely appressed-pubescent above and/or below. *Inflorescences* spicate, fascicled, on short lateral branchlets, or occasionally solitary, sometimes crowded into a terminal raceme. *Flowers* yellowish-white, sessile; spikes 1–10,2 cm long; peduncles 0,3–2,4 cm long, glabrous or subglabrous, occasionally pubescent. *Calyx* often tinged with pink or distinctly pinkish-red, glabrous, tube 0,7–1,75 mm long, lobes 0,3–0,8 mm long. *Corolla* glabrous, tube 1,5–2 mm long, lobes up to 0,75 mm long. *Stamen-filaments* free, up to 6 mm long; anthers 0,1 mm across, with a deciduous apical gland. *Ovary* glabrous, 0,6–1,5 mm long, very shortly stipitate. *Pods* olive- or dark-brown to blackish, 6,1–17,8 × 1,4–2,4(2,7) cm, oblong, straight or nearly so, acuminate apically, longitudinally dehiscent, glabrous, coriaceous, brittle, scarcely venose. *Seeds* olive to olive-brown, 10–13 × 10–13 mm, subcircular-lenticular, compressed; areole 6–8 × 6–8 mm, horse-shoe shaped.

Found from Tanzania southwards to South West Africa, Botswana, the Transvaal, Swaziland and Natal (Zululand). Occurs on a variety of soil types in woodland, wooded grassland and bushveld, sometimes near rivers and pans.

S.W.A.—1720 (Sambio): between Sambio and Masari, *De Winter 4078*. 1820 (Tarikora): Nyangana, *Maguire 1659*.

TRANSVAAL.—2229 (Waterpoort): northern entrance to Wyllie's Poort, *De Winter 7749*. 2230 (Messina): Messina, *Rogers 21842*. 2231 (Pafuri): Kruger National Park, 8 km E. of Punda Milia, *Codd 5385*. 2329 (Pietersburg): Vivo, 67,2 km W. of Louis Trichardt, *Schlieben 7363*. 2330 (Tzaneen): Hans Merensky Nature Reserve, *Oates 29*. 2331 (Phalaborwa): Kruger National Park, Letaba Rest Camp, *Lang sub TRV 30939*. 2430 (Pilgrim's Rest): Strydom Tunnel east, *Strey 7896*. 2431 (Acornhoek): between Klaserie and Acornhoek, *Rauh & Schlieben 9708*. 2531 (Komatipoort): 12 km N. of Barberton, *Codd 1627*. 2731 (Louwsburg): 1,6 km S. of Gollel on Candover road, *Ross 1554* (NH, NU).

SWAZILAND.—2631 (Mbabane): Stegi district, Ranches, *Compton 27023*. 2731 (Louwsburg): 32 km from Gollel on Hluti road, *Ross 1429* (K, NH, NU).

NATAL.—2632 (Bela Vista): Ndumu Game Reserve, western area, *Ross 667* (NU). 2731 (Louwsburg): Mkuzana River, ± 22 km from Magudu on Nongoma road, *Ross 1090* (K, NU). 2732 (Ubombo): Mkuzi Game Reserve, *Ward 2393*. 2831 (Nkandla): 9,6 km N. of Mahlabatini, *Acocks 11665*.

A. nigrescens is usually an easily recognized species on account of its 1–2 (rarely 3–4) pairs of large leaflets and glabrous calyces. However, plants are found with leaflets intermediate in size, shape and number between those of *A. nigrescens* and those of *A. burkei* Benth. As these plants have pubescent calyces, their relationship appears to be with *A. burkei* rather than with *A. nigrescens* and they are discussed in more detail under *A. burkei*.

A. nigrescens is usually glabrous throughout, but it is occasionally puberulous or even quite densely pubescent. The characteristic raised knobs on the trunk and the larger branches are variable in their occurrence and are, at times, even absent. It is these knobs which have given the species the common names Knob-Thorn or Knoppiesdoring.

A. nigrescens sometimes flowers when quite leafless and may then be difficult to distinguish from *A. galpinii* which sometimes does the same. *A. galpinii* may be distinguished by its shorter (0,75–1,25 mm long) usually ± puberulous calyx and by the corolla-lobes which are usually puberulous outside. In *A. nigrescens* the calyx is 1,5–2,2 mm long and glabrous, as are the corolla-lobes.

The wood of *A. nigrescens* is hard, heavy and durable.

9. Acacia burkei *Benth*. in Hook., Lond. J. Bot. 5 : 98 (1846); Harv. in F.C. 2 : 282 (1862); Benth. in Trans. Linn. Soc. Lond. 30 : 518 (1875) pro parte excl. specim. *Meller 9 et Kirk*; Burtt Davy in Kew Bull. 1908 : 156 (1908) pro parte excl. specim. *Meller 9*; Sim, For. Fl. P.E. Afr. 56 (1909) pro parte excl. specim. *Meller 9 et specim. Zambesia*; Glover in Ann. Bolus Herb. 1 : 146, t.18/4 (1915); Burtt Davy in Kew Bull. 1922 : 325 (1922); Burtt Davy Fl. Transv. 2 : 337, fig. 56 (1932); O. B. Miller, Checklist Bech. Prot. 17 (1948) pro parte excl. syn. *A. mossambicensis*; in J.S. Afr. Bot. 18 : 19 (1952) pro parte ut praec.; Codd, Trees & Shrubs Kruger Nat. Park 41, fig. 34b (1951); Young in Candollea 15 : 115 (1955); Palmer & Pitman, Trees S. Afr. 150 (1961); Von Breitenbach, Indig. Trees S. Afr. 2 : 280 (1965); De Winter et al, 66 Transv. Trees 42 (1966); Ross in Bol. Soc. Brot. Sér 2, 42 : 275 (1968); Brenan in F.Z. 3,1 : 76 (1970); Ross, Acacia Spp. Natal 24, fig. 1/4 (1971); Van Wyk, Trees Kruger Nat. Park 1 : 129 (1972); Ross, Fl. Natal 193 (1973): Palmer & Pitman, Trees S. Afr. 2 : 753 (1973). Type: Transvaal, Magaliesberg, *Burke* (K, holo.!; BM!, PRE!).

A. ferox Benth. in Hook., Lond. J. Bot. 5 : 97 (1846) pro parte, nom. illegit., non *A. ferox* Mart. & Gal. (1843); Harv. in F.C. 2 : 282 (1862) pro parte; Benth. in Trans. Linn. Soc. Lond. 30 : 517 (1875) pro parte; Glover in Ann. Bolus Herb. 1 : 151 (1915). Type: Transvaal, Magaliesberg, *Burke* (K, holo.!).

The holotype of *A. ferox* is a mixed gathering consisting of a vegetative twig of *A. burkei* and a fruiting specimen of *A. mellifera* (Vahl) Benth. subsp. *detinens* (Burch.) Brenan. *A. mossambicensis* sensu Henkel, Woody Pl. Natal 233 (1934); Henkel, Ecol. Hluhluwe Game Res. 17, t.vi (1937); non Bolle in Peters, Reise Mossamb. Bot. 1 : 5 (1861).

Tree to 27 m high with a rounded, flattened and somewhat spreading or irregularly open crown. *Bark* pale or dark greyish-yellow to brown or almost black, rough, irregularly fissured, flaking, often with persistent prickles scattered over the surface; young branchlets pale or dark greyish-yellow or reddish-brown to black, flaking, often minutely, subglabrous to densely pubescent. *Stipules* not spinescent, in pairs, ± linear, 1–3,5 × 0,2–0,6 mm, densely pubescent, soon deciduous. *Prickles* in pairs below the nodes, strongly recurved, often broad-based, grey to reddish-brown or black, 3–9 mm long. *Leaves*: petiole (0,4)1–2,3(3,5) cm long, adaxial gland often present, variable in position, 0,3–0,8 mm in diam.; rhachis (0)3,4–7,2(9,1) cm long, subglabrous to densely pubescent, abaxial surface usually without recurved prickles, eglandular or with a gland at the junction of the top pinna pair or top 1–3 pairs; pinnae (1)3–13 pairs; rhachillae 0,6–5,7 cm long, subglabrous to densely pubescent; leaflets (1)4–19 pairs per pinna, 1,2–20,2 × 0,8–13,1 mm, very variable in shape, linear to linear-oblong or obovate, obovate-oblong to ± orbicular, apex acute to rounded, usually markedly asymmetric basally, veins often prominent below, varying from glabrous above and/or below to sparingly or densely pubescent above and/or below, typically pubescent and with a small basal tuft of hairs on the lower surface. *Inflorescences* spicate, fascicled, often crowded into a terminal raceme, or occasionally solitary. *Flowers* yellowish-white, sessile or almost so; spikes 1,4–8,5(14,6) cm long; peduncles 0,4–2 cm long, sparingly to densely pubescent, rarely subglabrous. *Calyx* campanulate, often tinged with pink or distinctly pinkish-red, sparingly to densely pubescent, tube 0,7–1,6 mm long, lobes 0,4–1,1 mm long. *Corolla* often tinged with pink or pinkish-red, glabrous or apices of lobes sparingly pubescent, tube 1,5–2,1 mm long, lobes up to 1,2 mm long. *Stamen-filaments* free, up to 6 mm long; anthers ± 0,15 mm across, with a deciduous apical gland. *Ovary* to 1,6 mm long, glabrous, shortly stipitate. *Pods* reddish- or purplish-brown, 4,1–16,9 × 0,9–2,4 cm, straight, linear-oblong, ± acuminate to mucronate apically, longitudinally dehiscent, coriaceous, obscurely venose, glabrous or sparsely pubescent near the margins and stipe. *Seeds* olive- to reddish-brown, subcircular-lenticular, 6–13 × 6–11 mm, compressed; central areole 4–8 × 3–8 mm, horse-shoe shaped.

Found in south-eastern Botswana, south-eastern Rhodesia, the Transvaal, Mozambique, Swaziland and Natal (almost confined to Zululand). Occurs on a variety of soil types and on boulder strewn slopes in dry river valley scrub, thornveld, mixed woodland and in scrub.

TRANSVAAL.—2231 (Pafuri): Kruger National Park, 8 km N.E. of Punda Milia, *Codd & Dyer 4572*. 2426 (Mochudi): 96 km N. of Zeerust, *Louw 1503*. 2427 (Thabazimbi): 18,7 km from Mabula on Warmbad–Thabazimbi road, *Ross 1499* (NH, NU). 2428 (Nylstroom): 7,3 km N. of Warmbad on Nylstroom road, *Ross 1494* (NH, NU). 2431 (Acornhoek): Kruger National Park, 1,6 km E. of Skukuza, *Codd & Verdoorn 5486*. 2527 (Rustenburg): near Hartebeespoort Dam, 4,8 km W. of dam wall, *De Winter 5954*. 2528 (Pretoria): Pienaarsrivier, *Bremekamp sub PRE 29161*. 2529 (Witbank): hills on E. side of Loskopdam, *Mogg 17307*. 2531 (Komatipoort): Kruger National Park, Pretoriuskop, *Van der Schijff 3894*. 2731 (Louwsburg): Pongola settlement, *Codd 10153*.

SWAZILAND.—2631 (Mbabane): 17,6 km W. of Stegi on road to Mbabane, *Codd & Dyer 2864*. 2731 (Louwsburg): 24 km from Gollel on Hluti road, *Ross 1526* (K, NH, NU).

NATAL.—2632 (Bela Vista): Ndumu Game Reserve, lower margins of Ndumu Hill, *Tinley 898* (NU). 2731 (Louwsburg): between Candover and Magudu, *Strey 4789*. 2732 (Ubombo): Mkuzi Game Reserve, *Ward 4444*. 2831 (Nkandla): 1,6 km S. of Enseleni River, 11,2 km N. of Empangeni, *Ross 1364* (K, NH, NU). 2832 (Mtubatuba): Hluhluwe Game Reserve, near Hluhluwe River, *Ward 1653*. 2931 (Stanger): 4,8 km S. of Mandini on old main road, Lower Tugela valley, *Ross 1359* (K, NH, NU).

A. burkei is an extremely variable species, particularly in leaflet size, shape and number, and forms part of a taxonomically difficult complex of species. Within our area *A. burkei* is most closely related to *A. nigrescens* and to *A. welwitschii* subsp. *delagoensis*. However, it differs from both species in having a ± densely pubescent calyx.

In their typical forms *A. nigrescens* and *A. burkei* are readily distinguishable: the former with its 1–2 (rarely 3 or 4) pairs of large leaflets and glabrous calyces, and the latter with more numerous pairs of smaller leaflets and pubescent calyces. However, there are numerous plants with leaflets intermediate in size, shape and in number between those of *A. nigrescens* and those of *A. burkei*. Leaflet size varies considerably and an entire range from those the size of *A. burkei* to those the size of *A. nigrescens* may occasionally be found on a single plant. As these plants have pubescent calyces, their relationship seems to be with *A. burkei* rather than with *A. nigrescens*. Although distinct from *A. nigrescens* in having pubescent calyces and bracts, the superficial resemblance of some of these plants to *A. nigrescens* is strengthened because they often grow with, or in close proximity to, *A. nigrescens*.

The range of morphological variation within *A. burkei* is often not readily apparent to a casual collector, or even from an examination of material in many herbaria. The extremes of the species look very different yet when the range of variation is inspected, it becomes clear that it is not possible to divide this range of variation satisfactorily. It had been customary in the past to distinguish loosely between "big leaflet" and "small leaflet" plants of *A. burkei*, the former typically having leaflets more than 3 mm wide and the latter leaflets less than 3 mm wide. A detailed study (Ross, 1.c., 1968) revealed that the characters typifying "big leaflet" and "small leaflet" plants varied independently, certain combinations of characters being commoner than others. Each combination, however, is frequently modified by the substitution of individual characters which show correlation, to varying degrees of imperfection, with other characters. Owing to the presence of so many intermediates and the inability to divide this range of variation satisfactorily, no infraspecific categories have been recognized within *A. burkei*. A numerical study of the species (Ross & Morris in Bothalia 10 : 437, 1971) supported this view.

10. **Acacia galpinii** *Burtt Davy* in Kew Bull. 1922 : 326 (1922); Fl. Transv. 2 : 337 (1932); Steedman, Trees etc. S. Rhod. 13 (1933); Stapleton, Common Transvaal Trees 5 (1937); O. B. Miller, Checklist Bech. Prot. : 18 (1948); in J.S. Afr. Bot. 18 : 20 (1952); Pardy in Rhod. Agric. J. 49 : 12 (1952); Young in Candollea 15 : 97 (1955); Palgrave, Trees Central Afr. 239 (1956); Brenan in F.T.E.A. Legum.–Mimos. : 87 (1959); Palmer & Pitman, Trees S. Afr. 151, t.16, 30, 33 (1961); F. White, For. Fl. N. Rhod. 83, fig. 17F (1962); Brenan in Kew Bull. 17 : 164 (1963); Von Breitenbach, Indig. Trees S. Afr. 2 : 280 (1965); De Winter et al, 66 Transv. Trees 46 (1966); Brenan in F.Z. 3,1; 68 (1970); Ross in Bothalia 10 : 547 (1972); Palmer & Pitman, Trees S. Afr. 2 : 767 (1973). Type: Transvaal, banks of Bad-zynloop River, Mosdene Estate, Naboomspruit, 19 Sept. 1920, *Galpin 483 M* (K, holo.!; BM!, GRA!, PRE!).

A. caffra sensu Oliv. in F.T.A. 2 : 345 (1871) tantum quoad specim. *McCabe* pro parte; Bak.f., Leg. Trop. Afr. 3 : 833 (1930) etiam pro parte ut praec. *A. dulcis* sensu Henkel, Woody Pl. Natal 233 (1934), non Marloth & Engl. *A. senegal* sensu O. B. Miller, Checklist Bech. Prot. 21 (1948).

Tree to 36 m high; trunk to 2 m diam.; crown often rounded. *Bark* pale to dark yellowish- or greyish-brown to blackish, corky when young, rough, often longitudinally fissured, sometimes with scattered persistent prickles arising from swollen bases; young branchlets yellowish- or olive-brown to dark reddish-brown or purplish, glabrous or subglabrous to pubescent. *Stipules* not spinescent, in pairs, 1–3,5 × 0,2–0,8 mm, soon deciduous. *Prickles* in pairs just below the nodes, recurved or almost straight, often broad-based, dark reddish-brown to black, up to 1,2 cm long, rarely absent. *Leaves*: petiole (0,6)1,4–3,6(4,8) cm long, adaxial gland usually present, variable in position, round or oval, 0,6–1,5 × 0,4–1 mm; rhachis (2,6)5,5–11(14,2) cm long, subglabrous to pubescent, abaxial surface sometimes with recurved prickles, with a gland at the junction of the top 1–6 pinnae pairs, occasionally at the junction of all or nearly all pinnae pairs or absent altogether; pinnae (4)7–14 pairs; rhachillae (2,5)3–6,5(9,2) cm long, glabrous to pubescent; leaflets 12–35(45) pairs per pinna, (3)4–10(12) × 0,8–3 (3,8) mm, oblong to linear-oblong, the upper sometimes obovate, often slightly falcate, apex rounded to subacute, veins not very prominent beneath, glabrous except for the frequent presence of a small basal tuft of hairs abaxially, rarely sparingly pubescent beneath, margins with or without cilia. *Inflorescences* spicate, fascicled or solitary, often borne on short lateral shoots. *Flowers* cream to yellowish-white, sessile; spikes (3,2)5–10 cm long; peduncles 0,3–1,3 cm long, glabrous to sparingly or densely pubescent. *Calyx* cupular, red or purple, 0,75–1,25 mm long, ±puberulous or sometimes subglabrous or glabrous. *Corolla* red or purplish, subglabrous to puberulous, especially towards the apices of the lobes, tube up to 1,6 mm long, lobes up to 1,4 mm long. *Stamen-filaments* yellowish-white, free, 4–6 mm long; anthers 0,15–0,2 mm across, with a deciduous apical gland. *Ovary* glabrous, 0,8–1,5 mm long, on a stipe up to 0,8 mm long. *Pods* reddish- to purplish-brown, 11–28 × 2,3–3,5 cm, straight, apex acute to mucronate, longitudinally dehiscent, valves thinly woody but brittle, venose, slightly umbonate over the seeds, glabrous or almost so. *Seeds* olive-brown, subcircular-lenticular, 8–12(15) × 8–12,5 mm, compressed; central areole 5–8 × 3,5–7 mm, horse-shoe shaped.

Found in central Tanzania, Zambia, Malawi, Mozambique, Botswana, Rhodesia and the Transvaal.

Occurs in woodland or bushveld, frequently, but by no means always, on river banks or where moisture is available.

TRANSVAAL.—2328 (Baltimore): Blauwberg, along river near Leipsig, *Strey & Schlieben 8575.* 2329 (Pietersburg): 49,6 km W. of Louis Trichardt, south side of Soutpansberg range, *Codd 4441.* 2427 (Thabazimbi): Thabazimbi, *Dyer & Verdoorn 4218.* 2428 (Nylstroom): banks of Bad-zyn-loop River, Mosdene estate, Naboomspruit, 19 Sept. 1920, *Galpin 483M.* 2430 (Pilgrim's Rest): Steelpoort, *Smuts & Gillett 3521.* 2526 (Zeerust): Leeuwfontein, *Reyneke 416.* 2528 (Pretoria): Rust de Winter, *Codd 5598.* 2529 (Witbank): 6,4 km E. of Groblersdal, *Codd 2702.* Grid ref. unknown: Rustenburg Distr., *Melle sub PRE 29616.*

A. galpinii, popularly known as the Monkeythorn or Apiesdoring, is one of our largest acacias.

The red or purple calyces and corollas are unusual among the African species of *Acacia*.

A. galpinii sometimes flowers when leafless and then specimens may be difficult to distinguish from *A. nigrescens* in the same condition. The differences between the two species are given under *A. nigrescens*.

11. Acacia polyacantha *Willd.*, Sp. Pl. 4 : 1079 (1806); DC., Prodr. 2 : 459 (1825); Brenan in Kew Bull. 11 : 195 (1956); in F.T.E.A. Legum.–Mimos. : 87 (1959); in F.Z. 3,1 : 71 (1970). Type: Eastern India, collector unknown, *Herb Willdenow 19166* (B, holo., K, fragm.!, K, PRE, photo.).

subsp. **campylacantha** (*Hochst. ex A. Rich.*) *Brenan* in Kew Bull. 11 : 195 (1956); Palgrave, Trees Cent. Afr. 235 (1956); Keay in F.W.T.A. ed.2,1 : 499 (1958); Brenan in F.T.E.A. Legum.–Mimos. : 88, fig. 14/12 (1959); F. White, For. Fl. N. Rhod. 84 (1962); Von Breitenbach, Indig. Trees S. Afr. 2 : 282 (1965); Brenan in F.Z. 3,1 : 71, t.15/4, t.18 (1970); Van Wyk, Trees Kruger Nat. Park 1 : 153 (1972); Palmer & Pitman, Trees S. Afr. 2 : 765 (1973). Syntypes: Ethiopia, Mai Dogale, *Schimper 639* (BM!, E!, FI!, K!, P!, Z!); Dscheladscheranne [Jelajeranne], *Schimper 893* (BM!, E!, FI!, K!, P!, Z!).

A. campylacantha Hochst. ex A. Rich., Tent. Fl. Abyss. 1 : 242 (1847); Harms in Engl., Pflanzenw. Afr. 3,1 : 385, fig. 223 (1915); Burtt Davy in Kew Bull. 1922 : 325 (1922); Fl. Transv. 2 : 337 (1932); Bak.f., Leg. Trop. Afr. 3 : 831 (1930); Codd, Trees & Shrubs Kruger Nat. Park 42, fig. 35, 37e,f (1951); O. B. Miller in J.S. Afr. Bot. 18 : 19 (1952); Young in Candollea 15 : 99 (1955): Torre in C.F.A. 2 : 276 (1956); F. White, For. Fl. N. Rhod. 83, fig. 17H (1962). Syntypes as above. *A. catechu* sensu Schweinf. in Linnaea 35 : 363 (1867-8); Oliv. in F.T.A. 2 : 344 (1871); Harms in Warb., Kunene–Samb. Exped. : 243 (1903); Burtt Davy in Kew Bull. 1908 : 156 (1908); Sim, For. Fl. P.E. Afr. 56 (1909), non (L.f.) Willd. *A.*

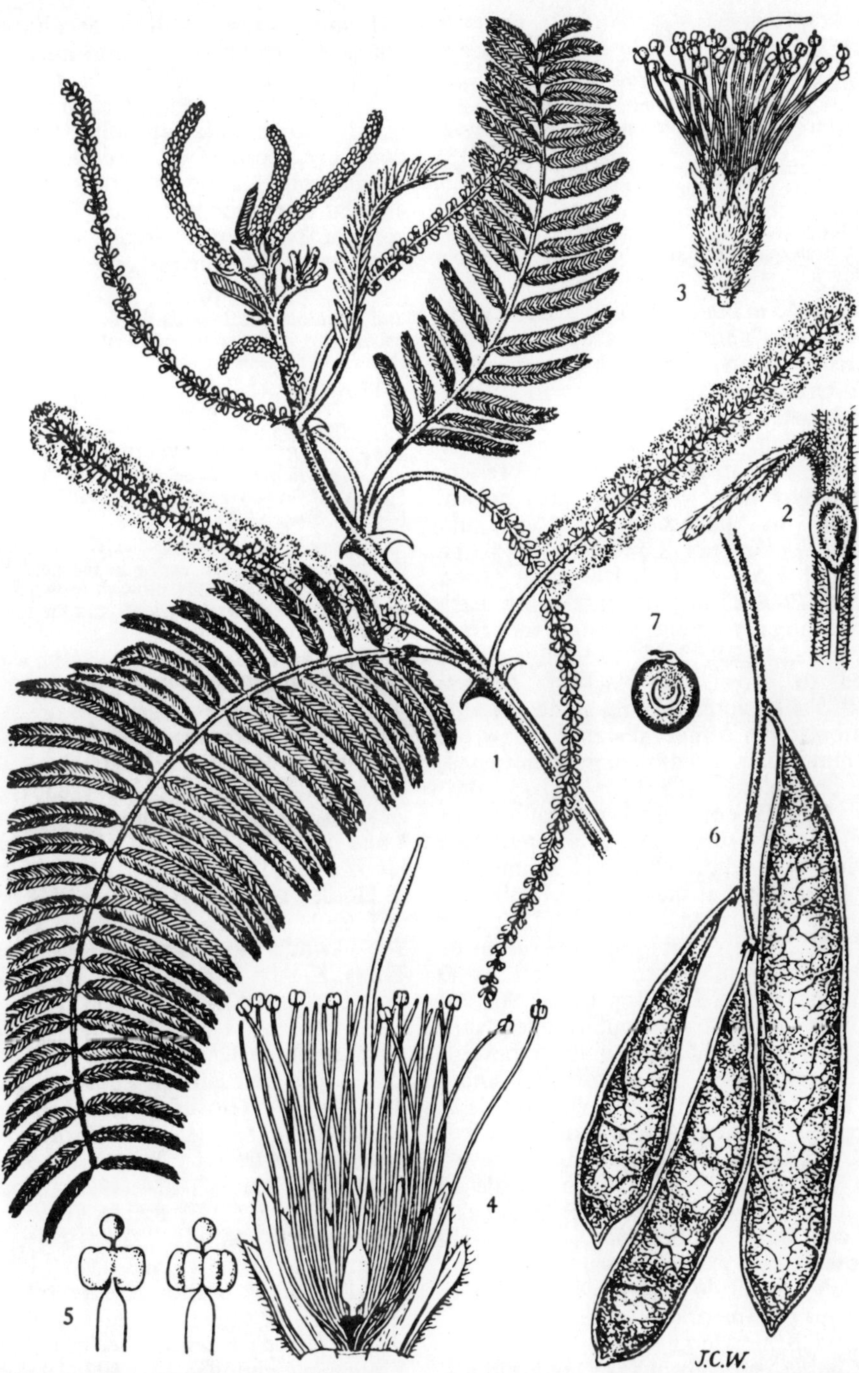

FIG. 8.—**Acacia polyacantha** subsp. **campylacantha**. 1, flowering branch, × ⅔; 2, gland on petiole, × 4; 3, flower, × 6; 4, flower, opened out to show ovary, × 6; 5, anthers, × 46, all from *Lusaka Natural History Club* 172; 6, pods × ⅔; 7, seed, × ⅔, all from *Gilliland* 199. Reproduced by permission of the Editorial Board of Flora Zambesiaca.

caffra sensu Oliv. in F.T.A. 2 : 345 (1871) pro parte quoad specim. Angola; Bak.f., Leg. Trop. Afr. 3 : 833 (1930) quoad specim. Angola; Eyles in Trans. Roy. Soc. S. Afr. 5 : 361 (1916). *A. suma* sensu Benth. in Trans. Linn. Soc. Lond. 30 : 519 (1875) pro parte, non sensu stricto. *A. catechu* (L.f.) Willd. subsp. *suma* (Roxb.) Roberty var. *campylacantha* (Hochst. ex A. Rich.) Roberty in Candollea 11 : 157 (1948). Syntypes as above. *A. caffra* (Thunb.) Willd. var. *campylacantha* (Hochst. ex A. Rich.) Aubrev., Fl. Forest. Soudano-Guin. 272, t.53/4, 53/5 (1950); Gilbert & Boutique in F.C.B. 3 : 150 (1952). Syntypes as above.

Tree to 15 m high; trunk to 0,6 m diam., often with persistent prickles up to 1,5 cm long arising from swollen bases (knobs); crown often somewhat flattened and spreading. *Bark* pale to dark yellow or yellowish-brown, fissured, flaking or peeling off to reveal a whitish inner layer; young branchlets creamy-white to light grey-brown, puberulous or pubescent, seldom subglabrous. *Stipules* not spinescent, in pairs, linear, 1,5–3,5 × 0,2–0,8 mm, soon deciduous. *Prickles* in pairs just below each node, occasionally absent, strongly recurved, usually broad-based, 3–15 mm long, straw-coloured to brown or blackish. *Leaves*: petiole 0,5–4 cm long, adaxial gland variable in position, often a short distance below the lowest pinna pair, usually slightly flattened, discoid or oblong, 1,5–4 × 1,5–3 mm; rhachis 5,6–20 cm long, puberulous or pubescent, rarely subglabrous, lower surface often with recurved prickles up to 3 mm long, a fairly large gland at the junction of the top 3–16 pinnae pairs; pinnae (6)14–35(45) pairs; rhachillae 1,2–5,5(7) cm long, puberulous or pubescent, rarely subglabrous; leaflets 20–60 (68) pairs per pinna, 2–5(6) × 0,4–0,9(1,25) mm, linear to linear-triangular, apex subacute to narrowly obtuse, usually pubescent on the margins only, the midrib and sometimes a few small basal nerves visible below. *Inflorescences* spicate, solitary or fascicled, usually produced with the new leaves. *Flowers* yellowish-white or cream, sessile or nearly so; spikes 3–12 cm long; peduncles 0,5–3 cm long, puberulous to densely tomentose, rarely subglabrous, often glandular. *Calyx* puberulous or pubescent, rarely subglabrous or puberulous on the lobes only, tube 0,7–1,8 mm long, lobes 0,2–0,8 mm long. *Corolla* subglabrous or puberulous, often only the lobes puberulous, tube up to 2,5 mm long, lobes up to 1 mm long. *Stamen-filaments* free, up to 6 mm long; anthers ±

0,1 mm across, with a deciduous apical gland. *Ovary* up to 1,2 mm long, glabrous, shortly stipitate. *Pods* brown, 6,5–12(15) × 0,9–1,7(2) cm, oblong, straight or almost so, usually acuminate apically, longitudinally dehiscent, coriaceous, venose, glabrous or subglabrous, rarely puberulous. *Seeds* subcircular to elliptic-lenticular, 7–9 × 6–8 mm, compressed; central areole 3–4 × 2,5–3,5 mm, horse-shoe shaped, not impressed.

Widespread in tropical Africa from the Gambia and Ethiopia southwards to the Transvaal. Occurs usually on colluvial or alluvial loams or clays by rivers and streams.

TRANSVAAL.—2230 (Messina): 66,4 km from Louis Trichardt on road to Punda Milia, *Grobbelaar 699*. 2231 (Pafuri): Kruger National Park, M'basa, *Lang sub TRV 32219*. 2329 (Pietersburg): 1,6 km E. of Louis Trichardt, *Codd 3024*. 2330 (Tzaneen): Elim (—AA), *Obermeyer sub TRV 29295*; Letaba River (—CA), *Strey 7935*.

A. polyacantha subsp. *campylacantha* may be distinguished from *A. caffra* in the field by its relatively straight trunk and yellowish-brown flaking bark, its large discoid petiolar glands and the tendency to a flattened spreading crown.

A. polyacantha subsp. *polyacantha*, with prickles straight or almost so, is known only from India and (probably) Sri Lanka (Ceylon).

12. **Acacia caffra** (*Thunb.*) *Willd.*, Sp. Pl 4 : 1078 (1806); DC., Prodr. 2 : 459 (1825); E. Mey., Comm. 1 : 169 (1836); Eckl. & Zeyh., Enum. 260 (1836); Benth. in Hook., Lond. J. Bot. 1 : 509 (1842); Meisn. in Hook., Lond. J. Bot. 2 : 105 (1843); Benth. in Hook., Lond. J. Bot. 5 : 98 (1846); Harv. in F.C. 2 : 282 (1862); Benth. in Trans. Linn. Soc. Lond. 30 : 520 (1875); Marloth in Trans. S. Afr. Phil. Soc. 5 : 269 (1889); Sim, For. Fl. Cape Col. 210, t.60 (1907) pro parte excl. specim. Lake Ngami, Mossamedes, Angola; Burtt Davy in Kew Bull. 1908: 156 (1908) pro parte excl. specim. *Lugard 93*, *McCabe 29*; Glover in Ann. Bolus Herb. 1 : 146, t.18/6 (1915); Harms in Engl., Pflanzenw. Afr. 3,1 : 387 (1915); Sim, Native Timbers S. Afr. 32, t.34 (1921); Burtt Davy, Fl. Transv. 2 : 337, fig. 55 (1932); Henkel, Woody Pl. Natal 232 (1934); Hutch., Botanist in S. Afr. 363, 366, 398 (1946); Codd, Trees & Shrubs Kruger Nat. Park 42 (1951); O. B. Miller in J. S. Afr. Bot. 18 : 19 (1952) pro parte excl. specim. *Curson 173*; Young in Candollea 15 : 102 (1955); Brenan in Kew Bull. 11 : 193 (1956); Palmer & Pitman, Trees S. Afr. : 150, t.IV, 32 (1961); Von Breitenbach, Indig. Trees S. Afr. 2 : 284

(1965); Ross in Ann. Natal Mus. 18 : 221 (1965); Ross & Gordon-Gray in Brittonia 18 : 267 (1966); De Winter et al, 66 Transv. Trees 44 (1966); Ross in Webbia 22 : 203 (1967); Brenan in F.Z. 3,1 : 72 (1970); Flow. Pl. Afr. 40 : t.1586 (1970); Ross, Acacia Spp. Natal 26, fig. 1/3 (1971); Van Wyk, Trees Kruger Nat. Park 1 : 131 (1972); Ross, Fl. Natal 193 (1973); in Bothalia 11 : 127 (1973); Palmer & Pitman, Trees S. Afr. 2 : 761 (1973). Type: Cape, Karoo near Slang River, *Herb. Willdenow* 19163 (B, holo., BOL, PRE, photo.).

Mimosa caffra Thunb., Prodr. 2 : 92 (1800); Fl. Cap., ed. Schult. 433 (1823). Type as above.

A. fallax E. Mey., Comm. 1 : 169 (1836); Meisn. in Hook., Lond. J. Bot. 2 : 105 (1843). Syntypes: Cape, Uitenhage Distr., Witrivier near Enon, *Drege* (BM!, K!, P!); Peddie Distr., Keiskamma, *Drege* (P!). *A. caffra* Willd. var. *namaquensis* Eckl. & Zeyh., Enum 260 (1836). Type: Cape, Clanwilliam Distr., Namaqualand, Olifants River, Clanwilliam, *Ecklon & Zeyher* 1694 (BOL!, K!). *A. catechu* sensu E. Mey., Comm. 1 : 170 (1836), non (L.f.) Willd. *A. multijuga* Meisn. in Hook., Lond. J. Bot. 2 : 105 (1843). Type: Natal, between Durban [Port Natal] and Tugela River, *Krauss* 112 (BM, iso.!). *A. caffra* var. *longa* Glover in Ann. Bolus Herb. 1 : 146 (1915). Syntypes: Cape, Umtata Distr., Umtata, *Convent of the Holy Cross* 233 (GRA!); King William's Town Distr., King William's Town, *Sim* 2137 (PRE!). *A. caffra* var. *tomentosa* Glover in Ann. Bolus Herb. 1 : 146 (1915); Bews, Fl. Natal 114 (1921); Burtt Davy, Fl. Transv. 2 : 337 (1932); Young in Candollea 15 : 107 (1955). Type: Cape, Komgha Distr., hillsides near Komgha, Oct. 1891, *Flanagan* 302 (BOL, lecto.!). *A. caffra* var. *transvaalensis* Glover in Ann. Bolus Herb. 1 : 146 (1915). Type: Transvaal, Pretoria Distr., Wonderboompoort, near Pretoria, *Rehmann* 4603 (PRE, lecto.!). *A. mellei* sensu O. B. Miller in J. S. Afr. Bot. 18 : 23 (1952) pro parte saltem quoad specim. *Miller* B/950.

Shrub or tree to 14 m high; trunk to 0,6 m diam.; crown often rounded. *Bark* reddish- or dark-brown to blackish, rough, frequently transversely and longitudinally fissured; young branchlets glabrous to sparingly or densely tomentose, eglandular or with minute reddish glands. *Stipules* not spinescent, in pairs, linear, 2,5–4 × 0,2–0,8 mm, soon deciduous. *Prickles* in pairs just below each node (rarely absent or with a few additional prickles scattered elsewhere on the stems), recurved or sometimes ± straight, dark-brown to blackish, up to 9 mm long. *Leaves:* petiole (0,5)0,8–3(4) cm long, adaxial gland variable in position, often slightly below the lowest pinna pair, 0,7–1,5 × 0,3–0,7 mm; rhachis (2)5–15(22,7) cm long, glabrous to densely tomentose, abaxial surface often with recurved prickles up to 3 mm long, with a gland at the junction of the top 1–3 pinnae pairs, occasionally between each pinna pair, or absent from some; pinnae (6)8–26(38) pairs; rhachillae (1,1)2,4–5(6,7) cm long, glabrous to tomentose; leaflets (13)21–50(64) pairs per pinna, 2–7(12,2) × 0,7–1,5(2,3) mm, linear to linear-oblong, apex rounded to subacute, glabrous to sparingly or densely appressed-pubescent below, margins with or without appressed or spreading cilia. *Inflorescences* spicate, solitary, fascicled or crowded into an irregular terminal panicle. *Flowers* yellowish-white, sessile; spikes (2)2,6–6,5(10) cm long; peduncles (0,2)0,8–2,5(4,1) cm long, subglabrous to densely tomentose, often glandular. *Calyx* campanulate, puberulous to pubescent or occasionally subglabrous or glabrous, tube 1–1,6 mm long, lobes 0,2–0,8 mm long. *Corolla* sometimes tinged with pink, puberulous to tomentose or occasionally glabrous, tube up to 2,5 mm long, lobes up to 1 mm long. *Stamen-filaments* free, up to 6 mm long; anthers ± 0,15 mm across, with a deciduous apical gland. *Ovary* 0,5–1,9 mm long, glabrous or subglabrous, sessile or shortly stipitate. *Pods* light or dark brown, 4,5–19,7 × 0,7–1,5 (2,7) cm, linear, straight or occasionally ± falcate, rounded to acuminate apically, longitudinally dehiscent, coriaceous, umbonate over the seeds, subglabrous or puberulous, seldom tomentose, with few to numerous reddish-brown glands scattered over the surface. *Seeds* olive- to light brown, subcircular to elliptic-lenticular, 6–12 × 4–8 mm, compressed; central areole 2–5 × 2–4 mm, horse-shoe shaped.

Found in south-eastern Botswana, the Transvaal, southern Mozambique, Swaziland, Natal and the Cape Province. Occupies a diverse range of habitats from coastal scrub, dry thornveld and river valley scrub, to mixed bushveld and tall grassland. Frequently occurs amongst boulders or near termite mounds which afford some protection from fire. A dwarf stunted from often occurs on rocky stream banks in the eastern Cape.

TRANSVAAL.—2329 (Pietersburg): 28,8 km E. of Pietersburg on Tzaneen road, *Van Vuuren 1292*. 2330 (Tzaneen): Westfalia Estate, Duiwelskloof, *Scheepers 1162*. 2427 (Thabazimbi): 18,7 km from Mabula store on Warmbad–Thabazimbi road, *Ross 1500* (NH, NU). 2428 (Nylstroom): 7,4 km N. of Warmbad on Nylstroom road, *Ross 1495* (K, NH, NU). 2429 (Zebediela): Chuniespoort, *Obermeyer sub PRE 34663*. 2430 (Pilgrim's Rest): Pilgrim's Rest,

Bushbuck Ridge, *Pritchard 46.* 2526 (Zeerust): Zeerust, *Thode A1405.* 2527 (Rustenburg): Rustenburg, *Galpin 11643.* 2528 (Pretoria): Plot 137, Willowglen, Pretoria, *Letty 345.* 2529 (Witbank): Loskop Dam Nature Reserve, *Mogg 30402.* 2530 (Lydenburg): Schoemanskloof, *Smuts 262.* 2531 (Komatipoort): Kruger National Park, Lebombo Mountains, Crocodile Bridge section, *Van der Schijff 3975.* 2627 (Potchefstroom): Potchefstroom, *Potts sub NU 29058.* 2628 (Johannesburg): Suikerbosrand, *Mogg 18113* (SRGH). 2726 (Odendaalsrus): Bezuidenhoudskraal, *Morris 1086.* 2730 (Vryheid): ridge above Assegaai River bridge on Moolman road, *Galpin 9621.*

SWAZILAND.—2631 (Mbabane): Mpisi, *Compton 28997.* 2731 (Louwsburg): 16 km from Gollel on Hluti road, *Ross 1426* (K, NH, NU).

NATAL.—2632 (Bela Vista): Ndumu Hill, *Ross 655* (NH, NU). 2730 (Vryheid): road crossing on Upper Blood River, on Kingsley–Viljoenspos road, *Edwards 2834.* 2731 (Louwsburg): 22,4 km from Nongoma on Magudu road, *Ross 1086* (NU). 2723 (Ubombo): Lebombo Mountains, 9,6 km from Jozini on road to Gwalaweni, *Strey 8142.* 2829 (Harrismith): Oliviers Hoek Pass, *Ross 143* (K, NH, NU). 2830 (Dundee): 8 km from Dundee on Wasbank road, *Ross 187* (NH, NU). 2831 (Nkandla): Umfolozi Game Reserve, *Ross 926* (NH, NU). 2832 (Mtubatuba): Hluhluwe Game Reserve, *Ward 1563.* 2929 (Underberg): Estcourt Hill, 3,2 km S. of Estcourt, *Ross 121* (NU). 2930 (Pietermaritzburg): Baynes Drift, *Ross 454.* 2931 (Stanger): Lower Tugela valley, below Bulwer farm, *Edwards 1927.* 3030 (Port Shepstone): Uvongo, *Ross 221* (K, NH, NU). 3130 (Port Edward): Port Edward, *Ross 225* (NH, NU).

CAPE.—2624 (Vryburg): Taungs, *Breuckner 597.* 3029 (Kokstad): bank of Umzimkulu River, near Clydesdale, *Tyson 2090.* 3118 (Vanrhynsdorp): on the Doorn River, *Drege s.n.* (K, PRE). 3126 (Queenstown): Queenstown, *Taylor s.n.* (GRA). 3128 (Umtata): Umtata River, *Pegler 1603.* 3129 (Port St. Johns): Port St. Johns, *G. Edwards 7892* (BM). 3218 (Clanwilliam): Olifants River, Clanwilliam, *Ecklon & Zeyher 1694* (BOL, K). 3226 (Fort Beaufort): bank of Fish River, near bridge on Fort Beaufort road, *Dyer 1178.* 3227 (Stutterheim): Prospect farm, near Komgha, *Flanagan 302.* 3228 (Butterworth): Kentani, *Pegler 49.* 3322 (Oudtshoorn): Oudtshoorn, *Burtt Davy 12624.* 3325 (Port Elizabeth): Bethelsdorp, near Uitenhage, *Burchell 4400* (BOL, GRA, K, P); on the banks of the Zwartkop and Sundays Rivers, *Ecklon & Zeyher 460* (BM, BOL, E, FI, GRA, K, TCD). 3326 (Grahamstown): bank of Kowie River, *R. Verdoorn 5.*

A. caffra is widespread throughout much of our area from the western Cape to the northern Transvaal. It is a very variable species, particularly in the degree of development of the indumentum and in the number of pinnae pairs. Of the four varieties recognized within *A. caffra*, only one, namely var. *tomentosa* Glover, remained in common use for some time. Var. *tomentosa* was distinguished from the other varieties on the number of pinnae pairs, the arrangement of the inflorescences and the degree of pubescence of the young branchlets, peduncles and calyces. In its typical form var. *tomentosa* is fairly distinctive but the characters typifying it vary independently. The extremes of the species look very different, yet when the range of variation is inspected, it becomes clear that intermediates are so numerous that infraspecific categories cannot be delimited satisfactorily. The many minutely differing forms in various parts of the species range, whose taxonomic significance seems unworthy of recognition, are therefore regarded as no more than part of the overall range of variation within the species.

However, mention must be made of some specimens from the Potgietersrust, Rustenburg and Waterberg districts of the western Transvaal, for example, *Codd 3741, 3996, 4005, 4424, Meeuse 10142, 10144.* The variation in leaflet size in these specimens is greater, and leaflets sometimes attain a larger size, than ever recorded in *A. caffra* in other parts of its distributional range. There is, however, a gradient in leaflet size from *A. caffra* to the largest leaflets on these specimens. The pods of these specimens differ in being broader than is usual for *A. caffra* and in frequently having irregularly constricted margins. The plants appear to represent a local variant of *A. caffra* but it has not been possible to establish whether they represent the response to some edaphic or other environmental condition, or whether they are the result of past introgression with an unknown parent which is now absent from the area. As no suitable means has been found of distinguishing these specimens from *A. caffra,* they are included in *A. caffra.*

The differences between *A. caffra* and *A. ataxacantha* and between *A. caffra* and *A. hereroensis* are given under *A. ataxacantha* and *A. hereroensis* respectively. Specimens of *A. caffra* in Namaqualand show a definite approach to specimens of *A. hereroensis.*

The wood of *A. caffra* is heavy and close-grained with a yellowish-white sapwood and a dark brown to blackish heart-wood. Although of no commercial value, it makes good fence poles and is an excellent firewood. The Zulu name of *A. caffra* is "um Thole".

13. **Acacia hereroensis** *Engl.* in Bot. Jahrb. 10 : 20 (1888); Schinz in Mém. Herb. Boiss. 1 : 112 (1900); Harms in Engl., Pflanzenw. Afr. 3,1 : 388 (1915); Dinter in Feddes Repert. 15 : 80 (1917); Bak. f., Leg. Trop. Afr. 3 : 835 (1930); O. B. Miller, Checklist Bech. Prot. 19 (1948); Ross in J. S. Afr. Bot. 31 : 220 (1965); in Webbia 22 : 213 (1967); Schreiber in F.S.W.A. 58 : 9 (1967); Brenan in F.Z. 3,1 : 73 (1970); Palmer & Pitman, Trees S. Afr. 2 : 761 (1973). Type: South West Africa, Otjimbingwe, *Marloth* 1331 (B, holo.†; GRA, pro parte!; PRE!).

A. gansbergensis Schinz in Mém. Herb. Boiss. 1 : 108 (1900); Dinter in Feddes Repert. 15 : 79 (1917); Bak.f., Leg. Trop. Afr. 3 : 833 (1930). Type: South West Africa, Gansberg, *Fleck 437a* (Z, holo.!). *A. caffra* sensu Schinz in Mém. Herb. Boiss. 1 : 105 (1900) pro parte quoad specim. *Fleck 494a;* sensu F. Bol. et al in Ann. Bolus Herb. 1 : 15 (1914), non (Thunb.) Willd. *A. mellei* Verdoorn in Flow. Pl. S.

Afr. 22 : t.860 (1942); O. B. Miller, Checklist Bech. Prot. 20 (1948); in J. S. Afr. Bot. 18 : 23 (1952) pro parte saltem excl. specim. *Miller B/950*; Young in Candollea 15 : 109 (1955); Brenan in Kew Bull. 11 : 197 (1956); Von Breitenbach, Indig. Trees S. Afr 2 : 286 (1965). Type: Transvaal, Pretoria Distr., Zwartkop, 9,6 km S. of Pretoria, *Melle sub PRE 26514* (PRE, holo.!; FHO!, K!).

Shrub or tree to 10 m high; trunk to 0,35 m in diam.; branches usually ascending. *Bark* pale to dark greyish-brown or brown, rough; young branchlets grey- or reddish-brown, densely puberulous to pubescent, frequently with minute reddish glands scattered in amongst the pubescence. *Stipules* not spinescent, in pairs, linear, 1,8–3,8 × 0,2–0,7 mm, soon deciduous. *Prickles* in pairs just below the nodes, usually strongly recurved, sometimes straightish, often broad-based, up to 8 mm long. *Leaves:* petiole 0,3–1,3(2,2) cm long, adaxial gland usually present, variable in position, slightly raised or occasionally stalked, 0,4–1,1 × 0,1–0,4 mm; rhachis 1,7–6(10) cm long, puberulous to densely pubescent, abaxial surface often with scattered recurved prickles up to 2 mm long, with a gland at the junction of the top 1–3(7) pinnae pairs and occasionally between the lowest 1–3 pairs; pinnae 8–14(26) pairs; rhachillae (0,5)1–2,4(3,4) cm long, puberulous to densely pubescent; leaflets grey-green, (16)20–36(48) pairs per pinna, 1–4 × (0,25)0,5–1,1 mm, linear to linear-oblong, apex rounded to subacute, glabrous or sparingly to densely appressed-pubescent below, margins with or without spreading cilia. *Inflorescences* spicate, solitary, fascicled or crowded into an irregular terminal panicle. *Flowers* yellowish–white, sessile; spikes (2,2)3–7,4(8,6) cm long; peduncles 0,3–2,8 cm long, puberulous to densely pubescent, often glandular. *Calyx* densely puberulous to pubescent, tube 1–2 mm long, lobes 0,4–1,1 mm long. *Corolla* subglabrous or appressed-pubescent especially towards the apices of the lobes, tube up to 3,4 mm long, lobes up to 1,1 mm long. *Stamen-filaments* free, up to 7,5 mm long; anthers ± 0,15 mm across, with a deciduous apical gland. *Ovary* 0,7–1,5 mm long, glabrous, shortly stipitate. *Pods* olive- to reddish-brown or brown, (5)6–11(14) × (0,9)1,2–2,3 cm, straight, acute to distinctly acuminate apically, longitudinally dehiscent, margins entire or frequently irregularly constricted between the seeds, umbonate over the seeds,

coriaceous, puberulous to densely pubescent, with numerous minute scattered reddish-brown glands. *Seeds* olive- to light brown, subcircular, 7–10 × 5–10 mm, compressed; central areole up to 4 × 3 mm, horse-shoe shaped.

Found in South West Africa, south-eastern Botswana, the western Transvaal, western Orange Free State and the northern Cape Province. Occurs in dry habitats; in grassland, woodland, thornveld, on rocky slopes and flats or, in drier areas, along shallow watercourses. In the Transvaal *A. hereroensis* is frequently associated with dolomite formations.

S.W.A.—1917 (Tsumeb): 17,6 km from Otavi on road to Grootfontein, *Tölken & Hardy 927*. 1918 (Grootfontein): 16 km N. of Grootfontein on road to Abenab Mine, *De Winter 3700*. 2017 (Waterberg): Waterberg plateau, *Boss sub TRV 35004*. 2116 (Okahandja): Waldau, *Dinter 370*. 2215 (Trekkopje): farm Wilsonfontein, *Hälbich 1308*. 2216 (Otjimbingwe): Auuanis, Khomas Hochland, *Merxmuller & Giess 1773*. 2217 (Windhoek): 8 km E. of Windhoek, *Codd 5801*. 2718 (Grunau): rocky banks of ravines in Great Karasberg, *Pearson 8072*. Grid ref. unknown: Gansberg, *Fleck 437a* (Z).

TRANSVAAL.—2425 (Gaberones): Lekkerlach, *Louw 604*. 2526 (Zeerust): Swartruggens, *Sutton 908*, 2527 (Rustenburg): farm Welgevonden, *Phillips & Mogg 14647*. 2528 (Pretoria): Pretoria, west end, Zeiler St. canal, *Repton 2743*. 2626 (Klerksdorp): 64 km S. of Lichtenburg on road to Ottosdal, *Morris 1118*.

O.F.S.—2925 (Jagersfontein): Fauresmith veld Reserve, *Henrici 4367*.

CAPE.—2525 (Mafeking): Mafeking Golf Course, *O. B. Miller B/499*. 2625 (Delareyville): 30,4 km from Setlagodi on road to Mafeking, *Codd 1323*. 2723 (Kuruman): Takoon, *Gerstner 6299*.

A. hereroensis is extremely closely related to *A. caffra*. It differs from *A. caffra* in the usually shorter and smaller leaves with more crowded pinnae pairs. These shorter leaves of *A. hereroensis* often tend to be held erect and do not droop as readily as the longer leaves of *A. caffra*. *A. hereroensis* has a more westerly distributional range than *A. caffra* although in the western Transvaal and south-eastern Botswana their ranges do overlap. However, in parts of the western Transvaal where the two species grow in close proximity, each species may usually be distinguished in the field without difficulty.

In view of the differences in distribution, and the ability of the two taxa to maintain their identity when they grow in proximity, it seems preferable to continue to recognize *A. hereroensis* and *A. caffra* as distinct species.

14. **Acacia fleckii** *Schinz* in Mém. Herb. Boiss. 1 : 108 (1900); Dinter in Feddes Repert. 15 : 79 (1917); Bak.f., Leg. Trop. Afr. 3 : 832 (1930); O. B. Miller, Checklist Bech. Prot. 18 (1948); Brenan in Kew Bull. 11 : 197 (1956); Torre in C. F. A. 2 : 277, t.54 (1956); Story in Mem. Bot. Surv. S. Afr. 30 : 22

(1958); F. White, For. Fl. N. Rhod. 84, fig.17i (1962); Von Breitenbach, Indig. Trees S. Afr. 2 : 285 (1965); Schreiber in F.S.W.A. 58 : 8 (1967); Brenan in F.Z. 3,1 : 74 (1970); Palmer & Pitman, Trees S. Afr. 2 : 758 (1973). Type: Botswana, Ghanzi [Chansis], *Fleck 412a* (Z, holo.!).

A. cinerea Schinz in Verh. Bot. Ver. Prov. Brandenb. 30 : 240 (1888), nom. illegit., non *A. cinerea* Spreng. (1826); Dinter in Feddes Repert. 15 : 78 (1917); Bak.f., Leg. Trop. Afr. 3 : 832 (1930); O. B. Miller, Checklist Bech. Prot. 17 (1948); in J. S. Afr. Bot. 18 : 19 (1952). Type: South West Africa, Amboland, "Omatope", *Schinz 252* (Z, holo.!). *A. catechu* sensu Harms in Warb., Kunene–Samb. Exped. 243 (1903), non (L.f.) Willd. *A. caffra* sensu N.E. Br. in Kew Bull. 1909 : 107 (1909) pro parte quoad specim. *Lugard 93. A. caffra* var. *tomentosa* sensu Bak.f., Leg. Trop. Afr. 3 : 833 (1930) pro parte quoad specim. *Lugard* 93; O. B. Miller, Checklist Bech. Prot. 17 (1948); in J. S. Afr. Bot. 18 : 19 (1952) saltem quoad specim. *Curson 173. A. catechu* subsp. *suma* (Roxb.) Roberty var. *baumii* Roberty in Candollea 11 : 157 (1948). Type: Angola, Cubango District, bank of the Cubango River, near Kalolo, *Baum 438* (G, holo.; BM!, E!, K!, Z!).

Shrub, often with many stems, or a tree to 7 m high; crown rounded or flattened and spreading. *Bark* pale to dark yellowish- or greyish-brown, rough, the outer layer papery and flaking off; young branchlets pale yellowish or grey to greyish-brown, puberulous to pubescent, with few to many small reddish glands intermixed. *Stipules* not spinescent, in pairs, linear, 2–4,5 × 0,2–0,8 mm, soon deciduous. *Prickles* in pairs below the nodes, usually strongly recurved, broad-based, grey-brown to reddish-brown or purplish, up to 1 cm long. *Leaves:* petiole 0,3–1(1,3) cm long, adaxial gland variable in position, elliptic or discoid, 0,8–2,2 × 0,4–1,1 mm, reddish-brown to black; rhachis (0,7)1,8–5,6(7,5) cm long, puberulous to densely pubescent, with few small reddish glands intermixed, without a conspicuous gland at the junction of any of the pinnae pairs; pinnae (4)8–16(20) pairs; rhachillae 0,7–2,9 cm long, puberulous to densely pubescent; leaflets 9–28(35) pairs per pinna, 2–5(6,5) × 0,5–1,2(1,6) mm, linear-oblong, straight or almost so, apex rounded to obtuse or subacute, veins often prominent beneath at first but becoming obscure with age, glabrous or sparingly to densely appressed-pubescent below, margins usually with spreading cilia. *Inflorescences* spicate, solitary or paired, sometimes forming a terminal panicle. *Flowers* yellowish-white, sessile;

spikes (1,8)2,6–5,4(6,5) cm long; peduncles 0,6–2,4 cm long, puberulous to densely pubescent. *Calyx* often olive-green, puberulous to densely pubescent, sometimes subglabrous, tube 1,3–2,5 mm long, lobes 0,6–1,2 mm long. *Corolla* glabrous or slightly puberulous, especially towards the apices of the lobes, tube up to 3,2 mm long, lobes up to 1,4 mm long. *Stamen-filaments* free, up to 9 mm long; anthers 0,2–0,25 mm across, with a deciduous apical gland. *Ovary* 0,7–1,5 mm long, glabrous or puberulous, stipitate. *Pods* fawn to dark brown, 4–13,5 × 1,1–2(2,3) cm, linear-oblong, straight, rounded to acute or sometimes rostrate apically, longitudinally dehiscent, venose, coriaceous, puberulous to subglabrous, with numerous small reddish glands. *Seeds* olive-brown, subcircular, 8–12,5 × 8–12 mm, compressed; areole 1–4 × 1–3 mm, horseshoe shaped.

Found in Angola, South West Africa, Botswana, Zambia, Rhodesia and the western Transvaal. Occurs in drier types of mixed deciduous woodland, thicket, bushland and scrub, sometimes on sandy flats adjacent to rivers. Frequent on Kalahari sand.

S.W.A.—1719 (Runtu): 12,8 km E. of Runtu, *De Winter 3778.* 1721 (Mbambi): 31,4 km E. of Nyangana Mission station, *De Winter & Marias 4870.* 1724 (Kátima Mulilo): Katima Mulilo, bank of Zambezi River, *Killick & Leistner 3090.* 1819 (Karakuwisa): near Karakuwisa, *Marsh sub PRE 29609.* 1821 (Andara): Andara, *Merxmuller & Giess 1357.* 1916 (Gobaub): farm Lazy Spade, 48 km N.E. of Outjo, *De Winter 3029.* 1920 (Tsumkwe): Tsumkwe, 251,2 km E. of Grootfontein, *Story 6209.* 2017 (Waterberg): Okakarara, Waterberg, *Liebenberg 4755.* 2219 (Sandfontein): farm Gemsbokfontein, near Botswana border, *Merxmuller 1179.* Grid ref. unknown: between Otjiwarongo and Otavi, *Werdermann & Oberdieck 2345.*

TRANSVAAL.—2326 (Mahalapye): Buffelsdrif, *Vahrmeijer 1291.* 2327 (Ellisras): 4 km N. of Ons Hoop Post Office, beside the Mogol River, *Codd 8483.* 2526 (Zeerust): 1,6 km S. of Nietverdiend Post Office, *Acocks 19184.*

A. fleckii is closely related to *A. erubescens* Welw. ex Oliv. and the two species are often confused. *A. fleckii* differs from *A. erubescens* in having:

1. shorter petioles 0,3–1 cm long, occasionally to 1,3 cm; in *A. erubescens* the petioles are normally 1,3–2,5 cm long. Occasionally petioles shorter or longer than these dimensions occur but then the extremes usually occur on shoots with at least some petioles of more usual length;
2. a larger petiolar gland which is elliptic to discoid, 0,8–2,2 mm long; in *A. erubescens* the gland is often slightly raised and is 0,3–1 mm long. As the dimensions for the two species overlap, gland size is a less useful character;

3. the leaf-rhachis without a gland at the junction of any of the pinnae pairs; in *A. erubescens* a gland is usually present at the junction of some of the pinnae pairs or between the top pair only;
4. (4)8–20 pinnae pairs; *A. erubescens* has only 3–7 pinnae pairs;
5. straight, smaller leaflets; in *A. erubescens* the leaflets are usually somewhat falcate and larger.

The above characters usually enable the two species to be readily distinguished. Another useful character is the tendency for *A. fleckii* to produce its inflorescences with or after the new leaves while *A. erubescens* often flowers when leafless.

15. Acacia erubescens *Welw. ex Oliv.* in F.T.A. 2 : 343 (1871); Benth. in Trans. Linn. Soc. Lond. 30 : 518 (1875); Harms in Engl., Pflanzenw. Afr. 3,1 : 385 (1915); Bak.f., Leg. Trop. Afr. 3 : 830 (1930);O. B. Miller in J. S. Afr. Bot. 18 : 20 (1952); Young in Candollea 15 : 111 (1955);Torre in C.F.A. 2 : 276, t.53B (1956); Brenan in F.T.E.A. Legum. -Mimos. : 88, fig. 14/13 (1959); F. White, For. Fl. N. Rhod. 83, fig. 17G (1962); Schreiber in F.S.W.A. 58 : 7 (1967); Brenan in F.Z. 3,1 : 73 (1970); Van Wyk, Trees Kruger Nat. Park 1 : 135 (1972); Palmer & Pitman, Trees S. Afr. 2 : 757 (1973); Schreiber in Mitt. Bot. Staatssamml. Munchen 11 : 116 (1973). Type: Angola, Mocamedes District, between Bumbo and Bruco, *Welwitsch* 1826 (LISU holo., BM!, K!)

A. caffra Willd. var. *pechuelii* Kuntze in Jahrb. K. Bot. Gart. Mus. Berl. 4 : 264 (1886). Type: South West Africa, Hereroland, *Pechuel-Loesche* (? B, holo. †). *A. dulcis* Marloth & Engl. in Bot. Jahrb. 10 : 24 (1888); Marloth in Trans. S. Afr. Phil. Soc. 5 : 269 (1889); Schinz in Mém. Herb. Boiss. 1 : 107 (1900); Harms in Engl., Pflanzenw. Afr. 3, 1 : 388 (1915); Dinter in Feddes Repert. 15 : 79 (1917); Bak.f., Leg. Trop. Afr. 3 : 830 (1930); Burtt Davy, Fl. Transv. 2 : 337 (1932); Codd, Trees & Shrubs Kruger Nat. Park 42, fig. 37a, b (1951): O. B. Miller in J. S. Afr. Bot. 18 : 20 (1952); Gilbert & Boutique in F.C.B. 3 : 151 (1952); Story, Mem. Bot. Surv. S. Afr. 30 : 22 (1958); Von Breitenbach, Indig. Trees S. Afr. 2 : 282 (1965). Type: South West Africa, Karibib Distr., Usakos, *Marloth* 1259 (B, holo. †; BOL!, GRA!, PRE!). *A. longipetiolata* Schinz in Mém. Herb. Boiss. 1 : 114 (1900). Syntypes: South West Africa, Hereroland, *Fleck* 491 (Z!); Kuiseb *Fleck* 492a (Z!), *Fleck* 493a (?Z). *A.* aff. *trispinosa* sensu Schinz in Mém. Herb. Boiss. 1 : 116 (1900). The specimen which Schinz referred to, namely, *Luderitz* 122 from Hereroland, is in the University of Zurich herbarium. This specimen is without leaves or fruits but is clearly referable to *A. erubescens*. However, it is as well to point out here that there is more than one Luderitz specimen from Hereroland with the collector's number 122 in the University of Zurich herbarium. Another specimen of *Luderitz* 122 is *A. karroo* Hayne. *A.* sp. sensu Schinz in Mém. Herb. Boiss. 1 : 116

(1900). *A. kwebensis* N.E. Br. in Kew Bull. 1909 : 108 (1909). Type: Botswana, Kwebe Hills, *Mrs. E. J. Lugard* 24 (K, holo.!).

Shrub, often with many stems, or a tree to 10 m high; trunk to 0,3 diam.; crown often flattened and spreading somewhat. *Bark* pale to dark yellowish- or greyish-brown, rough, the outer layer papery and flaking or peeling off; young branchlets yellowish- or greyish-brown to purplish, puberulous to pubescent, sometimes glabrous, indumentum usually slightly golden. *Stipules* not spinescent, in pairs, linear, 2–4,2 × 0,2–0,8 mm, soon deciduous. *Prickles* in pairs below the nodes (very rarely in threes but then all three prickles point downward), usually strongly recurved, broad-based, grey-brown to blackish, up to 7 mm long. *Leaves:* petiole (0,7)1,3–2,5(4) cm long, adaxial gland present or absent, variable in position, often slightly raised, 0,3–0,7(1) × 0,1–0,5 mm; rhachis (1)1,5–2,7(5) cm long, usually puberulous to densely pubescent, glands variable, either at the junction of some of the pinnae pairs, between the top pair only or sometimes absent; pinnae 3–7 pairs; rhachillae (1)1,2–3(3,9) cm long, usually puberulous to densely pubescent; leaflets 10–27 pairs per pinna, 3–7,5(10) × (0,75)1–1,6(2,8) mm, obliquely oblong, often slightly falcate or the upper somewhat obovate, apex usually oblique, acute or subacute, sometimes ± rounded, veins often prominent beneath at first but becoming obscure with age, glabrous below or occasionally sparingly pubescent, margins with or without cilia. *Inflorescences* spicate, fascicled or crowded into an irregular terminal panicle, sometimes solitary. *Flowers* yellowish-white, sessile; spikes (1,8)2,2–4,5(6) cm long, peduncles (0,6)1,2–2,5 cm long, pubescent. *Calyx* campanulate, puberulous to densely pubescent, tube 1–2,2 mm long, lobes 0,6–1,8 mm long. *Corolla* tubular, appressed-pubescent, especially towards apices of lobes, tube up to 3 mm long, lobes up to 1,5 mm long. *Stamen-filaments* free, up to 8 mm long; anthers 0,2–0,25 mm across, with a deciduous apical gland. *Ovary* 0,5–1,9 mm long, glabrous, sessile or shortly stipitate. *Pods* fawn to dark brown, 3–13 × 1,1–1,8(2,3) cm, linear-oblong, straight, rounded to acute or acuminate apically, longitudinally dehiscent, venose, coriaceous, subglabrous except for pubescence on the margins and stipe, glands absent or few and inconspicuous.

Seeds olive-brown, usually subcircular, 7–11 × 7–11 mm, compressed; areole 1–3 × 2–4,5 mm, horse-shoe shaped.

Found in Zaire, Tanzania, Zambia, Malawi, Rhodesia, Mozambique, Angola, South West Africa, Botswana and the Transvaal. Occurs in drier types of mixed deciduous woodland or scrub, often on rocky outcrops, in open bush on sand or along the sandy banks of dry watercourses.

S.W.A.—1713 (Swartbooisdrif): 17,6 km W. of Otjiwero, sandy course of Okahalalana Ndjala River, *De Winter & Leistner 5405*. 1718 (Kuring-Kuru): 12 km E. of Kuring-Kuru on road to Runtu, *De Winter 3950*. 1719 (Runtu): 4,8 km S. of Runtu, *De Winter 3775*. 1821 (Andara): road from Andara to Bagani, *Merxmuller & Giess 1994*. 1917 (Tsumeb): ± 11 km N. of Tsumeb, *Basson 41*. 1918 (Grootfontein): Kalkfontein, *Liebenberg 4897*. 2016 (Otjiwarongo): Omatjenne, *Liebenberg 4825*. 2115 (Karibib): Usakos, *Marloth 1259*. 2117 (Otjosondu): Quickborn flats, *Bradfield 22*. 2216 (Otjimbingwe): farm Otjiseva, near Teufelsbach, *Kinges & Wiss 788* (PRE) pro parte. 2217 (Windhoek): near Windhoek, *Codd 5797*.

TRANSVAAL.—2229 (Waterpoort): Dongola reserve, *Pole Evans 3532*. 2230 (Messina): Messina, *Gerstner 5711*. 2231 (Pafuri): Kruger National Park, near Punda Milia, *Codd & Dyer 4528*. 2328 (Baltimore): farm Koppermyn, 128 km N.N.W. of Potgietersrust, *Galpin 9218*. 2426 (Mochudi): 96 km N. of Zeerust, *Louw 1499*. 2427 (Thabazimbi): Thabazimbi, *Dyer & Verdoorn 4219*. 2430 (Pilgrim's Rest): 12 km S. of Olifants River bridge on Acornhoek–Leydsdorp road, *Codd 1668*. 2527 (Rustenburg): farm Welgevonden, *Mogg 14635*. 2528 (Pretoria): near Flinks Drift on Pienaars River flats, *Smuts 356*.

In *A. erubescens* the inflorescences are often produced when the plants are leafless. The flowers are usually sweet-smelling.

The differences between *A. erubescens* and *A. fleckii* are given under *A. fleckii*.

16. **Acacia kraussiana** *Meisn. ex Benth.* in Hook., Lond. J. Bot. 1 : 515 (1842); Meisn. in Hook., Lond. J. Bot. 2 : 103 (1843); Harv. in F.C. 2 : 283 (1862); Benth. in Trans. Linn. Soc. Lond. 30 : 530 (1875); Wood & Evans, Natal Plants 3,2 : t.245 (1901); Glover in Ann. Bolus Herb. 1 : 147, t.18/8 (1915); Harms in Engl., Pflanzenw. Afr. 3,1 : 390 (1915); Bews, Fl. Natal 114 (1921); Henkel, Woody Pl. Natal 235 (1934); Brenan & Exell in Bol. Soc. Brot. Sér. 2,31 : 103, t.1 fig. C (1957); Mogg in Macnae & Kalk, Nat. Hist. Inhaca Island 9, 145 (1958); Brenan in F.Z. 3,1 : 80, t.19C (1970); Ross, Acacia Spp. Natal 33, fig. 2/11 (1971); Fl. Natal 193 (1973). Type: Natal, Umlaas River, Durban [Port Natal], *Krauss 198* (K, holo.!; BM!, FI!, OXF!, PRE!, TCD!, Z!).

Scandent shrub or climber to 15 m high. *Bark* pale to dark greyish-brown or black; young branchlets grey-brown, usually pube-rulous and inconspicuously glandular but sometimes pubescent, angular, often hexagonal, some branchlets usually modified into tendrils. *Stipules* not spinescent, in pairs, ± linear, up to 4,5 mm long and 1,1 mm wide, soon deciduous. *Prickles* irregularly scattered along the internodes, up to 2 mm long, occasionally minute and apparently absent, recurved, arising from longitudinal bands along the stem which are usually concolorous with the intervening lenticellate bands. *Leaves:* petiole 0,8–2,7 cm long, subglabrous to densely puberulous, adaxial gland usually immediately above the pulvinus, raised, up to 2,8 mm long; rhachis 2–6,2 cm long, subglabrous to densely puberulous or pubescent, lower surface usually with scattered recurved prickles up to 3 mm long, a small gland usually at the junction of the top pinna pair only; pinnae (1)3–6 pairs; rhachillae 0,9–5(6,9) cm long, subglabrous to densely puberulous; leaflets 6–17 pairs per pinna, 5–15(23) × (2)3–6(8) mm, linear-oblong to obovate-oblong, often slightly falcate, the lowest pair very much reduced and bract-like, apex acute or shortly mucronate, midrib excentric basally, lateral veins conspicuous beneath, glabrous or nearly so, rarely puberulous or pubescent, marginal cilia usually absent. *Inflorescences* capitate, forming a terminal panicle. *Flowers* yellowish-white, very shortly pedicellate; peduncles 0,7–2,8 cm long, densely puberulous or pubescent. *Calyx* glabrous to sparingly or densely puberulous, tube 1,6–2,2 mm long, lobes 0,6–1,1 mm long. *Corolla* glabrous, tube 2,6–4,2 mm long, lobes 0,5–1,5 mm long. *Stamen-filaments* free, up to 8,5 mm long; anthers with a deciduous apical gland. *Ovary* pilose, up to 2 mm long, on a stipe longer than itself. *Pods* pale or dark brown, 5,9–16,2 × (0,9)1,4–2,5 cm, linear-oblong, subcoriaceous, markedly umbonate over the seeds, sometimes irregularly constricted between some of the seeds, venose, tardily longitudinally dehiscent or at times indehiscent, apex rounded to mucronate, margins not strongly thickened, mostly glabrous or subglabrous and with minute reddish glands, especially when young. *Seeds* olive or dark-brown to blackish, 5,5–9 × 4–6 mm, ellipsoid, somewhat compressed; areole 4–7 × 2,5–4 mm.

Found in Natal and Mozambique. Essentially a coastal species and usually found below 110 m, often growing near the shore. Occurs on forest margins, in forest clearings, in scrub, bushland and, in Tongaland, in dry sand forest. Often forms fairly dense impenetrable thickets.

NATAL.—2632 (Bela Vista): Ndumu Game Reserve, Matini forest, *Tinley 944* (NH, NU). 2731 (Louwsburg): between Vryheid and Nongoma, *Pole Evans 3581*. 2732 (Ubombo): False Bay Park, *Ross 2328*. 2832 (Mtubatuba): Hluhluwe Game Reserve, *Ward 3965*. 2930 (Pietermaritzburg): Isipingo, *Acocks 10873*. 2931 (Stanger): Hawaan forest, S. bank of Umhlanga River, *Ross & Moll 2261*. 3030 (Port Shepstone): Doonside, *Galpin 9534*.

A. kraussiana is a very distinct species which is readily distinguished by its few pinnae pairs and large leaflets from both of the other *Acacia* species with scattered recurved prickles and capitate inflorescences occurring in our area. However, *A. kraussiana* is sometimes confused with *Entada spicata* (E. Mey.) Druce. *E. spicata* differs in having a spicate inflorescence and pods in which the valves at maturity (but not the margins) split transversely into one-seeded segments, the segments falling away from the margins which persist as a continuous but empty frame. Unlike *A. kraussiana* in which the lowest pair of leaflets on each pinna are bract-like, the lowest pair of leaflets in *E. spicata* are well developed.

A specimen of *A. kraussiana*, *Mrs. Hutton* 106 (BM, GRA), carries the locality "Shafton, Howick, Natal". As *A. kraussiana* is not known to occur more than a few kilometres inland from the coast south of the Tugela River, it seems unlikely that this specimen was collected near Howick. Perhaps the label does not belong with the specimen.

17. **Acacia brevispica** *Harms* in Notizbl. Bot. Gart. Berl. 8 : 370 (1923); Bak.f., Leg. Trop. Afr. 3 : 853 (1930); Brenan, Checklist Tang. Terr. 332 (1949); Brenan & Exell in C.F.A. 2 : 287 (1956); in Bol. Soc. Brot., Sér. 2, 31 : 108, t.1 fig. B (1957); Brenan in F.T.E.A. Legum.-Mimos. : 96, fig. 15/22 (1959); in Kew Bull. 21 : 477 (1968); in F.Z. 3,1 : 81 (1970); Ross in Bothalia 10 : 419 (1971). Type: Tanzania, Lushoto District, Kitivo, *Holst 606* (B, holo.†).

subsp. **dregeana** (*Benth.*) *Brenan* in Kew Bull. 21 : 479 (1968); in F.Z. 3,1: 81 (1970); Ross, Acacia Spp. Natal 22, fig. 2/10 (1971); in Bothalia 10 : 419 (1971); Fl. Natal 193 (1973). Type: Cape, Pondoland, *Drege s.n.* (K, holo.!; P!).

A. pennata var. *dregeana* Benth. in Hook., Lond. J. Bot. 1 : 516 (1842). Type as above. *A. pennata* sensu E. Mey., Comm. 1 : 169 (1836); Harv. in F.C. 2 : 283 (1862) pro parte quoad specim. *Drege*; Glover in Ann. Bolus Herb. 1 : 147 (1915) pro parte quoad specim. *Wood* 4469, excl. ref. Wood & Evans, Natal Pl. 3, 2 : t.244 (1901); Bews, Fl. Natal 114 (1921) pro parte; Henkel, Woody Pl. Natal 234 (1934) pro parte, non (L.) Willd. sensu stricto. *A. brevispica* sensu Brenan &

Exell in Bol. Soc. Brot., Sér. 2, 31 : 114 (1957) quoad pl. Mozamb., Natal, Cape. *A. brevispica* var. *dregeana* (Benth.) Ross & Gordon-Gray in Brittonia 18 : 63 (1966). Type as above.

Scandent shrub up to 12 m high, a non-climbing shrub or sometimes a small tree to 3 m high; young branchlets grey-brown, very shortly puberulous or rarely glabrous, with numerous minute reddish glands. *Stipules* not spinescent, in pairs, ± linear, up to 4,5 mm long, 0,3–1,2 mm wide, soon deciduous. *Prickles* scattered along the internodes, recurved or spreading, arising from longitudinal bands which are usually paler than the intervening lenticellate bands, 0,5–4 mm long. *Leaves:* petiole 0,5–3,5 cm long, adaxial gland often present, variable in position; rhachis 3,2–10 cm long, shortly puberulous to subglabrous, with or without recurved prickles abaxially, a gland often present at the junction of the top 1–3 pinnae pairs; pinnae 5–20 pairs; rhachillae 1,4–3,8 cm long; leaflets 24–62 pairs per pinna, 2,5–5 × 0,5–1(1,2) mm, linear-oblong, lower surface usually appressed-pubescent throughout, at times on portion of surface only or occasionally leaflet entirely glabrous, margins usually with short inconspicuous cilia, sometimes absent. *Inflorescences* capitate, on axillary peduncles, racemosely arranged or aggregated into an irregular terminal panicle. *Flowers* yellowish-white, sessile or very shortly pedicellate; peduncles 1–2,5 cm long, puberulous, with numerous minute reddish glands. *Calyx* campanulate, puberulous to almost glabrous, tube 0,7–1,8 mm long, lobes 0,3–0,8 mm long. *Corolla* campanulate, tube up to 2,5 mm long, lobes up to 1 mm long. *Stamen-filaments* free, up to 4 mm long; anthers with a deciduous apical gland. *Ovary* densely pubescent, up to 1,2 mm long, on a stipe longer than itself. *Pods* brown, 5,7–15(21) × (1,1)1,8–2,8 cm, oblong to linear-oblong, straight or almost so, rarely slightly falcate, longitudinally dehiscent, subcoriaceous to coriaceous, umbonate over the seeds, usually densely puberulous and with numerous minute reddish glands. *Seeds* olive-brown, 8–12 × 6–10 mm, elliptic, compressed; areole 6–8 × 3–5 mm.

Occurs in southern Mozambique, the eastern Transvaal, Swaziland, Natal and the eastern Cape (Pondoland). Found in dry thornveld and in dry river valley scrub; often forms dense impenetrable thickets.

TRANSVAAL.—2431 (Acornhoek): Kruger National Park, Munweni River, *Brynard & Pienaar 4327*.

SWAZILAND.—2631 (Mbabane): Stegi Hill, *Compton 28418*. 2731 (Louwsburg): Ingwavuma Poort, *Compton 28629*.

NATAL.—2731 (Louwsburg): 1 km from Pongola bridge on Magudu road, *Edwards 3187*. 2732 (Ubombo): ± 5 km S. of Pongola Poort, *Ward 3917*. 2829 (Harrismith): Estcourt Pasture Research Station, *Acocks 9893*. 2830 (Dundee): 4,8 km from Weenen on Estcourt road, *Ross 766* (K, NH, NU). 2831 (Nkandla): Umfolozi Game Reserve, bank of White Umfolozi River, *Ross 2029*. 2930 (Pietermaritzburg): Ashburton, near Pietermaritzburg, *Ross 443*. 2931 (Stanger): Lower Tugela valley, opposite confluence of Nembe and Tugela Rivers, *Edwards 3045*.

CAPE.—3128 (Umtata): Mqanduli, *Pegler 559* (GRA). 3129 (Port St. Johns): Msikaba drift, *Strey 8478*. Grid ref. unknown: Pondoland, *Drege s.n* (K, P); Umzimvubu, 1450 m, *Schlechter 6429* (GRA).

Typical *A. brevispica* is widespread in tropical Africa, but occurs nowhere closer to our area than Angola and central Tanzania. In tropical Africa typical *A. brevispica* is relatively uniform and no difficulty is experienced in distinguishing it from *A. schweinfurthii*. *A. brevispica* subsp. *dregeana*, however, is characteristically very variable and in our area it bridges many of the discontinuities which exist between typical *A. brevispica* and *A. schweinfurthii* in the tropics. Consequently difficulty is sometimes experienced in distinguishing specimens of *A. brevispica* subsp. *dregeana* from *A. schweinfurthii*.

A. schweinfurthii var. *schweinfurthii* differs from *A. brevispica* subsp. *dregeana* in having:

1. typically longer petioles (1,5)2,6–5,5 cm long;

2. a distinctly humped petiolar gland situated immediately above the pulvinus, sometimes two glands are present or, on occasions, the gland is absent;

3. leaflets usually larger, 2,5–8 × 0,8–2 mm;

4. leaflets glabrous on the lower surface apart from appressed marginal cilia. In *A. brevispica* subsp. *dregeana* when the lower leaflet surface is entirely glabrous the marginal cilia are usually lacking too so that the leaflets are completely devoid of hairs;

5. pods glabrous or almost so except for the glands; lacking the dense puberulence of *A. brevispica* subsp. *dregeana*.

There is an overlap in petiole length and in leaflet width between *A. brevispica* subsp. *dregeana* and *A. schweinfurthii* var. *schweinfurthii*, but in combination the above characters usually enable *A. schweinfurthii* var. *schweinfurthii* to be distinguished from subsp. *dregeana*. Certainly, most specimens can be readily sorted. Some specimens do, however, undoubtedly create difficulties and it is sometimes debatable whether they are robust specimens of *A. brevispica* subsp. *dregeana* or depauperate specimens of *A. schweinfurthii*. *Ross 874* (NU) from 3 km S. of Mandini in the Lower Tugela valley is difficult to place as it has long petioles and the general facies of *A. schweinfurthii*, but the petiolar

glands are not humped, the leaflets are appressed pubescent beneath and the pods are ± densely puberulous. The specimen seems best regarded as a robust specimen of *A. brevispica* subsp. *dregeana*.

Ross 764 (NU) from the Weenen district of Natal is unusual in that some of the pinnae are themselves bipinnate.

True *A. pennata* (L.) Willd. does not occur in Africa, but until fairly recently the name was used in a wide sense to cover *A. brevispica*, *A. schweinfurthii* and several other species which occur in tropical and subtropical Africa.

18. **Acacia schweinfurthii** *Brenan & Exell* in Bol. Soc. Brot., Sér. 2, 31 : 128, t.1 fig. E (1957); Brenan in F.T.E.A. Legum.-Mimos. : 99, fig. 15/27 (1959); White, For. Fl. N. Rhod. 84 (1962); Von Breitenbach, Indig. Trees S. Afr. 2 : 286 (1965); Brenan in Kew Bull. 21 : 477 (1968); in F.Z. 3,1 : 83, t.19 fig. E (1970); Palmer & Pitman, Trees S. Afr. 2 : 813 (1973). Type: the Sudan, Gubbiki, *Schweinfurth 2206* (BM, holo.!; K!, P!, Z!).

A. brevispica Harms var. *schweinfurthii* (Brenan & Exell) Ross & Gordon-Gray in Brittonia 18 : 62 (1966). Type as above.

var. **schweinfurthii.**

Brenan & Exell l.c. : 130 (1957); Brenan in F.T.E.A. Legum-Mimos. : 99 (1959); in F.Z. 3,1 : 83 (1970); Ross, Acacia Spp. Natal 38, fig.2/9 (1971); in Bothalia 10 : 419 (1971); Fl. Natal 193 (1973).

A. pennata sensu Harv. in F.C. 2 : 283 (1862) pro parte quoad specim. *Gueinzius*; Wood & Evans, Natal Plants 3, 2 : t.244 (1901); Bews, Fl. Natal 114 (1921) pro parte; Glover in Ann. Bolus Herb. 1 : 147 (1915) pro parte quoad ref. Wood & Evans Natal Plants t.244; Henkel, Woody Pl. Natal 234 (1934) pro parte; Hutch., Botanist in S. Afr. 308, 390, 468, 664 (1946); Codd, Trees & Shrubs Kruger Nat. Park 48 (1951), non (L.) Willd. sensu stricto.

Scandent shrub up to 12 m high or a non-climbing shrub to 3 m high; young branchlets olive-green to olive-brown or grey-brown, puberulous when young but often becoming glabrous, with numerous minute reddish glands. *Stipules* not spinescent, in pairs, ± linear, up to 5 mm long, up to 1,2 mm wide, soon deciduous. *Prickles* irregularly scattered along the internodes, recurved or spreading, arising from longitudinal bands which are often darker than the intervening lenticellate bands, 1–4,5 mm long. *Leaves:* petiole (1,5)2,6–5,5 cm long, with a humped gland 1–2 × 0,5–1 mm usually situated immediately above the pulvinus, sometimes two glands present or gland absent; rhachis (3)6–16 cm long, sparingly to densely puberulous, with or without recurved prickles abaxially, a gland often present at the junction of the top 1–2 pinnae pairs; pinnae (5)9–17 pairs; rhachillae (1,8)3–7,6 cm long, sparingly to densely puberulous; leaflets 17–60 pairs per pinna, 2,5–8 × (0,8)1–2 mm, linear or linear-oblong, midrib nearer one margin basally, margins with whitish ± conspicuous appressed cilia, invariably glabrous beneath except for the marginal cilia. *Inflorescences* capitate, on axillary peduncles, usually arranged in ± pyramidal panicles. *Flowers* yellowish-white, sessile or very shortly pedicellate; peduncles 0,5–2,4 cm long, sparingly to densely puberulous, with numerous, minute reddish glands. *Calyx* puberulous or rarely subglabrous, tube 1–1,8 mm long, lobes up to 0,8 mm long. *Corolla* subglabrous to puberulous, especially towards the apices of the lobes, tube up to 2,2 mm long, lobes up to 0,8 mm long. *Stamen-filaments* free, up to 5 mm long; anthers with a deciduous apical gland. *Ovary* densely pubescent, up to 1,5 mm long, on a stipe longer than itself. *Pods* olive-brown to brown, 8–18 × 1,4–2,5 cm, oblong, straight or almost so, tardily longitudinally dehiscent or at times indehiscent, coriaceous or subcoriaceous, ± transversely plicate and umbonate over the seeds, glabrous or almost so, usually with numerous minute reddish glands, especially towards the margins and stipe. *Seeds* dark reddish-brown, 8–12 × 6–8 mm, elliptic, compressed; areole 6–8 × 3–5 mm.

Found from the Sudan southwards to the Transvaal and Natal. Occurs most frequently in riverine fringing vegetation where it usually forms dense impenetrable thickets; also occurs in woodland, scrub and on forest margins away from rivers.

TRANSVAAL.—2229 (Waterpoort): Wyllie's Poort, *Hutchinson 2095*. 2230 (Messina): Nwanedzi, [Vanetzi] River, *Gerstner 6083*. 2328 (Baltimore): Wagon Drift, Mogalakwena [Magalakwin] River, *Hutchinson 2674*. 2330 (Tzaneen): 80 km N. or Gravelotte station, on bank of Great Letaba River, *Galpin 13532*—the precise locality of this specimen is uncertain because the Great Letaba River is only 32 km N. of Gravelotte by road; the Small Letaba River, however, is ± 80 km N. of Gravelotte. 2331 (Phalaborwa): Kruger National Park, Letaba, *Lang s.n.* 2430 (Pilgrim's Rest): Blyde River Poort, *Van der Schijff 5933*. 2431 (Acornhoek): Kruger National Park, near Skukuza Camp, *Codd & De Winter 5113*. 2531 (Komatipoort): Crocodile River Poort, *Rogers 22233*.

NATAL.—2632 (Bela Vista): Ndumu Game Reserve, near Banzi Pan, *Ross 697* (K, NH, NU). 2732 (Ubombo): Mkuzi bridge, 3 km N. of Mkuze on Candover road, *Ross 1022* (K, NH, NU). 2831 (Nkandla): Umfolozi Game Reserve, Matshamshlope, *Downing 561* (NH, NU). 2832 (Mtubatuba): Hluhluwe Game Reserve, *Ward 1835*. 2930 (Pietermaritzburg): Durban district, Clairmont, *Wood 8022* (E, GRA). 2931 (Stanger): 40 km from Kranskop on Mapumulo valley road, *Moll 921*. Grid ref. uncertain: Durban [Port Natal], *Gueinzius s.n.* (K, TCD).

Difficulty is sometimes experienced in distinguishing specimens of *A. schweinfurthii* var. *schweinfurthii* from specimens of *A. brevispica* subsp. *dregeana*. The main differences between these two species are discussed under *A. brevispica* subsp. *dregeana*.

So far only var. *schweinfurthii* has been recorded from our area. Variety *sericea* Brenan & Exell differs from var. *schweinfurthii* in having the leaflets ± appressed silky-pubescent beneath. Var. *sericea* is known from relatively few gatherings in Tanzania, Zambia, Rhodesia and Mozambique.

The specimen, *Ross 874*, discussed under *A. brevispica* subsp. *dregeana* differs from var. *sericea* in having ± densely puberulous pods.

True *A. pennata* (L.) Willd. does not occur in Africa, but until fairly recently the name was used in a wide sense to cover *A. schweinfurthii*, *A. brevispica* and several other species which occur in tropical and subtropical Africa. Often *A. schweinfurthii* is likely to be the species referred to, but there is usually no certainty.

19. **Acacia karroo** *Hayne*, Arzneyk. Gebr. Gewächse 10 : t.33 (1827); Glover in Ann. Bolus Herb. 1 : 150, t.18/10 (1915); Bews, Fl. Natal 115 (1921): Burtt Davy in Kew Bull. 1922 : 328 (1922); Marloth, Fl. S. Afr. 2 : 51, t.18A, 21 (1925); Bak.f., Leg. Trop. Afr. 3 : 843 (1930); Burtt Davy, Fl. Transv. 2 : 346 (1932); Steedman, Trees etc. S. Rhod. 13 (1933); Henkel, Woody Pl. Natal 229 (1934); Hutch., Botanist in S. Afr. 138, 260, 411, 472, 512, 543, 547, 550, 552, 664 (1946); Gerstn. in J.S. Afr. Bot. 14 : 19 (1948); Codd, Trees & Shrubs Kruger Nat. Park 44, fig. 38h & i (1951); O. B. Miller in J.S. Afr. Bot. 18 : 22 (1952); Story, Mem. Bot. Surv. S. Afr. 27 : 26 (1952); Verdoorn in Bothalia 6 : 409 (1954); Flow. Pl. S. Afr. 31 : t.1220 (1956); Palgrave, Trees Cent. Afr. 242 (1956); Palmer & Pitman, Trees S. Afr. 157, t.36, 37 (1961); F. White, For. Fl. N. Rhod. 85, fig. 18D (1962); Von Breitenbach, Indig. Trees S. Afr. 2 : 298 (1965); De Winter et al, 66 Transv. Trees 50 (1966); Leistner, Mem. Bot. Surv. S. Afr. 38 : 123 (1967); Schreiber in F.S.W.A. 58 : 9 (1967); Brenan in F.Z. 3,1 : 87, t.15/7 (1970); Ross in Bothalia 10(2) : 385 (1971); Acacia Spp. Natal 30, fig. 2/14, 2/15 (1971); in Bothalia 10(3) : 427 (1971); Van Wyk, Trees Kruger Nat. Park 1 : 142 (1972); Ross, Fl. Natal

193 (1973); Palmer & Pitman, Trees S. Afr. 2 : 795 (1973); Ross in Bothalia 11(4) : 445 (1975). Type: Cape Province, precise locality unknown, *Herb. Willdenow* 19184 fol. 2 (B, lecto.; PRE, photo.).

Mimosa nilotica sensu Burm. f., Prodr. Fl. Cap. 27 (1768), non L. *M. capensis* Burm. f., Prodr. Fl. Cap. 31 [sphalm 27] (1768) pro parte. *M. leucacantha* Jacq., Hort. Schoenbr. 3 : 75, t.393 (1798), non *Acacia leucacantha* Bert. ex Spreng., Syst. Veg. 3 : 144 (1826). *M. nilotica* Thunb., Fl. Cap. ed. Schult. 432 (1823), non L. *M. eburnea* sensu Boj., Hort. Maurit. 115 (1837), non L.

A. horrida Willd., Sp. Pl. 4 : 1082 (1806) pro parte quoad fig. Jacq.; sensu auct. mult. : E. Mey., Comm. 1 : 166 (1836); Eckl. & Zeyh., Enum. 260 (1836); Meisn. in Hook., Lond. J. Bot. 2 : 103 (1843); Pappe, Silv. Cap. 14 (1853); Harv. in F.C. 2 : 281 (1862); Benth. in Trans. Linn. Soc. Lond. 30 : 507 (1875); Engl. in Bot. Jahrb. 10 : 23 (1888); Marloth in Trans. S. Afr. Phil. Soc. 5 : 270 (1889); Fourcade, Report on Natal Forests 106 (1889); Schinz in Mém. Herb. Boiss. 1 : 113 (1900); Sim, For. Fl. Cape Col. 211, t.61 (1907); Burtt Davy in Kew Bull. 1908 : 158 (1908); Sim, For. Fl. P.E. Afr. 57 (1909); Dinter, Veg. Veldkost Deutsch-Sudwest-Afrikas 36 (1912); Harms in Engl., Pflanzenw. Afr. 3, 1 : 364 (1915); Dinter in Feddes Repert. 15 : 80 (1917); Hutch., Botanist in S. Afr. 541, 542 (1946). *A. reticulata* (L.) Willd., Sp. Pl. 4 : 1056 (1806) pro parte, nom. rejec.; Willd., Enum. Hort. Berol. 1051 (1809) pro parte. *A. capensis* (Burm. f.) Burch., Trav. 1 : 114, 189 (1822); Sw., Hort. Britt. 1 : 103 (1826); Colla in Mem. Acad. Torin 35 : 175 (1831); Zeyher & Burke in Hook., Lond. J. Bot. 5 : 111, 113, 116, 119, 120, 125 (1846). *A. hirtella* E. Mey., Comm. 1 : 167 (1836); Harv. in F.C. 2 : 281 (1862); Benth. in Trans. Linn. Soc. Lond. 30 : 513 (1875);* Glover in Ann. Bolus Herb. 1 : 150, t.19/21 (1915); Bews, Fl. Natal 115 (1921). Type: Natal between Umkomaas [Omcomas] and Umlazi [Omblas], *Drege* (K!, P! iso). *A. natalitia* E. Mey., Comm. 1 : 167 (1836); Meisn. in Hook., Lond. J. Bot. 2 : 103 (1843); Benth. in Hook., Lond. J. Bot. 5 : 97 (1846) quoad specim. *Krauss* 66; Harv. in F.C. 2 : 281 (1862); Benth. in Trans. Linn. Soc. Lond. 30 : 508

---

*non Sim in Agric. J. 19 (1900); For. Fl. Cape Col. 211, t.59 (1907). Sim was apparently referring to *A. sieberana* DC. var. *woodii* (Burtt Davy) Keay & Brenan as evidenced by his description of the bark as "yellowish white flaky" and of the pod as "4 inches long, ¾–1 inch wide, solid, indehiscent, tomentose". However, *A. sieberana* var. *woodii* seldom has only 4 pinnae pairs as described by Sim and the involucel is in the upper half of the peduncle or apical and not in the lower third as illustrated by Sim in t.59. Sim described the inflorescence as "light yellow or nearly white" which is in contrast to the bright yellow inflorescence of *A. karroo*. *A. robusta* Burch., however, which often has only 4 pinnae pairs, has a whitish inflorescence and has the involucel in the lower third of the peduncle so it appears as though Sim's description of *A. hirtella* may possibly have been taken from *A. sieberana* var. *woodii* and from *A. robusta*. Sim, For. Fl. P.E. Afr. 57, t.35A (1909), was clearly referring to *A. robusta*.

(1875); Burtt Davy in Kew Bull. 1908 : 159 (1908;) Sim, For. Fl. P.E. Afr. 57 (1909); Glover in Ann. Bolus Herb. 1 : 150, t.19/18 (1915); Bews, Fl. Natal 115 (1921); Burtt Davy in Kew Bull. 1922 : 329 (1922); Fl. Transv. 2 : 347 (1932); Gerstn. in J. S. Afr. Bot. 14 : 22 (1948). Syntypes: Natal, Durban [Port Natal] and Umgeni, 91,44 m alt., *Drege* (K!, P!); Cape, Port St. Johns Distr., between Umgazana [Omgaziana] and Umzimvubu [Omsamwubo], *Drege* (P!). *A. hirtella* Willd. var. *inermis* Walp. in Linnaea 13 : 542 (1839). Type: Cape Province, without locality, *Mund* (whereabouts unknown). There is a Mund specimen in the Kew Herbarium, but, as the flowering branchlets are armed with spines it is assumed that it cannot be an isotype of var. *inermis*. As flowering twigs of *A. karroo* are fairly often devoid of spines, no justification is seen for upholding var. *inermis*. *A. sp. nov.* sensu Schinz in Mém. Herb. Boiss. 1 : 116 (1900). *A. seyal* sensu Sim., For. Fl. P.E. Afr. 57, t.35B (1909), non Del. *A. horrida* var. *transvaalensis* Burtt Davy in Kew Bull. 1908 : 158 (1908). Syntypes: Transvaal, Pretoria Distr., Groenkloof, near Pretoria, *Burtt Davy* 2468 (BOL!, FHO!, K!, PRE!); Arcadia, Pretoria, *Burtt Davy* 2807 (FHO!, K!, PRE!). *A. karroo* var. *transvaalensis* (Burtt Davy) Burtt Davy in Kew Bull. 1922 : 328 (1922); Fl. Transv. 2 : 347 (1932). Types as for *A. horrida* var. *transvaalensis*. *A. inconflagrabilis* Gerstn. in J. S. Afr. Bot. 14 : 24 (1948). Syntypes: Natal, Nongoma Distr., Nongoma township, *Gerstner* 4562 (K!, NBG!, NH!, PRE!); *Gerstner* 4635 (NBG!, NH!, PRE!); *Gerstner* 4637 (NBG!); *Gerstner* 5258 (whereabouts unknown).

Shrub, often several stemmed, or a tree to 22 m high, sometimes very slender, spindle-like, and sparsely branched; crown rounded, often irregularly so, or flattened; trunk to 0,75 m diam. *Bark* dark brown, reddish-brown, brownish-black to black, rough, often fissured, or white to pale greyish-white or greyish-brown and smooth, the latter often with scattered persistent paired spines; young branchlets reddish- to purplish- or blackish-brown, often flaking to expose a rusty-red inner layer, sometimes white to yellowish- or greyish-brown and smooth, glabrous or sometimes sparingly to densely pubescent, eglandular or with small inconspicuous reddish sessile glands. *Stipules* spinescent, in pairs, usually 0,4–7(10) cm long, sometimes greatly elongated to 25 cm long, the latter usually slightly inflated and up to ± 1 cm in diam., remaining distinct to the base and not confluent, straight or sometimes ± deflexed, whitish or the same colour as the stem, entire plant frequently exceedingly spinescent; other prickles absent. *Leaves*; petiole 0,5–1,8 cm long, adaxial gland usually present, variable in position, usually rounded or oval, up to 1,5 × 1,5 mm; rhachis (0)1–4,6(9) cm long, glabrous or sometimes sparingly to densely pubescent, with a yellowish- or reddish-brown to black gland at the junction of each of the top 1–3 pinna pairs, between every pinna pair or absent from some; pinnae (1)2–6(13) pairs; rhachillae (1)1,5–3,8(7,2) cm long, glabrous or sometimes densely pubescent; leaflets 5–15(27) pairs per pinna, (2,8)3,5–8(12,5) × 1–2,5(5) mm, linear, linear-oblong to obovate-oblong, eglandular, apex rounded to subacute but not spinulose-mucronate, usually glabrous but sometimes fairly densely pubescent beneath, margins usually without cilia but sometimes spreading cilia present. *Inflorescences* capitate, on axillary peduncles, fascicled or sometimes solitary, forming terminal racemes, sometimes on lateral axillary branchlets, the entire inflorescence forming an irregular terminal panicle. *Flowers* bright yellow, sessile; peduncles 0,7–2,4(4) cm long, glabrous or occasionally densely pubescent, sometimes glandular; involucel $\frac{1}{3}$–$\frac{3}{4}$ way up the peduncle (when the flowers are young the involucel often appears apical, but as the peduncle lengthens the involucel soon assumes its true position), ± 2 mm long. *Calyx* glabrous throughout or apices of lobes sparingly or sometimes ± densely pubescent, tube 1,2–1,8 mm long, lobes up to 0,5 mm long. *Corolla* glabrous or almost so, tube 1,5–2,3 mm long, lobes up to 0,8 mm long, reflexed. *Stamen-filaments* free, up to 5 mm long; anthers with a deciduous apical gland. *Ovary* glabrous, up to 1,5 mm long. *Pods* yellowish- or reddish-brown to brown, (4)5–10,5(21) × 0,5–0,7(1,1) cm, linear, slightly to strongly falcate or sometimes straightish, usually constricted between the seeds, often distinctly moniliform, apex rounded to acuminate, sometimes attenuate at both ends, longitudinally dehiscent, usually longitudinally venose, mostly glabrous but at times densely tomentellous, sometimes inconspicuously glandular. *Seeds* olive-brown or brown, (3,5)4,5–6,5(9) × (2)3–4(7) mm, elliptic or lenticular, sometimes ± quadrate, compressed; areole 3–5,5(7) × 2–3,5(4,5) mm.

Found in southern Angola, Botswana, Zambia, Rhodesia, Malawi, Mozambique and throughout our area. *A. karroo* is the most widespread *Acacia* in southern Africa and it occupies a diverse range of habitats including dry thornveld, river valley scrub, bushveld, woodland, grassland, the banks of dry watercourses, riverbanks, coastal dunes and coastal scrub.

S.W.A.—1917 (Tsumeb): Gaub, *Mrs Borle 36*. 1918 (Grootfontein): Magistrates office, Grootfontein, *Le Roux 240*. 2017 (Waterberg): Waterberg plateau, *Boss sub TRV 35006*. 2117 (Otjosondu): farm Hummelshain, 97,6 km W.S.W. of Steinhausen, *De Winter 2422*. 2216 (Otjimbingwe): Windhoek Swakop River at farm Otjiseva, *Wiss & Kinges 746*. 2217 (Windhoek): near Windhoek, *Codd 5798*. 2218 (Gobabis): Breitenberg, *Seydel 2483*. 2316 (Nauchas): Nauchas Mountains, *Keet 1687*. 2416 (Maltahöhe): Bullspoort, base of Naukloof Mountains, *Rodin 3948*. 2616 (Aus): Kuibis, *Van Son sub TRV 31818*. 2618 (Keetmanshoop): Keetmanshoop, *Rogers 29787* (GRA, K). 2718 (Grunau): farm Noachabeb, Great Karas Mountains, *Ortendahl 337* (UPS). 2817 (Vioolsdrif): farm Aussenkehr, *Ortendahl 297* (UPS).

TRANSVAAL.—2229 (Waterpoort): Wyllie's Poort, *Strey 7939*. 2230 (Messina): Tshipise, *Gerstner 6225*. 2326 (Mahalapye): Buffelsdrif, *Vahrmeijer 1285*. 2329 (Pietersburg): Louis Trichardt, *Gerstner 5800*. 2330 (Tzaneen): Westfalia Estate, Duiwelskloof, banks of Merensky Dam, *Scheepers 1095*. 2428 (Nylstroom): 1,6 km S. of Warmbaths, *Codd 2880*. 2430 (Pilgrim's Rest): Abel Erasmus Forest Reserve, *Schlieben & Strey 8416*. 2431 (Acornhoek): Bushbuck Ridge, *Pritchard 24*. 2526 (Zeerust): Zeerust, *Thode A1403*. 2527 (Rustenburg): Skeerpoort, *Prosser 1171*. 2528 (Pretoria): Fountains valley, Pretoria, *Verdoorn 595*. 2529 (Witbank): Laersdrif, 10 km from Stoffberg on Roossenekal road, *Ross 2090*. 2530 (Lydenburg): 6,4 km N. of Lydenburg, *Young A451*. 2531 (Komatipoort): near Sheba siding, *Codd 9536*. 2626 (Klerksdorp): Hendriksrus, *Morris 1151*. 2627 (Potchefstroom): Boskop, *Louw 592*. 2628 (Johannesburg): Heidelberg, *Thode A1314*. 2629 (Bethal): N. of Amersfoort, *Strey 7881*. 2725 (Bloemhof): Boskuil, *Sutton 120*. 2726 (Odendaalsrus): Kommandodrif, banks of Vaal River, *Morris 1048*. 2730 (Vryheid): Oliemyne, 28,8 km from Wakkerstroom on Piet Retief road, *Devenish 327*.

O.F.S.—2726 (Odendaalsrus): Bothaville, *Goossens 1193*. 2727 (Kroonstad): ± 10 km N. of Kroonstad, *Scheepers 1344*. 2827 (Senekal): Allemanskraal, *Rycroft 2737* (NBG). 2924 (Hopetown): Luckhoff, *C. A. Smith 5324*. 2925 (Jagersfontein): near Fauresmith at the Garings drift on the Luckhoff road, *C. A. Smith 5278*. 2926 (Bloemfontein): Bloemfontein, *Sim s.n.*

SWAZILAND.—2631 (Mbabane): Palata, *Compton 30816*. 2632 (Bela Vista): Abercorn Pont, eastern border of Swaziland over Usutu River, *Codd & Dyer 2860*.

NATAL.—2730 (Vryheid): road crossing on Upper Blood River on Kingsley–Viljoenspos road, *Edwards 2836*. 2731 (Louwsburg): Nongoma commonage, *Ward 3036*. 2732 (Ubombo): Mpangazi Lake, *Strey 4960*. 2829 (Harrismith): Oliviershoek Pass, *Ross 538* (NH, NU). 2830 (Dundee): 8 km from Dundee on Wasbank road, *Ross 1258* (NH, NU). 2831 (Nkandla): 4,8 km N. of Empangeni, *Codd 9642*. 2832 (Mtubatuba): Hluhluwe Game Reserve, *Ward 2146*. 2929 (Underberg): Estcourt Hill, 1,6 km S. of Estcourt, *Ross 745* (NH, NU). 2930 (Pietermaritzburg): Bisley, near Pietermaritzburg, *Ross 1602*. 2931 (Stanger): Inyoni, *Gerstner 4526*. 3030 (Port Shepstone): Uvongo, *Ross 802*. 3130 (Port Edward): Port Edward, *Ross 806*.

LESOTHO.—2828 (Bethlehem): Leribe, *Dieterlen 185* (NH).

CAPE.—2524 (Vergelee): 9,6 km W. of Tshidila-molomo, banks of Setlagodi River, *Leistner 558.* 2624 (Vryburg): near Taungs, *Rodin 3418.* 2723 (Kuruman): near Kuruman, *Gerstner 6280.* 2816 (Oranjemund): S. bank of Orange River at Arris Drift, *Pillans 5260* (BOL). 2817 (Vioolsdrif): Kuboos, *Pillans 5399* (BOL, K). 2823 (Griekwastad): Klaarwater, Griqua-town, *Burchell 1953–1* (BOL, GRA, K). 2824 (Kim-berley): Barkly West, *Breuckner 54.* 2922 (Prieska): Prieska, *Bryant J172.* 2923 (Douglas): Maselsfontein, *E. Anderson 633* (GRA). 3025 (Colesberg): Orange River, S. bank, *Barker 8852* (NBG). 3118 (Van-rhynsdorp): Doorn River, ± 1,6 km S. of Klawer, *Wilman 237* (BOL.) 3125 (Steynsburg): Ebenezer Weir, De Wet's farm, Vlekpoort River, *Archibald 3201.* 3126 (Queenstown): 25,6 km N. of Queenstown, *De Winter 8357.* 3129 (Port St. Johns): Mbotyi, *Strey 8602.* 3222 (Beaufort West): Acacia, *J. G. Ross 1.* 3224 (Graaff-Reinet): near Graaff-Reinet, *H. Bolus 374* (BOL). 3225 (Somerset East): Bergzebra National Park, *Brynard 168.* 3226 (Fort Beaufort): Alice, *Barker 898* (BOL, NBG). 3227 (Stutterheim): 9,6 km from Berlin on road to East London, *Comins 1436.* 3228 (Butterworth): ± 5 km from coast on Mr Miles' farm, adjacent to Kei River, *Dyer 4502.* 3318 (Cape Town): Langrietvlei, 8 km from Hopefield, *Letty 340.* 3319 (Worcester): Robertson, *Garside 61* (K). 3320 (Montagu): Cogmans Kloof, *Barker 8839* (NBG). 3321 (Ladismith): near eastern foot of Bosluiskloof Pass, *Edwards 3365.* 3322 (Oudtshoorn): between Calitzdorp and Oudtshoorn, *Barker 506* (NBG). 3323 (Willowmore): between Uniondale and Georgida, *Fourcade 4512* (K). 3324 (Steytlerville): Armmands-vriend, Baviaanskloof, *Bayliss 3811* (NBG). 3325 (Port Elizabeth): near the Swartkop River, and on hills of Addo, *Ecklon & Zeyher 605* (GRA, SAM). 3326 (Grahamstown): between Southwell and Port Alfred, *Strey 10372.* 3327 (Peddie): East London, *Batten sub NBG 69945.*

A. karroo, the Sweet Thorn or Soetdoring, is apparently one of the least exacting *Acacia* species in regard to habitat preference and has therefore been able to inhabit a diverse range of habitats. Con-sequently, *A. karroo* is an exceedingly variable species. There is strong evidence that the variation within *A. karroo* is regional; plants in various parts of the species geographical range often having a different "look". This variation is considered in more detail in Bothalia 10(2) : 385–402 and in Bothalia 10(3) : 427–430 (1971).

"Typical" *A. karroo* grows in the Karoo and in the drier parts of the Cape Province as a shrub or relatively small tree with dark rough bark, usually (1)2–3(5) pinnae pairs per leaf, and 6–12 pairs of leaflets per pinna which are 4–8 × 1,5–2,5 mm. A narrow fringe of larger trees often occurs along the banks of rivers and these frequently enable the course of a river to be detected from afar.

Apart from the "typical" form of *A. karroo*, a number of other entities are recognizable within the species. Some of the more important are enumerated below:

1. The white-barked trees or shrubs with short spines, 4–7(13) pinnae pairs per leaf and 12–18(27) pairs of smaller, narrower leaflets per pinna (*A. natalitia*)

which are found chiefly in the eastern Cape, Natal, Swaziland and the eastern Transvaal. *A. hirtella* from the Natal coast is similar but differs in having pubescent young branchlets, leaves, leaflets and peduncles.

2. The small slender shrubs up to 1 m high found in the eastern Cape in the vicinity of the Kei River mouth. A very local entity of which *Dyer 4502* is a typical specimen.

3. The "fire-resistant" shrubs found in the Nongoma district of Zululand (*A. inconflagrabilis*).

4. The slender, sparingly branched trees up to 6 m high found in Zululand, particularly in the Hluhluwe and Umfolozi Game Reserves and in the corridor linking the two reserves (popularly termed "spindle *A. karroo*"). A "spindle" growth form also occurs near the Loskop Dam in the Transvaal. The plants typically have bright reddish-brown minutely flaking bark, glaucous foliage, large flattened ± discoid petiolar glands and a large gland at the junction of each or almost every pinna pair. *Ward* 2123 and *Codd* 9616 will serve to establish the identity of this entity.

5. The large trees with greyish-white bark, long spines up to 25 cm long and long moniliform pods found along the Zululand coast from about the mouth of the Tugela River and northwards to Mozambique. Plants are confined to a fairly narrow belt along the coast which is sometimes only a few kilometres wide. Plants grow on the coastal plain, among the coast dunes, in the mouths of many river estuaries, for example, the Amatikulu and around the shore of the fresh water Lake Sibayi. The plants, which usually form very dense pure stands and are often dominant to the exclusion of other trees, often act as pioneers in stabilising loose sand dunes, especially in disturbed areas and in patches of regenerating coast dune forest. Unlike in "typical" *A. karroo*, the paired spines often persist on the trunk in these plants. *Gerstner* 4526 and *Strey* 4960 will serve to establish the identity of these plants.

6. On the Transvaal highveld from Pretoria east-wards there appears to be a local tendency for the production of a sparse indumentum on the young branchlets, leaves, peduncles and pods (*A. karroo* var. *transvaalensis*). However, occasionally this tendency is so extreme, for example, at Steelpoort, as to alter the general appearance of the plants completely. Indeed, the latter, for example *Codd* 6702 and *Ross* 2089, 2094, bear a strong superficial resemblance to *A. gerrardii*. The differences between these plants and *A. gerrardii* are given under the latter. These pubescent Transvaal specimens also differ in many respects from the pubescent specimens of *A. karroo* found on the Natal south coast.

7. On the Springbok Flats north of Pretoria small shrubby plants occur which often can be dis-tinguished from *A. tenuispina* only with difficulty. Some of the plants have a similar growth form to *A. tenuispina* but as they lack spinulose-mucronate leaflet apices and glandular pods they are referred to *A. karroo*. It has been suggested by some collectors that the plants may be hybrids between *A. karroo* and *A. tenuispina. Burtt Davy* 4075 and *Codd* 7040 are examples of this entity.

Within the species numerous biotypes are recognizable, each of which varies independently but usually within certain limits, the limits of each falling within the range of variation that is accepted as *A. karroo*. The extremes of each of the variants are usually quite distinctive and naturally it is these extremes that attract immediate attention. However, it is found that the extremes of each variant are linked to the "central *A. karroo* gene-pool" by numerous and varied intermediate stages that become progressively less and less distinct until a stage is reached where it is difficult to assign a specimen to a particular entity with any degree of certainty. It therefore becomes difficult to delimit each entity clearly. Consequently, it seems preferable to regard *A. karroo* as an inherently variable species in which no formal infraspecific categories are recognized rather than to fragment the species into a number of somewhat arbitrary infra-specific taxa.

A. karroo has the ability to encroach rapidly into grassland grazing areas, particularly in over-grazed areas, and is consequently considered a serious menace in parts of its range. Attempts to eradicate plants by chopping often result in a vigorous coppice growth.

A good quality gum is exuded from the stems and was at one time exported as "Cape Gum".

A tendency of *A. karroo*, shared also by *A. nilotica* (L.) Willd. ex Del. and sometimes also by *A. davyi* N.E. Br., is for a few flowers to develop in the involucel on the peduncle, sometimes giving the appearance of a smaller secondary capitulum below the main one. The flowers in this secondary capitulum in *A. karroo* often develop before those in the main capitulum. Most of these flowers appear to be sterile, but this needs further investigation.

Of all the indigenous *Acacia* species, *A. karroo* appears to be subjected to the severest attacks by the wattle bagworm, *Kotochalia junodii* (Heyl.). The degree of infestation is often sufficient to kill fairly large trees.

20. Acacia tenuispina *Verdoorn*

in Bothalia 6 : 156, fig. 5 (1951); Brenan in F.Z. 3,1 : 90 (1970); Ross in Bothalia 10 : 351 (1971). Type: Transvaal, Hoogbult Farm, Naboomspruit, *Galpin* 475 M (PRE, holo.!; K!).

A. permixta var. *glabra* Burtt Davy in Kew Bull. 1922 : 330 (1922); Fl. Transv. 2 : 340 (1932); Hutch., Botanist in S. Afr. 664 (1946). Type as above.

Stoloniferous slender shrub 0,3–1(2) m high, many stemmed and often forming dense thickets; young branchlets grey- or reddish-brown to purplish-black, with numerous small scattered glands, often glutinous, glabrous or subglabrous. *Stipules* spinescent, in pairs, 0,4–5,6 cm long, straight or slightly deflexed, slender, whitish, glabrous; other prickles absent. *Leaves:* petiole 0,2–1,1 cm long, glabrous or subglabrous, adaxial gland absent; rhachis 0–3,8 cm long, glabrous or subglabrous, with a sessile to shortly stipitate gland at the junction of the top 1–2 pinnae pairs or of the only pair of pinna, otherwise with small scattered glands; pinnae 1–6 pairs; rhachillae 0,3–1,8 cm long, glabrous or sub-glabrous, usually with small scattered glands; leaflets (3)4–9 pairs per pinna, (2,1)3–4,8 × 0,8–1,5 mm, linear or linear-oblong to obovate-oblong, glabrous, entire, eglandular, lateral nerves invisible beneath, apex spinulose-mucronate. *Inflorescences* capitate, on axillary peduncles, solitary or fascicled, scattered along shoots of the current or previous season. *Flowers* bright yellow, sessile or very shortly pedicellate; peduncles 0,8–3 cm long, glabrous or subglabrous, sparingly to densely glandular; involucel at or above the middle of the peduncle, 1–2 mm long. *Calyx* glabrous, tube 1,4–1,8 mm long, lobes up to 0,5 mm long. *Corolla* glabrous, tube 1,9–2,8 mm long, lobes up to 1 mm long, often reflexed. *Stamen-filaments* free, up to 5 mm long; anthers with a deciduous apical gland. *Ovary* shortly stipitate, glabrous, up to 2 mm long. *Pods* pale to dark yellowish- or chestnut-brown, 1,9–4(7,4) × 0,4–0,8 cm, slightly to strongly falcate, longitudinally dehiscent, venose, not or only slightly constricted between the seeds, with numerous dark sessile pustular glands scattered over the surface, slightly glutinous. *Seeds* olive to olive-brown, 4–7 × 3,5–5 mm, elliptic, compressed; areole 2–4 × 1,5–3 mm.

Found in south-eastern Botswana and the western Transvaal. Forms extensive low thickets, usually on black cotton soil.

TRANSVAAL.—2328 (Baltimore): near Villa Nora, *Acocks 8819*. 2426 (Mochudi): 12,8 km S.W. of Rooibokkraal, *Codd 8655*. 2427 (Thabazimbi): Rooiberg, *Mogg s.n.* 2428 (Nylstroom): Mosdene, Naboomspruit, farm Doornshoek, *Galpin M541*. 2527 (Rustenburg): 8 km N.E. of P.O. Bospoort on road to Beestekraal, *Codd 6370*. 2528 (Pretoria): 16 km N. of Pienaars River station, 3,2 km E. of turning to Mackenzie, *Codd 850*.

On the Springbok Flats north of Pretoria small shrubby plants occur which often can be distinguished from *A. tenuispina* only with some difficulty. Some of the plants have a similar growth form to *A. tenuispina*, but, as they lack the spinulose-mucronate leaflet apices and glandular pods, they are referred to *A. karroo*. It has been suggested by some collectors that the plants may be hybrids between *A. karroo* and *A. tenuispina*.

21. Acacia exuvialis *Verdoorn*

in Bothalia 6 : 154, fig. 2 (1951); Codd, Trees & Shrubs Kruger Nat. Park : 52 (1951); Von Breitenbach, Indig. Trees S. Afr. 2 : 288 (1965); Brenan in F.Z. 3,1 : 90 (1970); Ross in Bothalia 10 : 351 (1971); Palmer & Pitman, Trees S. Afr. 2 : 791 (1973). Type: Transvaal, Nelspruit Distr., Kruger National Park, 25,6 km W. of Skukuza, *Codd & Verdoorn 5464* (PRE, holo.!).

Many stemmed shrub or small tree with slender ascending branches to 5 m high. *Bark* pale to dark yellowish- or greyish-brown, often oily in appearance, peeling in long strips; young branchlets reddish-brown to purplish-black, sometimes flaking, with some scattered inconspicuous dark sessile glands, often glutinous, glabrous or subglabrous. *Stipules* spinescent, in pairs, 0,4–7,7 cm long, straight or ± deflexed, sometimes slightly enlarged and swollen, whitish, glabrous; other prickles absent. *Leaves:* petiole (0,3)0,8–1,5(2,4) cm long, adaxial gland usually absent; rhachis (0)0,7–2,4(4,7) cm long, glabrous or subglabrous, with a small sessile to shortly stipitate gland at the junction of each pinna pair, otherwise eglandular or with a few scattered inconspicuous glands; pinnae (1)2–4(6) pairs; rhachillae 0,6–2(3,3) cm long, glabrous or subglabrous; leaflets 3–6 pairs per pinna, (2,4)4–7(10) × 1,5–3(4,5) mm, linear-oblong to ovate- or obovate-oblong, margins entire, eglandular or almost so, glabrous, lateral nerves inconspicuous beneath, apex ± spinulose-mucronate. *Inflorescences* capitate, on axillary peduncles, solitary or fascicled along shoots of the current or previous season. *Flowers* bright yellow, sessile; peduncles (1,3)2–3(3,9) cm long, glabrous or subglabrous, glandular; involucel at or above the middle of the peduncle, 2–4 mm long. *Calyx* glabrous or subglabrous, tube 1,2–1,8 mm long, lobes up to 0,6 mm long. *Corolla* glabrous, tube 1,5–2,5 mm long, lobes up to 0,7 mm long, often reflexed. *Stamen-filaments* free, up to 4,5 mm long; anthers with a deciduous apical gland. *Ovary* shortly stipitate, up to 1,8 mm long, glabrous. *Pods* pale to dark yellowish- or reddish-brown, 1,5–3,6(6,5) × 0,4–0,9 cm, slightly to strongly falcate, somewhat torulose, subcoriaceous, venose, longitudinally dehiscent, eglandular or with few scattered glands, slightly glutinous, glabrous. *Seeds* olive-green to olive-brown, 5–8 × 3,5–6 mm, elliptic, compressed; areole 3–5 × 2,5–3 mm.

Found in south-eastern Rhodesia and the eastern Transvaal. Occurs in mixed deciduous bush or woodland, often with *Colophospermum mopane*.

TRANSVAAL.—2331 (Phalaborwa): Kruger National Park, 5,6 km N.W. of Shingwedzi Camp, *Codd & De Winter 5568*. 2430 (Pilgrim's Rest): 3,2 km S. of Steelpoort, *Morris 1183*. 2431 (Acornhoek): Klaserie, *Strey 7902*. 2531 (Komatipoort): Kruger National Park, 15,2 km S. of Skukuza, *Codd & Verdoorn 5502*.

22. **Acacia swazica** *Burtt Davy* in Kew Bull. 1922 : 332 (1922); Fl. Transv. 2 : 342 (1932); Hutch., Botanist in S. Afr. 365, 370 (1946); Verdoorn in Bothalia 6 : 156, fig. 6 (1951); Codd, Trees and Shrubs Kruger Nat. Park 51 (1951); Von Breitenbach, Indig. Trees S. Afr. 2 : 290 (1965); Brenan in F.Z. 3,1 : 92 (1970); Ross, Acacia Spp. Natal 41, fig. 2/18 (1971); in Bothalia 10 : 351 (1971); Fl. Natal 193 (1973); Palmer & Pitman, Trees S. Afr. 2 : 793 (1973). Type: Swaziland, near Manzini [Bremersdorp], *Burtt Davy 3045* (PRE, holo.!, K, fragm.!; BM, iso.!).

A. glandulifera sensu Burtt Davy in Kew Bull. 1908 : 158 (1908) pro parte quoad specim. *Burtt Davy 3045*, prope Bremersdorp, Swaziland; Henkel, Woody Pl. Natal 228 (1934), non Schinz sensu stricto. *A. nebrownii* sensu Burtt Davy in Kew Bull. 1921 : 50 (1921) pro parte quoad *Burtt Davy 3045*, prope Bremersdorp, Swaziland, non Burtt Davy sensu stricto.

Slender shrub, often several stemmed, or small slender tree with short ascending branches up to 3 m high. *Bark* grey- to yellowish- or reddish-brown, sometimes flaking to reveal a yellowish inner layer; young branchlets grey- or reddish-brown to purplish, sometimes flaking minutely, glabrous except for scattered conspicuous reddish sessile pustular glands, often glutinous. *Stipules* spinescent, in pairs, 0,6–7,4 cm long, straight or slightly deflexed, slender, whitish, glabrous; other prickles absent. *Leaves:* petiole 0,3–1,9 cm long, glabrous or subglabrous, adaxial gland absent; rhachis 0–1,8(4,4) cm long, glabrous or subglabrous, often with a small sessile to shortly stipitate gland at the junction of each pinna pair, otherwise with small scattered glands; pinnae 1–3(5) pairs; rhachillae 0,6–2,9 cm long, glabrous, usually with small scattered glands; leaflets 3–7(9) pairs per pinna, (2)4–9(13) × (1)1,5–5,1 mm, lanceolate to obovate-oblong or broadly obovate, glabrous, margins entire, eglandular or with few very inconspicuous glands towards the apex, lateral nerves ± prominent and conspicuous beneath, apex spinulose-mucronate. *Inflorescences* capitate, on axillary peduncles, solitary or fascicled along shoots of the current or previous season. *Flowers* bright yellow, sessile or very shortly pedicellate; peduncles 1,4–4,7 cm long, glabrous or subglabrous, sparingly to densely glandular; involucel at or above the middle of the peduncle, large, 2–5 mm long. *Calyx* glabrous, tube 1,4–1,8 mm long, lobes up to 0,5 mm long. *Corolla* glabrous, tube 1,8–2,6 mm long, lobes up to 0,6 mm long. *Stamen-filaments* free, up to 4,8 mm long; anthers with a deciduous apical gland. *Ovary* shortly stipitate, glabrous, up to 1,8 mm long. *Pods* pale to dark yellowish- or chestnut-brown, (1,5)2–6 × 0,7–1,2 cm, slightly to strongly falcate, subcoriaceous, venose, longitudinally dehiscent, with numerous conspicuous dark sessile pustular glands, slightly glutinous, glabrous. *Seeds* olive to olive-brown, 4–7 × 4–6 mm, elliptic to subcircular, compressed; areole 2,5–4 × 2,5–4 mm.

Found in southern Mozambique, the eastern Transvaal, Swaziland and the north-western corner of Tongaland in the vicinity of Ndumu. Occurs in dry bushveld and mixed scrub, usually on boulder strewn slopes or in rocky situations.

TRANSVAAL.—2531 (Komatipoort): Kruger National Park, 1 km S.E. of Pretorius Kop (—AB), *Codd 5682*; 8 km E. of Kaapmuiden (—CB), *Buitendag 783*; Lebombo Flats, 24 km S. of Komatipoort (—DB), *Strey 4025*.

SWAZILAND.—2631 (Mbabane): near Manzini [Bremersdorp] (—AD), *Burtt Davy 3045*; Tulwane (—BC), *Compton 31932*; Grand Valley (—CD), *Compton 29518*. 2731 (Louwsburg): near Gollel, *Rodin 4211* (K).

NATAL.—2632 (Bela Vista): 4,8 km S. of Abercorn Pont on road to Ndumu, *Ross 1920*.

A. swazica may sometimes be confused with *A. exuvialis*. However, *A. swazica* differs from *A. exuvialis*, and from all of the other species in the complex with glandular glutinous pods, by the conspicuous venation on the lower leaflet surfaces.

23. Acacia nebrownii *Burtt Davy* in Kew Bull. 1921 : 50 (1921) pro parte excl. specim. *Burtt Davy 3045*, Swaziland, et *Burtt Davy 5230*, Potgietersrust, Transvaal; Bak.f., Leg. Trop. Afr. 3 : 851 (1930); O. B. Miller, Checklist Bech. Prot. 20 (1948); in J. S. Afr. Bot. 18 : 23 (1952); Verdoorn in Bothalia 6 : 156, fig. 4 (1951); Schreiber in Mitt. Bot. Staatssamml. Munchen 2 : 284 (1957); Von Breitenbach, Indig. Trees S. Afr. 2 : 289 (1965); Leistner, Mem. Bot. Surv. S. Afr. 38 : 123 (1967); Schreiber in F.S.W.A. 58 : 10 (1967); Brenan in F.Z. 3,1 : 92,

t.15/9 (1970); Ross in Bothalia 10 : 351 (1971); Palmer & Pitman, Trees S. Afr. 2 : 793 (1973). Syntypes: Botswana, Kwebe Hills, *Mrs. E. J. Lugard* 14 (K!) and 16 (K!).

A. glandulifera Schinz in Mém. Herb. Boiss. 1 : 111 (1900) nom. illegit., non *A. glandulifera* S. Wats. in Proc. Am. Acad. 25 : 147 (1890); Burtt Davy in Kew Bull. 1908 : 158 (1908) pro parte quoad *Mrs. E. J. Lugard* 14 & 16, excl. specim. *Burtt Davy 3045 & 5230*; Glover in Ann. Bolus Herb. 1 : 149, t.18/9 (1915); Dinter in Feddes Repert. 15 : 79 (1917). Syntypes: South West Africa, Great Namaqualand, *Fleck* 484a (Z!); Hereroland, Tsoachaub, *Fleck* 480a (Z!). *A. rogersii* Burtt Davy in Kew Bull. 1922 : 331 (1922); in Fl. Transv. 2 : 342 (1932); Bak.f., Leg. Trop. Afr. 3 : 851 (1930); O. B. Miller, Checklist Bech. Prot. 21 (1948). Type: Transvaal, Soutpansberg Distr., Messina, *Rogers 21843* (PRE, holo.!; K!, NH!, Z!). *A. walteri* Suesseng. in Mitt. Bot. Staatssamml. Munchen 1 : 333 (1953). Syntypes as for *A. glandulifera* Schinz.

Many stemmed shrub with slender ascending branches or occasionally a slender tree to 4 m high. *Bark* reddish-brown to purplish-black; young branchlets yellowish- or reddish-brown to purplish, sometimes as though whitewashed over a purple background, with numerous dark sessile pustular glands, sometimes glutinous, glabrous or subglabrous. *Stipules* spinescent, in pairs, 0,5–7 cm long, straight or slightly arcuate and deflexed, slender, whitish, tips usually reddish-brown; other prickles absent. *Leaves*: petiole 0,3–1,4 cm long, glabrous or subglabrous; rhachis 0–0,9 cm long, glabrous or subglabrous, a shortly columnar gland at the junction of the pinna pair; pinnae mostly 1 pair, rarely 3 pairs; rhachillae 0,5–1,4 cm long, glabrous or subglabrous, glandular; leaflets 3–5 pairs per pinna, 2,1–5,7 × 0,9–3(4) mm, linear-oblong to obovate or ovate, eglandular or with some small pale inconspicuous glands on the margin and sometimes the surface, glabrous, margins entire, apex usually shortly mucronate, lateral nerves inconspicuous beneath. *Inflorescences* capitate, on axillary peduncles, solitary or fascicled. *Flowers* bright yellow, sessile; peduncles 0,6–2,4 cm long, glabrous or subglabrous, glandular; involucel basal or up to $\frac{1}{3}$-way up the peduncle. *Calyx* glabrous, tube 1–1,6 mm long, lobes up to 0,6 mm long. *Corolla* glabrous, tube 1,8–2,5 mm long, lobes up to 0,8 mm long, often reflexed. *Stamen-filaments* free, up to 5,5 mm long; anthers with a deciduous apical gland. *Ovary* shortly stipitate, up to 1,8 mm long, glabrous. *Pods* pale to dark yellowish-brown or chestnut, 2–4(5,6) × 0,6–1,3 cm, slightly to strongly falcate, not or scarcely constricted between the seeds, with numerous conspicuous dark sessile pustular glands scattered over the surface, otherwise glabrous, longitudinally dehiscent, venose, apex obtuse, acute or mucronate. *Seeds* olive to olive-brown, 6–10 × 5–7 mm, elliptic, compressed; areole 4–6 × 2,5–3,5 mm.

Found in South West Africa, Botswana, the western portion of Rhodesia and north of the Soutpansberg in the Transvaal. Occurs in low lying sandy flats, around pans or along river banks; in dry bush, thornveld or woodland, sometimes with *Colophospermum mopane*. Frequently forms thickets.

S.W.A.—1914 (Kamanjab): Kamanjab, *Story 5649*. 2218 (Gobabis): 24 km W. of Gobabis, *Codd 5824*. 2416 (Maltahöhe): farm Bullsport, Bullsport flats, *Strey 2018*. 2717 (Chamaites): Holoog, common on plateau and banks of Great Fish River, *Pearson 9814* (K). 2718 (Grunau): 18,4 km S.S.E. of Narubis, *Acocks 15563*. Grid ref. unknown: Grootfontein District, farm Kumkauas, *Kinges 2850*; between Keetmanshoop and Aus, *Gerstner 6291*; Gurinaris, *Pearson 9256* (K); without locality, *Keet 1677*.

TRANSVAAL.—2229 (Waterpoort): farm Little Muck 604, Dongola area, *Codd 4326*; farm "Zoutpan 193", *Obermeyer, Schweickerdt & Verdoorn 60*. 2230 (Messina): Messina, *Pole Evans 2037*.

24. **Acacia permixta** *Burtt Davy* in Kew Bull. 1922 : 330 (1922) pro parte excl. var. *glabra*; Fl. Transv. 2 : 340 (1932); Verdoorn in Bothalia 6 : 155, fig. 3 (1951); Von Breitenbach, Indig. Trees S. Afr. 2 : 289 (1965); Brenan in F.Z. 3,1 : 91 (1970); Ross in Bothalia 10 : 351 (1971); Palmer & Pitman, Trees S. Afr. 2 : 791 (1973). Type: Transvaal, Potgietersrust, *Burtt Davy 5230* (PRE, holo.!, K, fragm.!).

A. glandulifera sensu Burtt Davy in Kew Bull. 1908 : 158 (1908) pro parte quoad specim. *Burtt Davy 5230*, Potgietersrust, non Schinz. *A. nebrownii* sensu Burtt Davy in Kew Bull. 1921 : 50 (1921) pro parte quoad specim. *Burtt Davy 5230*, Potgietersrust, non Burtt Davy sensu stricto.

Many stemmed shrub or a small tree up to 4 m high with slender weakly ascending branches. *Bark* pale to dark chestnut or reddish-brown; young branchlets sparingly to densely hairy or tomentose with spreading grey to whitish hairs 0,75–2 mm long among which some conspicuous reddish sessile glands are scattered, epidermis sometimes splitting to reveal a reddish-brown inner layer. *Stipules* spinescent, in pairs, up to 6,5 cm long, straight or often slightly deflexed, slender, whitish, tips often reddish-brown;

other prickles absent. *Leaves*: petiole 0,3–1,7 cm long, sparingly to densely clothed with spreading hairs, adaxial gland absent; rhachis 0–4,2 cm long, sparingly to densely clothed with spreading hairs, with scattered conspicuous sessile reddish glands, a small raised columnar gland at the junction of the top 1–3 pinnae pairs or absent; pinnae 1–5(7) pairs; rhachillae 0,4–2,8 cm long, with irregularly scattered sessile reddish glands in amongst the spreading pubescence; leaflets 3–10 pairs per pinna, 2,2–7 × 0,9–2(2,9) mm, linear- to obovate-oblong or obovate, eglandular or sometimes with a few glands on the margins towards the apex but margins not crenulate, glabrous or sometimes sparingly pubescent below, margins with few to many spreading cilia, lateral nerves inconspicuous beneath, at least some leaflets spinulose-mucronate apically. *Inflorescences* capitate, on axillary peduncles, solitary or fascicled. *Flowers* bright yellow, sessile; peduncles 1,2–3(4,2) cm long, sparingly to densely clothed with spreading hairs, with scattered sessile reddish glands; involucel at or above the middle of the peduncle, 2,5–3,5 mm long. *Calyx* glabrous or sparingly pubescent especially on apices of the lobes, tube 1,4–2,2 mm long, lobes up to 0,7 mm long. *Corolla* glabrous or subglabrous, tube 2–3,2 mm long, lobes up to 0,9 mm long. *Stamen-filaments* free, up to 5 mm long; anthers with a deciduous apical gland. *Ovary* up to 1,8 mm long, shortly stipitate, glabrous. *Pods* pale to dark yellowish- or reddish-brown, (2)2,7–5(8,9) × 0,6–1,4 cm, slightly to strongly falcate, not or scarcely constricted between the seeds, with few to many conspicuous dark reddish-brown ± raised glands scattered over the surface, otherwise glabrous or subglabrous, longitudinally dehiscent, venose, apex obtuse, acute or sometimes mucronate. *Seeds* olive to olive-brown, 5–8 × 4–6,5 mm, subcircular to elliptic, compressed; areole 2–4 × 1,5–2,5 mm.

Found in south-western Rhodesia and the western Transvaal. Occurs in woodland, thorn scrub and grassland; usually on sandy or coarse gritty soils, often derived from granitic formations, particularly along the edges of the Pietersburg plateau.

TRANSVAAL.—2328 (Baltimore): 24 km S.E. of Villa Nora, *Codd 6581*. 2329 (Pietersburg): 40 km N. of Pietersburg on Kalkbank road, *Story 1558*. 2427 (Thabazimbi): 11,2 km N. of Maraheki, *Codd 5916*.

2527 (Rustenburg): 8 km N. of Beestekraal, *Codd 2981*.

    A. permixta is readily distinguished from all of the other species in the complex with glandular glutinous pods by the spreading hairs on the young branchlets, leaf-rhachides and peduncles.

    25. **Acacia borleae** *Burtt Davy* in Kew Bull. 1922 : 325 (1922); Verdoorn in Bothalia 6 : 154, fig. 1 (1951); Codd, Trees & Shrubs Kruger Nat. Park 41, fig. 34a (1951); Von Breitenbach, Indig. Trees S. Afr. 2 : 287 (1965); Brenan in F.Z. 3, 1 : 90, t.15/8 (1970); Ross, Acacia Spp. Natal 23, fig. 2/17 (1971); in Bothalia 10 : 351 (1971); Ross, Fl. Natal 192 (1973); Palmer & Pitman, Trees S. Afr. 2 : 789 (1973). Type: Mozambique, Lourenco Marques, *Borle* 271 (PRE, holo.!; FHO!).

    A. sp., Henkel, Woody Pl. Natal 229 (1934). *A. barbertonensis* Schweick. in Kew Bull. 1937 : 445 (1937); Gerstner in J. S. Afr. Bot. 4 : 57, fig. 2 (1938). Type: Transvaal, Barberton Distr., Komatipoort, *Cotton Experimental Station, Barberton* (K, holo.!; PRE, fragm.!).

Many stemmed shrub with slender ascending branches or sometimes a slender tree, 1–5 m high. *Bark* dark reddish-brown to purplish-black or black, rough; young branchlets dark reddish-brown to blackish, flaking minutely, with numerous sessile glands, often glutinous, glabrous or subglabrous. *Stipules* spinescent, in pairs, 0,5–6,5 cm long, straight or slightly reflexed, slender, whitish, tips usually reddish-brown; other prickles absent. *Leaves*: petiole 0,5–1,8 cm long, adaxial gland often absent, sometimes a rather large sessile gland at or below the junction of the lowest pinna pair; rhachis (0) 2,5–4,7(6,8) cm long, glabrous or subglabrous, with a small sessile gland at the junction of the top 1–3 pinnae pairs, smaller scattered sessile glands present and sometimes numerous; pinnae (1)5–10(17) pairs; rhachillae 0,6–3,4 cm long, glabrous or subglabrous, glandular; leaflets (5)8–15 (18) pairs per pinna, 1,5–5 × 0,8–2,2 mm, linear-oblong, margins clearly crenulate-glandular, surface glandular-punctate, glabrous, apex rounded to subacute or shortly spinulose-mucronate. *Inflorescences* capitate, on axillary peduncles, solitary or fascicled, forming terminal racemes. *Flowers* bright yellow, sessile; peduncles 1,6–3,4 cm long, glabrous or subglabrous, glandular; involucel at or above the middle of the peduncle. *Calyx* glabrous or subglabrous, tube 1,2–1,8 mm long, lobes up to 0,5 mm long. *Corolla* tubular, glabrous, tube 2–2,5 mm long, lobes up to 0,8 mm long, spreading slightly, often reflexed. *Stamen-filaments* free, up to 4,5 mm long; anthers with a deciduous apical gland. *Ovary* up to 1,4 mm long, shortly stipitate, glabrous. *Pods* dark brown to reddish-brown or blackish, 2,8–7,5(10) × 0,5–0,8 cm, slightly to strongly falcate, often curled into an almost complete circle, ± moniliform, irregularly constricted between the seeds, with numerous sessile pustular glands on the surface, glutinous, longitudinally dehiscent, glabrous or sub-glabrous. *Seeds* olive-brown, 4,5–6 × 3,5 — 5 mm, elliptic to subcircular, compressed; areole 2,5–4 × 2,5–3,5 mm.

    Occurs in south-eastern Rhodesia, the eastern Transvaal, southern Mozambique, Swaziland and Zululand. Found in mixed woodland and thornveld, usually in low lying areas on heavy soils or on rocky outcrops. Often gregarious and forming thickets.

    TRANSVAAL.—2331 (Phalaborwa): Kruger National Park, Shingwedzi, *Lamont 53*. 2431 (Acornhoek): Kruger National Park, Skukuza, *Van der Schijff 3420*. 2531 (Komatipoort): Kruger National Park, 22,4 km N.E. of Pretorius Kop Camp, *Codd 4312*.

    SWAZILAND.—2631 (Mbabane): 22,4 km W. of Stegi on road to Mbabane, *Codd & Dyer 2865*. 2731 (Louwsburg): Gollel, *Bayer sub NH 31431*.

    NATAL.—2632 (Bela Vista): Ndumu Game Reserve, western area, *Ross 661* (K, NH, NU). 2731 (Louwsburg): bridge over Mona River on Mkuze-Nongoma road, *Ross 1058* (NU). 2732 (Ubombo): 14,4 km E. of Mhlosinga, farm Sutton, *Strey 5671*. 2831 (Nkandla): Umfolozi Game Reserve, road from Tobothi to Ngoloti, *Ross 2019*. 2832 (Mtubatuba): Somkele, *Ward 2958*.

    A. borleae is easily distinguished from all of the other species in our area by the numerous sessile glands on the surface and margins of the leaflets.

    In Rhodesia and Mozambique the young branchlets, leaflets and pods are frequently shortly puberulous, but in our area the puberulence is generally absent.

    26. **Acacia davyi** *N.E.Br.* in Kew Bull. 1908 : 161 (1908); Glover in Ann. Bolus Herb. 1 : 150 (1915); Burtt Davy, Fl. Transv. 2 : 346 (1932); Henkel, Woody Pl. Natal 227 (1934); Hutch., Botanist in S. Afr. 361 (1946); Von Breitenbach, Indig. Trees S. Afr. 2 : 306 (1965); Brenan in F.Z. 3, 1 : 89 (1970); Ross, Acacia Spp. Natal 27, fig. 2/13 (1971); Fl. Natal 193 (1973); in Bothalia 11 : 127 (1973); Palmer & Pitman, Trees S. Afr. 2 : 803 (1973). Syntypes: Transvaal, Houtbosch (Woodbush), *Rehmann 6276* (BM!, K!, Z!); *Burtt Davy 5132* (T.D.A. Herb. No. 1211) (PRE!);

Soutpansberg, *Junod sub T.D.A. Herb. No.* 1323 (PRE!). Swaziland, near Manzini [Bremersdorp], *Burtt Davy* 3024 (BM!, FHO!).

Shrub or small tree up to 5 m high. *Bark* yellow to yellowish-brown or brown, soft, corky, sometimes slightly papery, fissured; young branchlets creamy-white or yellowish to pale grey-brown or brown, glabrous to puberulous. *Stipules* spinescent, in pairs, 0,4–3 cm long, straight, slender, mostly ascending, glabrous or densely puberulous; other prickles absent. *Leaves*: petiole 0,3–0,9 cm long, glabrous to densely puberulous, adaxial gland often absent, when present usually just below the lowest pinna pair; rhachis (3,6)8–14(18) cm long, glabrous to densely puberulous, with a gland at the junction of the top 1–6 pinnae pairs; pinnae of well-developed leaves in 12–27 pairs (reduced leaves with as few as 8 pairs of pinnae sometimes also present); rhachillae 1,8–4,4(5,6) cm long; leaflets (17) 20–36(44) pairs per pinna, 1,9–6 × 0,6–1 mm, linear to linear-oblong, apex obtuse to subacute but not spinulose-mucronate, eglandular, margins glabrous or occasionally with few short cilia, usually glabrous below. *Inflorescences* capitate, on axillary peduncles, fascicled, borne along shoots of the current season and often aggregated into ± elongate terminal "racemes". *Flowers* bright-yellow, sessile; peduncles 0,8–3(6) cm long, glabrous to densely puberulous; involucel ⅓–⅔-way up the peduncle. *Calyx* glabrous except for pubescence on the lobes, especially apically, or sparingly puberulous throughout, tube 0,9–1,4 mm long, lobes up to 0,4 mm long. *Corolla* glabrous, tube 1,4–1,9 mm long, lobes up to 0,6 mm long. *Stamen-filaments* free, up to 4 mm long; anthers with a deciduous apical gland. *Ovary* up to 1,75 mm long, glabrous. *Pods* pale to dark yellowish-brown or brown, (5)7–12(17) × 0,5–0,8 cm, linear, straight to slightly falcate, longitudinally dehiscent, not or somewhat constricted between the seeds, glabrous to sparingly puberulous, eglandular, flat, inconspicuously venose, sometimes irregularly swollen and pustulate from insect attack. *Seeds* olive- to dark-brown, 6–7,5 × 3–5,5 mm, elliptic, compressed; areole 4–5 × 2,5–3,5 mm.

Found in the Transvaal, southern Mozambique, Swaziland and Natal. Occurs in grassland, woodland, thornveld, bushveld and scrub; often on rocky slopes. *A. davyi* is seldom dominant to the exclusion of all other *Acacia* species except very locally.

TRANSVAAL.—2230 (Messina); Makonde, *Van Warmelo 5115/20.* 2329 (Pietersburg): slopes of Lajuma, *Strey 7988.* 2330 (Tzaneen): Duiwelskloof, *Galpin 9407.* 2430 (Pilgrim's Rest): Mariepskop, *Van der Schijff 5016.* 2431 (Acornhoek): Lothian, *Strey 3326.* 2530 (Lydenburg): 4,8 km S. of Nelspruit, *Leach 11573.* 2531 (Komatipoort): hillside adjoining Barberton, *Codd 1589.* 2731 (Louwsburg): 24 km from Pongola on road to Piet Retief, *Grobbelaar 522.*

SWAZILAND.—2631 (Mbabane): Mlilwane, *Compton 31157.* 2731 (Louwsburg): 3,2 km E. of Goedgegun on Hlatikulu road, farm Buckwood, *Ross 1666.*

NATAL.—2730 (Vryheid): road crossing Upper Blood River on Kingsley–Viljoenspos road, *Edwards 2835.* 2731 (Louwsburg): Ceza, *Strey 9825.* 2732 (Ubombo): 2,4 km from Ubombo on Ubombo–Mkuze road, summit of Lebombo Mts., *Ross 264* (K, NH, NU). 2830 (Dundee): De Jagersdrif, Buffalo River valley between Dundee and Vryheid, *Edwards 2838.* 2832 (Mtubatuba): Hluhluwe Game Reserve, *Ward 2472.*

A tendency of *A. davyi* is for a few flowers to develop in the involucel on the peduncle, sometimes giving the appearance of a smaller secondary capitulum below the main one.

A. davyi bears a strong superficial resemblance to *A. arenaria* Schinz. *A. arenaria* differs, however, in having:

1. white or pale pink flowers;
2. the corolla usually 2–4 times as long as the small cupular calyx;
3. dark reddish-brown arcuate pods;
4. a very different distribution.

27. **Acacia erioloba** *E. Mey.*, Comm. 1 : 171 (1836), non *A. erioloba* Edgw. in J. Asiat. Soc. Beng. 16 : 1215 (1847); Harv. in F.C. 2 : 280 (1862); Engl. in Bot. Jahrb. 10 : 22 (1888); Ross in Bothalia 11(4) : 444 (1975). Type from Namaqualand (whereabouts unknown); Transvaal, Wolmaransstad Distr., between Kommandodrif and Makwassie, *J. W. Morris* 1042 (K, neo.!).

A. giraffae sensu auct. mult., non *A. giraffae* Willd., Enum. Hort. Berol. 1054 (1809) sensu stricto: Burch., Trav. 2 : 240 (1824); DC., Prodr. 2 : 472 (1825); Harv. in F.C. 2 : 280 (1862); Benth. in Trans. Linn. Soc. Lond. 30 : 503 (1875); Marloth in Trans. S. Afr. Phil. Soc. 5 : 271 (1889); Schinz in Mém. Herb. Boiss. 1 : 108 (1900); Sim, For. Fl. Cape Col. 213, t.58 (1907); Burtt Davy in Kew Bull. 1908 : 157 (1908); Glover in Ann. Bolus Herb. 1 : 148, t.18/1 (1915); Harms in Engl., Pflanzenw. Afr. 3, 1 : 352 (1915); Dinter in Feddes Repert. 15 : 79 (1917); Pole Evans in S. Afr. J. Sci. 17 : figs. 35, 36 (1920); Burtt Davy in Kew Bull. 1922 : 327 (1922); Marloth, Fl. S. Afr. 2 : 54, tt. 18D, 19 (1925); Bak.f., Leg. Trop. Afr.

3 : 835 (1930); Burtt Davy, Fl. Transv. 2 : 340, fig. 59 (1932); Hutch., Botanist in S. Afr. 178, 341, 386, 412, 418, 424, 425, 481, 543, 547, cum photogr. (1946); West in Rhod. Agric. J. 47 : 206 (1950); O. B. Miller in J. S. Afr. Bot. 18 : 21 (1952); Pardy in Rhod. Agric. J. 50 : 4 (1953); Torre in C.F.A. 2 : 281 (1956); Story, Mem. Bot. Surv. S. Afr. 30 : 23 (1958); Leistner in Koedoe 4 : 101 (1961); Palmer & Pitman, Trees S. Afr. 153, tt. vi, 34, 35 (1961); F. White, For. Fl. N. Rhod. 84, fig. 17L (1962); Von Breitenbach, Indig. Trees S. Afr. 2 : 292 (1965); De Winter et al, 66 Transv. Trees 46 (1966); Leistner, Mem. Bot. Surv. S. Afr. 38 : 67, 123, tt. 21, 23, 25, 28, 30, 36, 38, 44, 48 (1967); Schreiber in F.S.W.A. 58 : 8 (1967); Brenan in F.Z. 3, 1 : 93, t.15/10 (1970); Ross in Bothalia 10(2) : 359 (1971); in Bothalia 10(4) : 547 (1972); Palmer & Pitman, Trees S. Afr. 2 : 769 (1973); Schreiber in Mitt. Bot. Staatssamml. Munchen 11 : 117 (1973).

Mimosa sp. sensu Paterson, Journeys into Country of Hottentots & Caffraria : 133, tt.16, 17 (1789).

A. giraffae var. *espinosa* Kuntze in Jahrb. K. Bot. Gart. Mus. Berl. 4 : 264 (1886). Type: South West Africa, Hereroland, *Pechuel-Loesche* (B, holo. †).

Tree up to 15 m high or less frequently a shrub to 4 m high; trunk to 1 m in diam.; crown rounded, the branches often drooping somewhat, or flattened and spreading. *Bark* dark greyish-brown to blackish, rough, fissured, often flaking off in thick ± woody sections when old; young branchlets pale to dark grey- or reddish-brown to purple, sometimes as though whitewashed over a purplish background, often flaking minutely, glabrous or subglabrous, seldom pubescent. *Stipules* spinescent, in pairs, 0,5–5(10) cm long, usually rather stout, often thickened below and fused together basally into an enlarged "ant-gall", 1,5–2 × 2–2,5 cm, sometimes furrowed down the middle, tapering to a sharp point apically; other prickles absent. *Leaves*: petiole 0,4–1,4 cm long, adaxial gland absent; rhachis (0)1–3,5(5,5) cm long, glabrous or subglabrous, a small gland at the junction of each pinna pair; pinnae (1)2–5(6) pairs; rhachillae (1,3)1,6–3,2(4,2) cm long, glabrous or subglabrous; leaflets (6)8–15(18) pairs per pinna, 4–11,5 × (0,7)1,4–2,4(4) mm, linear-oblong to narrowly obovate, oblique basally, apex rounded to subacute, glabrous throughout or sometimes marginal cilia present, rarely pubescent below, lateral nerves prominent and conspicuous above and beneath. *Inflorescences* capitate, on axillary peduncles, solitary or fascicled, scattered along the shoots. *Flowers* bright golden-yellow, sessile or shortly pedicellate, (sometimes some flowers are male only); peduncles (1,8)2,3–4(5,5) cm long,

glabrous or subglabrous, eglandular; involucel apical. *Calyx* glabrous, sometimes apices of lobes with few glandular hairs, tube 1,5–2,2 mm long, lobes up to 0,6 mm long. *Corolla* glabrous or apices of lobes with glandular hairs, 2,7–3,6 mm long, lobes sometimes free for most of their length. *Stamen-filaments* free or connate into groups basally, up to 7,5 mm long; anthers with a deciduous apical gland. *Ovary* 1,2–2 mm long, sessile or shortly stipitate, glabrous at first but soon becoming pubescent. *Pods* densely grey-velutinous all over, with numerous minute dark reddish-brown to purplish glands particularly when young, (4)6–13 cm long, 1,8–5 cm wide, 0,8–2 cm thick, indehiscent, semi-woody, frequently semilunate to suborbicular, sometimes curled almost into a circle, apex rounded to acute or beaked, spongy within, seeds irregularly scattered but separated by transverse partitions. *Seeds* dark reddish-brown, 8–14 × 7–10 mm, lenticular to elliptic, sometimes scarcely compressed; areole 3–9 × 2–5,5 mm.

Found in Angola, South West Africa, Botswana, Zambia, south-western Rhodesia, the Transvaal, western Orange Free State and northern Cape Province. Occurs frequently on the Kalahari sands and in other areas where sandy soils are prevalent; in dry woodland, bush or thornveld. Often the dominant species in the Kalahari thornveld. In very dry areas it occurs along watercourses and in other situations where underground water is available.

S.W.A.—1713 (Swartbooisdrif): bank of Ososouu River at Otjivero, *De Winter & Leistner 5361*. 1724 (Katima Mulilo): 80 km from Katima Mulilo on road to Linyanti, *Killick & Leistner 3141*. 1816 (Namutoni): Ondonga, *Rautanen 645* (GRA). 1819 (Karakuwisa): Omuramba bed, 27,2 km S. of Runtu on road to Karakuwisa, *De Winter 3771*. 1914 (Kamanjab): farm Beulah, near Kamanjab, *De Winter 3105*. 1917 (Tsumeb): 31 km S.W. of Otavi on road to Otjiwarongo, *De Winter 2843*. 2116 (Okahandja): Okahandja, *Dinter 267* (GRA, K). 2117 (Otjosondu): Quickborn, *Bradfield 11*. 2214 (Swakopmund): near Goanikontes, *Rodin 2154*. 2216 (Otjimbingwe): farm Friedenau, 24,8 km in Khomas Hochland south west of Windhoek, *De Winter 2590*. 2415 (Sossusvlei): Sesriem, *Strey 2292* (BOL). 2616 (Aus): Aus, *Dinter 6140* (K). 2618 (Keetmanshoop): Gobas, *Pillans 5907* (BOL).

TRANSVAAL.—2326 (Mahalapye): Buffelsdrif, *Vahrmeijer 1290*. 2328 (Baltimore): Swerwerskraal, 54,4 km N.W. of Potgietersrust, *Hutchinson 2633*. 2428 (Nylstroom): 1,6 km from Potgietersrust on road to Moorddrif, *Meeuse 10143*. 2429 (Zebediela): Lowveld Fishery station, near Marble Hall, on banks of Elands River, *Marais 1098*. 2526 (Zeerust): Swartruggens, *Sutton 1180*. 2528 (Pretoria): Rust de Winter, *Gerstner 5534*. 2529 (Witbank): N. side of Loskop Dam, 9,6 km from dam wall, *Mogg 17311*.

2725 (Bloemhof): Schweizer-Reneke to Wolmaransstad, *Burtt Davy 1685*. 2726 (Odendaalsrus): Kommandodrif, *Morris 1042*. Grid ref. unknown: Sekukuniland, Eersterecht, *Mogg & Barnard 1163*.

O.F.S.—2725 (Bloemhof): between Bloemhof and Hoopstad, *Hutchinson 2988* (K). 2825 (Boshof): between Sandfontein and farm Boshof, *Schweickerdt 1105*.

CAPE.—2620 (Twee Rivieren): Kalahari Gemsbok National Park, 1,6 km S.E. of Rooibrak in river bed of Auob, *Leistner 1492*. 2624 (Vryburg): Taungs, *Pole Evans sub PRE 15833*. 2722 (Olifantshoek): Olifantshoek, 100,8 km S.E. of Kuruman, *Lang sub TRV 31705*. 2816 (Oranjemund): Numees, *Werdermann & Oberdieck 564*. 2820 (Kakamas): 40 km W. by N. of Aughrabies Falls Hotel, *Barclay, Acocks & Tainton 977*. 2823 (Griekwastad): Klaarwater, Griquatown, *Burchell 1952*. 2824 (Kimberley): near Schmidtsdrif, *Acocks 735*. 2919 (Pofadder): 1,6 km E. of Pofadder, *Hutchinson 942*. 2923 (Douglas): near Douglas, *Kotze 793*. 3017 (Hondeklipbaai): bed of Spoeg River, *Acocks 14943*.

Unfortunately the familiar name *A. giraffae* Willd. can no longer be applied to this species (see Ross in Bothalia 11, 4 : 443, 1975).

The combination of stout spines which are often enlarged into "ant-galls", leaflets with prominent venation, bright yellow flowers, glabrous peduncles, apical involucels and densely grey-velutinous pods enable *A. erioloba* to be easily distinguished from all other indigenous species.

Strey 2292 from the Rehoboth district of South West Africa is unusual in having distinctly coiled pods.

A. erioloba, commonly known as the "Camelthorn" or "Kameeldoring", was formerly much more abundant than it is today. In Burchell's time numerous large trees adorned the country in the northern Cape and north of the Orange River. However, with the opening of the diamond mines and railways there was a tremendous demand for the wood which, being hard and heavy, is an excellent fuel. Almost all of the large specimens within a radius of several hundred kilometres of Kimberley disappeared to provide fuel. Marloth, Fl. S. Afr. 2 : 54 (1925), reports that in one year alone 10 000 tons of the wood were forwarded to Kimberley from the Vryburg area. This large-scale destruction of *A. erioloba* from areas which are otherwise devoid of large trees is much regretted. *A. erioloba* is now protected in the northern Cape and in the Jacobsdal district of the Orange Free State.

In parts of the northern Cape, South West Africa and Botswana *A. erioloba* is still the only tree of any size to be seen for kilometres on end and it forms a very conspicuous feature of the landscape. In these areas the sociable weavers, *Philetairus socius* (Latham), frequently build their immense nests in trees of *A. erioloba*. These nests, which are used year after year and are continually added to, sometimes become so heavy that even the largest branches break under their weight.

Roots, probably of *A. erioloba*, have been reported at a depth of 45,72 m in a borehole in South West Africa (Mem. Bot. Surv. S. Afr. 27 : 117, 1952).

Coetzee in S. Afr. J. Sci. 52 : 23 (1955) reports that the pollen-grains of *A. erioloba* are anomalous in consisting of 32 cells as opposed to the 16 cells in all other species studied.

A. erioloba hybridizes with *A. haematoxylon*.

28. **Acacia haematoxylon** *Willd.*, Enum. Hort. Berol. 1056 (1809); DC., Prodr. 2 : 462 (1825); Harv. in F.C. 2 : 280 (1862); Benth. in Trans. Linn. Soc. Lond. 30 : 504 (1875); Engl. in Bot. Jahrb. 10 : 23 (1888); Marloth in Trans. S. Afr. Phil. Soc. 5 : 269 (1889); Schinz in Mém. Herb. Boiss. 1 : 112 (1900); Sim, For. Fl. Cape Col. 211 (1907); Dinter, Veg. Veldkost Deutsch-Sudwest-Afrikas 32 (1912); Glover in Ann. Bolus Herb. 1 : 148 (1915); Harms in Engl., Pflanzenw. Afr. 3,1 : 354 (1915); Dinter in Feddes Repert. 15 : 79 (1917); Pole Evans in S. Afr. J. Sci. 17 : fig. 29 (1920); Bak.f., Leg. Trop. Afr. 3 : 835 (1930); Hutch., Botanist in S. Afr. 413 (1946); O. B. Miller in J. S. Afr. Bot. 18 : 22 (1952); Palmer & Pitman, Trees S. Afr. 156, t.vii (1961); Von Breitenbach, Indig. Trees S. Afr. 2 : 296 (1965); Volk in J. S.W. Afr. Wiss. Ges. 20 : 43, fig.7 (1966); Leistner, Mem. Bot. Surv. S. Afr. 38 : 67, 123, t.21 (1967); Schreiber in F.S.W.A. 58 : 8 (1967); Brenan in F.Z. 3,1 : 93, t.15/11 (1970); Ross in Bothalia 10 : 359 (1971); in Bothalia 10 : 548 (1972); Palmer & Pitman, Trees S. Afr. 2 : 773 (1973). Type: Interior of the Cape Province, *Lichtenstein sub Herb. Willdenow 19186* (B, holo.).

A. atomiphylla Burch., Trav. 1 : 341 (1822). Type: Cape, Hopetown Distr., Asbestos mountains, at Kloof village, *Burchell 1685* (K, holo.!).

Shrub or small tree to 10 m high; trunk to 0,3 m in diam.; crown often irregularly rounded, narrow, branches drooping somewhat. *Bark* dark greyish-brown to blackish, rough; young branchlets pale to dark grey or reddish-brown to purple, often appearing as though whitewashed over a purplish background, flaking minutely, slender, sparingly to densely grey-puberulous or tomentellous, seldom subglabrous. *Stipules* spinescent, in pairs, 0,5–5,7 cm long, slender, never inflated, straight, greyish-white to reddish-brown; other prickles absent. *Leaves* usually densely grey-tomentellous, bipinnate but the leaflets so small and laterally compressed that the pinnae resemble single linear crenulate leaflets: petiole 1–5(7) mm long, adaxial gland often absent; rhachis

0,8–5,1(8,2) cm long, with minute reddish glands scattered amongst the grey indumentum, a small yellowish-brown gland often present at the junction of the top 1–6 and the lowest 1–3 pinnae pairs or between each or most pairs; pinnae 6–26 pairs; rhachillae (0,3)0,5–1(1,5) cm long, with minute reddish glands scattered amongst the grey indumentum; leaflets grey, 12–24(35) pairs per pinna, 0,25–0,8 × 0,2–0,5 mm, oblong, tightly laterally compressed, superficially appearing simply pinnate, densely puberulous. *Inflorescences* capitate, on axillary peduncles, solitary or fascicled. *Flowers* bright golden-yellow, grey in bud, sessile, sometimes some flowers are male only; peduncles 1–2,4 cm long, densely grey-tomentellous, somewhat glandular; involucel at or above the middle or at the apex of the peduncle. *Calyx* 1,4–2 mm long, apices of lobes densely tomentellous, lobes shallow or free for most of their length. *Corolla* 1,8–3 mm long, apices of lobes sparingly to densely tomentellous, lobes free almost to the base. *Stamen-filaments* free or connate into groups basally, up to 4,5 mm long; anthers with a deciduous apical gland. *Ovary* up to 1,5 mm long, sessile, glabrous at first but soon becoming pubescent. *Pods* densely grey-velutinous all over, with numerous minute dark reddish-brown to purplish glands particularly when young, 8–21 cm long, 0,6–1,4 cm wide, up to 0,9 cm thick, indehiscent, falcate or curled into a complete circle, seldom straightish, margin entire or irregularly constricted between the seeds and ± moniliform, slightly spongy within. *Seeds* dark reddish- or purplish-brown, 8,5–11,5 × 6–9 mm, lenticular to elliptic, sometimes scarcely compressed; areole 5–7 × 3,5–5 mm, almost closed.

Restricted to South West Africa, south-western Botswana and the northern Cape Province. Occurs mostly on the Kalahari sands and in other areas where loose sandy soils are prevalent; in dry woodland, bush or thornveld. Often found on the sandy flats between the dunes or along dry watercourses.

S.W.A.—2217 (Windhoek): Smalhoek, *Merxmuller 1045* (K). 2317 (Rehoboth): Rehoboth, *Fleck 486a* (Z). 2318 (Leonardville): 19,2 km S.E. of Pretorius Post Office, between Nossob River and Botswana, *Codd 5847.* 2416 (Maltahöhe): farm Bullsport, *Strey 2510.* 2619 (Aroab): 14,4 km S.S.E. of Aroab on road to Klipdam, *De Winter 3433.* 2819 (Ariamsvlei): Ariamsvlei, farm Walzersbrunn, *Ortendahl 309.* Grid ref. unknown: 160 km N.E. of Mariental, *Basson 143*; Gründorn, *Dinter 5047.*

CAPE.—2520 (Mata Mata): western border of Kalahari Gemsbok National Park, *Story 5572.* 2620 (Twee Rivieren): Kalahari Gemsbok Park, 32 km N. of Twee Rivieren, *Werger 1497.* 2622 (Tsabong): Duffield, *Breuckner 1317.* 2623 (Morokweng): 240 km N.W. of Vryburg near Heuningvlei, *Rodin 3574.* 2723 (Kuruman): near Kuruman, *Marloth 1056.* 2821 (Upington): 94,4 km from Olifantshoek on road to Upington, *Tölken & Schlieben 1194.* 2822 (Glen Lyon): 30,4 km S.S.W. of Olifantshoek, *Leistner & Joynt 2782.* 2823 (Griekwastad): Klaarwater, Griquatown, *Burchell 1900* (K). 2922 (Prieska): 9,6 km W.S.W. of Abrahams Dam, *Acocks 13191.* 2923 (Douglas): Asbestos Mts., Kloof village, *Burchell 1685* (K). 2924 (Hopetown): Hopetown commonage, *Schweickerdt 1178.* Grid ref. unknown: Vryburg Distr., along Molopo River, *De Winter 7831.*

A very distinctive species which is easily recognized by its compact densely grey-tomentellous leaves with minute leaflets.

A. haematoxylon hybridizes with *A. erioloba.*

Lichtenstein, in his Travels in Southern Africa (1815), states that the wood of *A. haematoxylon* is of fine quality and suitable for the manufacture of musical instruments.

29. Acacia erioloba *E. Mey.* × Acacia haematoxylon *Willd.* Ross in Bothalia 11(4) : 444 (1975).

A. giraffae Willd., Enum. Hort. Berol. 1054 (1809). Type: Interior of the Cape Province, *Herb. Willdenow* 19171 (B, holo.).

A. giraffae Willd. × *A. haematoxylon* Willd., Leistner, Mem. Bot. Surv. S. Afr. 38 : 67, 123, t.24 (1967); Ross in Bothalia 10(2) : 359 (1971); Robbertse in Proc. Electron Micros. Soc. S. Afr. 3 : 29 (1973).

Tree to 7 m high; crown rounded, spreading, branches usually drooping somewhat; habit resembling that of *A. erioloba.* *Bark* dark greyish-brown to blackish, rough; young branchlets grey or reddish-brown to purplish, sometimes appearing as though whitewashed over a purplish background, glabrous to densely grey-tomentellous, glandular or eglandular. *Stipules* spinescent, in pairs, 0,3–5 cm long, slender but usually stouter than in *A. haematoxylon*, never inflated as in *A. erioloba.* *Leaves* distinctly bipinnate, sparingly to densely grey-tomentellous: petiole 2–9 mm long, adaxial gland absent; rhachis 0,9–4,8 cm long, with or without minute scattered glands, a small yellowish-brown gland often present at the junction of each pinna pair; pinnae 3–12 pairs; rhachillae 0,4–2,2 cm long, glandular or eglandular; leaflets greyish, 11–25 pairs per pinna, 1–4 × 0,4–1,1 mm, linear to linear-oblong, often slightly falcate, apex rounded to subacute, sparingly to densely puberulous above and below, lateral nerves not prominent. *Inflorescences* capitate, on axillary peduncles, solitary or sometimes fascicled. *Flowers* bright golden-yellow, grey in bud, sessile; peduncles 1–3 cm long, densely grey-tomentellous, glandular or eglandular; involucel apical or a short distance below the apex of the peduncle. *Calyx* 1,8–2,4 mm long, apices of lobes sparingly to densely tomentellous, shallowly lobed or lobes free for ± half their length. *Corolla* 2,2–3 mm long, apices of lobes sparingly to densely tomentellous, lobes free almost to the base. *Stamen-filaments* free or connate basally into groups of usually 3–6, up to 4,5 mm long. *Ovary* up to 2 mm long, sessile or shortly stipitate, glabrous at first but soon becoming pubescent. *Pods* densely grey-velutinous all over, with numerous minute dark reddish-brown to purplish glands particularly when young, 7–14 cm long, 1,2–2,3 cm wide, up to 1 cm thick, falcate or curled into a complete circle, margin irregular, often constricted between the seeds and ± moniliform, slightly spongy within, each seed separated by a thin transverse septum. *Seeds* dark reddish-brown, 9–12 × 6–8 mm, lenticular-elliptic, sometimes subcircular, scarcely compressed; areole 6–8 × 2,5–3,5 mm, almost closed.

Restricted to the northern Cape Province. Occurs on the Kalahari sands, often on the flats of loose sand between the hills; in dry woodland, bush or thornveld. Although specimens are relatively widespread in the northern Cape, they are nowhere common. Usually only a solitary plant is found or, at most, five or six individuals.

CAPE.—2520 (Mata Mata): Kalahari Gemsbok National Park, 14,4 km N. of Mata Mata, *Leistner 1494*. 2620 (Twee Rivieren): Kalahari Gemsbok National Park, 3,2 km S.E. of Kamkwa along Auob River, *Leistner 3151* (KMG). 2722 (Olifantshoek): Moeswal Post Office, Langeberg Mts., *Leistner 1728*. 2922 (Prieska): Bloubosfontein, 59 km N.N.W. of Prieska, *Leistner 1340*; 9,6 km W.S.W. of Abrahams Dam, *Acocks 13190*; *Acocks 12689*; *Codd 1261*. 2923 (Douglas): 14,4 km W.N.W. of Douglas, *Leistner 1197*.

This convincing hybrid is of great interest and a fuller account of it appears in Bothalia 10 : 359 (1971).

Some of the characters displayed by the hybrid, for example, number of pinnae pairs and leaflet size, are intermediate between the values of *A. erioloba* and those of *A. haematoxylon*, while other characters, for example, the degree of pubescence and the presence of glands, are those exhibited by a single parent, namely, *A. haematoxylon*. The young branchlets, leaf-rhachides and peduncles in the hybrid are usually as densely pubescent as in *A. haematoxylon* and are not only

sparingly pubescent as an intermediate state between the glabrous *A. erioloba* and the densely pubescent *A. haematoxylon*. Recombination of the characters of the two parent species apparently does not take place at random, but there is a marked tendency for characters associated together in one species to remain associated in the hybrid.

Although the parentage of the hybrid is known, it is not known which species functions as the male parent and which as the female parent. Furthermore, it is not known whether the same species is always, for example, the male parent or whether the same species may sometimes serve as the female parent. Consequently, there is at present no understanding of differences arising in the progeny as a result of this. As the hybrid is fertile it should be possible to find all stages of back-crossing with the parents. Careful field studies are required.

30. **Acacia nilotica** (*L.*) *Willd. ex Del.*, Fl. Egypt I11. 79 (1813); A.F. Hill in Bot. Mus. Leafl. Harvard Univ. 8 : 97 (1940); Brenan in Kew Bull. 12 : 83 (1957); in F.T.E.A. Legum.-Mimos. : 109 (1959); in F.Z. 3,1 : 96 (1970). Type: Egypt, *Herb. Linnaeus* 1228.28 (LINN, lecto.!).

Mimosa nilotica L., Sp. Pl. 521 (1753). Type as above.

subsp. **kraussiana** (*Benth.*) *Brenan* in Kew Bull. 12 : 84 (1957); in F.T.E.A. Legum.–Mimos. : 110 (1959); Palmer & Pitman, Trees S. Afr. 161, t.43 (1961); F. White, For. Fl. N. Rhod. 86, fig. 18G (1962); Von Breitenbach, Indig. Trees S. Afr. 2 : 302 (1965); Brenan in F.Z. 3,1 : 97, t.16/13, t.21 (1970); Ross, Acacia Spp. Natal 36, fig. 2/6 (1971); Flow. Pl. Afr. 41 : t.1636 (1971); Van Wyk, Trees Kruger Nat. Park 1 : 151 (1972); Ross, Fl. Natal 193 (1973); Palmer & Pitman, Trees S. Afr. 2 : 787 (1973); Schreiber in Mitt. Bot. Staatssamml. Munchen 11 : 120 (1973). Type: Natal, Durban [Port Natal], *Krauss 69* (K, holo.!; FI!, TCD!).

Mimosa nilotica, Thunb., Prodr. Pl. Cap. 2 : 92 (1800).

Acacia arabica (Lam.) Willd. var. *kraussiana* Benth. in Hook., Lond. J. Bot. 1 : 500 (1842), non *A. kraussiana* Meisn. ex Benth. in Hook., Lond. J. Bot. 1 : 515 (1842); Meisn. in Hook., Lond. J. Bot. 2 : 103 (1843); Harv. in F.C. 2 : 281 (1862); Benth. in Trans. Linn. Soc. Lond. 30 : 506 (1875); Burtt Davy in Kew Bull. 1908 : 156 (1908); Fl. Transv. 2 : 343, fig. 60 (1932); N.E. Br. in Kew Bull. 1909 : 106 (1909); Codd, Trees & Shrubs Kruger Nat. Park 38, fig. 33b (1951); O. B. Miller in J. S. Afr. Bot. 18 : 18 (1952). Type as above. *A. arabica* sensu E. Mey., Comm. 1 : 168 (1836); Sim, For. Fl. P.E. Afr. 57, t.36B (1909); Henkel, Woody Pl. Natal 231 (1934). *A. benthamii* Rochebr., Toxicol. Afr. 2 : 192 (1898), non *A. benthamii* Meisn. (1844); Glover in Ann. Bolus Herb. 1 : 149, t.19/15 (1915); Bews, Fl. Natal 115 (1921); Burtt Davy in Kew Bull.

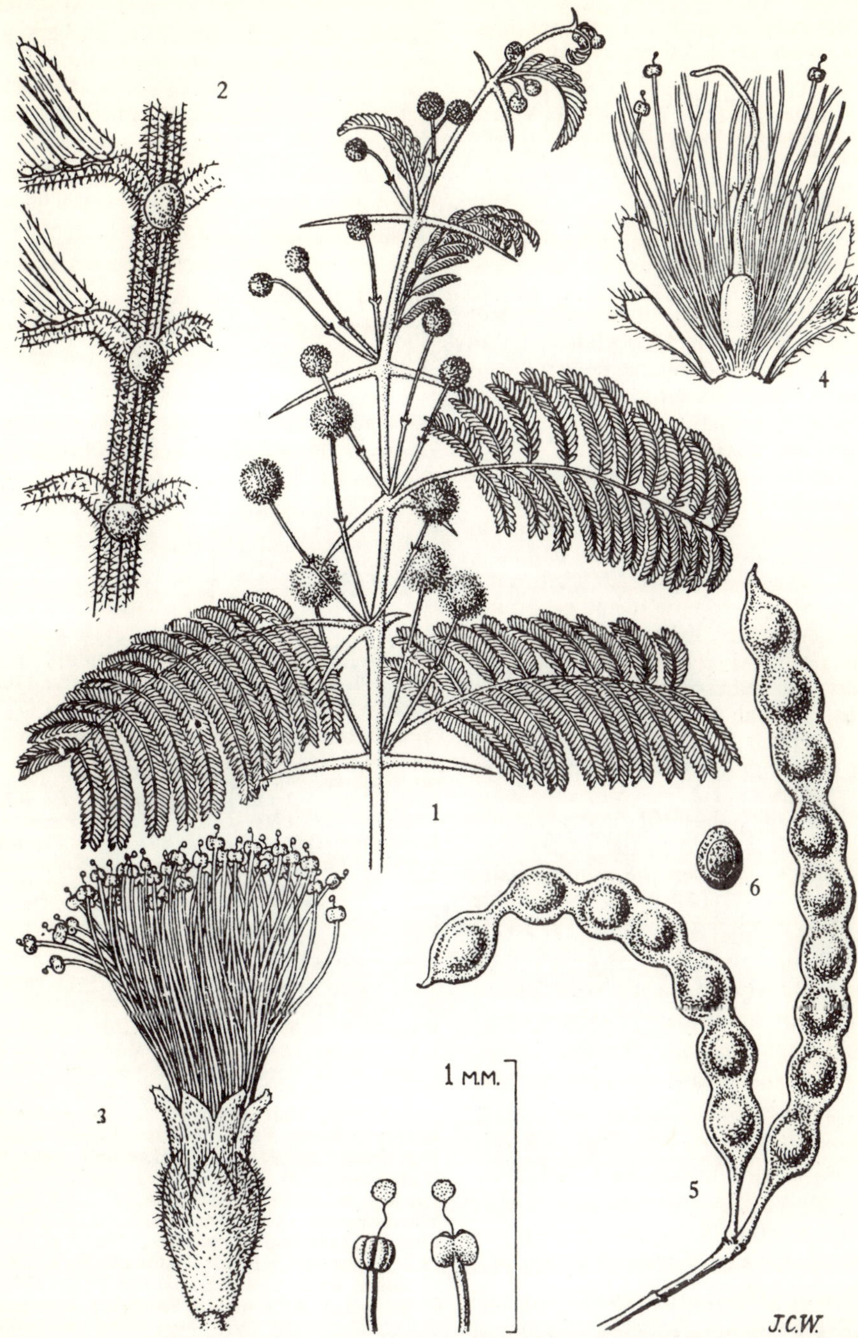

FIG. 9.—**Acacia nilotica** subsp. **kraussiana.** 1, flowering branch, × ⅔; 2, part of leaf-rhachis showing glands, × 6; 3, flower, × 10 with enlargement of anthers to show glands, all from *Purves* 215; 4, flower, opened out to show ovary, × 10, *Robson* 49; 5, pods, × ⅔, *White* 2279; 6, seed, × 1, *Milne-Redhead* 413. Reproduced by permission of the Editorial Board of Flora Zambesiaca.

1922 : 324 (1922); Bak.f., Leg. Trop. Afr. 3 : 850 (1930); Hutch., Botanist in S. Afr. 268 ("benthamiana"), 664 (1946); O. B. Miller, Checklist Bech. Prot. 16 (1948). Type as for *A. nilotica* subsp. *kraussiana. A. nilotica* var. *kraussiana* (Benth.) A. F. Hill in Bot. Mus. Leafl. Harvard Univ. 8 : 98 (1940). Type as above. *A. subalata* sensu Brenan, Checklist Tang. Terr. 333 (1949) pro parte; West in Rhod. Agric. J. 47 : 209 (1950); Pardy in Rhod. Agric. J. 51 : 489 (1954); Torre in C.F.A. 2 : 286 (1956), non Vatke sensu stricto. *A. nilotica* subsp. *subalata* sensu Boughey in J. S. Afr. Bot. 30 : 158 (1964); Schreiber in F.S.W.A. 58 : 11 (1967), non (Vatke) Brenan sensu stricto.

Tree to 10 m high; crown flattened, sometimes irregularly so, or rounded; trunk to 0,35 m in diam. *Bark* dark brownish-black to black, rough, longitudinally fissured; young branchlets greyish- or reddish-brown to purplish-black, often pinkish when young, sparingly to densely pubescent or occasionally subglabrous, glands inconspicuous or absent. *Stipules* spinescent, in pairs, 0,4–5,5(15) cm long, straight or almost so, often deflexed, whitish to greyish- or reddish-brown, sparingly to densely pubescent but becoming glabrescent with age; other prickles absent. *Leaves*: petiole 0,4–1,2(2,4) cm long, sparingly to densely pubescent or occasionally subglabrous, adaxial gland usually present, sometimes two present, often a short distance below the lowest pinna pair, up to 1,8 × 0,8 mm; rhachis (1,2)2,7–5,8(8) cm long, sparingly to densely pubescent or occasionally subglabrous, with a gland at the junction of each pinna pair or between the top few pairs only; pinnae (3)5–11(14) pairs; rhachillae 0,9–2, 8 (4,4) cm long; leaflets 12–27(36) pairs per pinna, 1,5–5,5 × 0,5–1,5 mm, linear to linear-oblong, apex rounded to subacute but not spinulose-mucronate, margins usually sparsely to densely ciliate, glabrous above and below or rarely puberulous below. *Inflorescences* capitate, on axillary peduncles, solitary or fascicled. *Flowers* bright-yellow, sessile; peduncles 1,2–4,5 cm long, sparingly to densely pubescent, rarely subglabrous; involucel from near the base to just over halfway up the peduncle. *Calyx* glabrous to densely pubescent, expecially apically, tube 0,8–1,6 mm long, lobes up to 0,4 mm long. *Corolla* glabrous to pubescent, especially apically, tube 2–2,6 mm long, lobes up to 1,1 mm long, often reflexed. *Stamen-filaments*

free or sometimes united into bundles basally, up to 5 mm long; anthers with a deciduous apical gland. *Ovary* up to 1,1 mm long, sessile or shortly stipitate, glabrous but soon becoming pubescent. *Pods* green and fleshy when young but shrivelling and turning black with age, (5,2)8–17(19,5) × 0,9–1,6(1,9) cm, oblong, straight or slightly falcate, margins shallowly to deeply crenate between each seed, each joint marked with a distinct raised bump which corresponds to the seed inside, sparingly to densely pubescent all over at first but the raised parts over the seeds becoming glabrescent and shining with age, rarely subglabrous throughout, sweet-smelling, indehiscent, breaking up transversely into segments on the ground. *Seeds* olive-brown, 6,5–9 × 5–8 mm, subcircular, scarcely compressed; areole 5–7 × 4–7 mm.

Found from Tanzania southwards to the Transvaal, Swaziland and Natal. Occurs in dry thornveld, river valley scrub, woodland, bushveld and scrub.

S.W.A.—1713 (Swartbooisdrift): 22,4 km N. of Otjihangasems (11,2 km N. of Omuhonga River), *Giess 9355* (M). 1716 (Enana): 19 km S.W. of Omafa on road to Ndola store, *De Winter 3626*. 1813 (Ohopoho): 3,2 km E. of Ohopoho, *De Winter & Leistner 5311*. 1916 (Gobaub): farm Pierre, Outjo–Otavi road, *Tölken & Hardy 895*. 1917 (Tsumeb): 7,5 km E. of Otavi on road to Grootfontein, *De Winter 2859*. 1918 (Grootfontein): farm Welgemoed, *Merxmuller & Giess 2165*. 2016 (Otjiwarongo): Omatjenne, *Volk 2937*.

TRANSVAAL.—2229 (Waterpoort): Dongola area, farm Breslau, *Codd 4834*. 2230 (Messina): Nwanedzi River, *Gerstner 6033a*. 2329 (Pietersburg): Vivo, 67,2 km W. of Louis Trichardt, *Schlieben 7550*. 2331 (Phalaborwa): Kruger National Park, Letaba Camp, *Codd 4266*. 2428 (Nylstroom): Mosdene, Naboomspruit, *Galpin M109*. 2429 (Zebediela): near Marble Hall, *Strey 8007*. 2431 (Acornhoek): Kruger National Park, Skukuza, Lower Sabie road, *Van der Schijff 3415*. 2526 (Zeerust): near Marico Dam, *Sutton 1135*. 2527 (Rustenburg): Krokodilpoort, 6,4 km N. of Brits, *Mogg 14604*. 2528 (Pretoria): Wonderboom, *C. A. Smith 6175*. 2529 (Witbank): Loskop Dam Nature Reserve, *Mogg 17512*. 2531 (Komatipoort): Kruger National Park, Nsikazi-Crocodile River junction, *Van der Schijff 3952*. 2731 (Louwsburg): 3,2 km N. of Pongola River on road to Gollel, *Ross 1708* (NH, NU). Grid ref. unknown: Sekukuniland, farm Korenvelden, *Barnard 74*.

SWAZILAND.—2631 (Mbabane): Nokwane, *Compton 31231*. 2731 (Louwsburg): Maloma, *Compton 29478*.

NATAL.—2632 (Bela Vista): Ndumu Game Reserve, *Ward 2018*. 2730 (Vryheid): Utrecht, *Ross 557* (NU). 2731 (Louwsburg): 6,4 km N. of Candover on Gollel road, *Ross 1418* (NH, NU). 2732 (Ubombo): 1,6 km along road to Ubombo, 3,2 km N. of Mkuze, *Ross 1695* (NH, NU). 2829 (Harrismith): 4,8 km from Ladysmith on Helpmekaar road, *Ross 547* (NU). 2830 (Dundee): 4,8 km from Dundee on Elandslaagte road, *Ross 550* (NU). 2831 (Nkandla): Middledrift, *Edwards 1429*. 2832 (Mtubatuba): Hluhluwe Game Reserve, Hluhluwe valley, *Ward 1604*. 2929 (Underberg): Estcourt Hill, 1,6 km S. of Estcourt, *Ross 748* (NU). 2930 (Pietermaritzburg): Bisley, near Pietermaritzburg, *Ross 1669* (NH, NU). 2931 (Stanger): near Stanger, *Wood 4011* (K). 3030 (Port Shepstone): Doonside, *Ross 791* (NU).

A. nilotica subsp. *kraussiana* occupies a wide range of habitats and is consequently rather variable. The pods, in particular, show considerable variation. Schreiber in F.S.W.A. 58 : 11 (1967) referred all of the South West African material of *A. nilotica* to subsp. *subalata* (Vatke) Brenan. However, after examining the specimens and comparing them with the Kew collections of typical subsp. *subalata* from East Africa, it is felt that the South West African specimens are better placed in subsp. *kraussiana*. It is extremely difficult to differentiate concisely in words between subsp. *kraussiana* and subsp. *subalata*, and some specimens of subsp. *kraussiana* certainly do approach subsp. *subalata*, particularly when the pods are immature.

In northern Zululand, and especially in Tongaland, the paired spines are frequently greatly elongated, so much so in some instances that they cause the young branchlets to droop. From the vicinity of Hluti in Swaziland to just north of Mkuze in Zululand there appears to be a tendency for the occurrence of plants with ±glabrous young branchlets and pods.

A tendency of *A. nilotica* is for a few flowers to develop in the involucel on the peduncle, sometimes giving the appearance of a smaller secondary capitulum below the main one. The flowers in this secondary capitulum often develop before those in the main capitulum. A specimen from the Skukuza area of the Kruger National Park (*Van der Schijff* 4009) is unusual in that a few of the peduncles are forked either at or some distance above the involucel and each limb bears a capitulum of flowers.

A good quality gum, at one time used for confectionery and for adhesive purposes, is exuded from the stems. The wood is reddish-brown with a darker heartwood, close-grained, very hard, durable and termite-proof. It is consequently useful for fencing posts and also provides a good fuel.

31. **Acacia xanthophloea** *Benth*. in Trans. Linn. Soc. Lond. 30 : 511 (1875); Burtt Davy in Kew Bull. 1908 : 160 (1908) pro parte excl. specim. *Elliott* 163; Sim, For. Fl. P.E. Afr. 58, t.41 (1909); Glover in Ann. Bolus Herb. 1 : 150 (1915); Eyles in Trans. Roy. Soc. S. Afr. 5 : 363 (1916); Bews, Fl. Natal 115 (1921); Bak. f., Leg. Trop. Afr. 3 : 851 (1930); Burtt Davy, Fl. Transv. 2 : 343 (1932); Henkel, Woody Pl. Natal 228 (1934); Codd, Trees & Shrubs Kruger Nat. Park 52, figs. 44c, d, 46 (1951): O.B. Miller in J. S. Afr. Bot. 18 : 26 (1952); Brenan in F.T.E.A. Legum.-Mimos. : 108, fig. 16/36 (1959); Palmer & Pitman, Trees S. Afr. 166, t.48 (1961); Von Breitenbach, Indig. Trees S. Afr. 2 : 308 (1965); De Winter et al, 66 Transv. Trees 60 (1966); Brenan in F.Z. 3,1 : 96, t.20 (1970); Ross, Acacia Spp. Natal 43, fig. 2/7 (1971); Flow. Pl. Afr. 41 : t.1637 (1971); Van Wyk, Trees Kruger Nat. Park 1 : 167 (1972); Ross, Fl. Natal 193 (1973); Palmer & Pitman, Trees S. Afr. 2 : 815 (1973). Syntypes: Malawi, E. end of Lake Shirwa [Chilwa], *Meller* (K!); Mozambique, Sena, *Kirk* (K!).

Tree up to 30 m high with a rounded or flattened and somewhat spreading crown. *Bark* on trunk lemon to greenish-yellow, flaking minutely, becoming powdery, dark brown to black where damaged; young branchlets lemon to greenish-yellow, powdery, glabrous or subglabrous, young extremities brown to plum-coloured but the outer layer soon flaking off to reveal the greenish-yellow inner layer. *Stipules* spinescent, in pairs, 0,9–8,5 cm long, straight or almost so; "ant-galls" and other prickles absent. *Leaves*: petiole 0,1–1,5 cm long, glabrous to sparingly pubescent, adaxial gland usually present, variable in position, up to 1,5 × 1 mm; rhachis (0)2,5–7 cm long, glabrous to sparingly pubescent, a gland often present at the junction of the top 1–2 pinna pairs; pinnae (1)3–6(8) pairs; rhachillae 0,3–3 cm long; leaflets 8–17 pairs per pinna, 2,5–6,5 × 0,75–1,75 mm, linear to linear-oblong, apex rounded to acute or mucronate, margins usually without cilia, glabrous beneath, lateral nerves invisible beneath. *Inflorescences* capitate, usually on abbreviated lateral shoots whose axes do not elongate and are represented by clustered scales, the peduncles thus appearing to be in lateral fascicles on the older yellow-barked twigs. *Flowers* bright yellow (at least in our area), sessile; peduncles 0,8–2,9 cm long, sparingly to densely pubescent or subglabrous, glandular chiefly

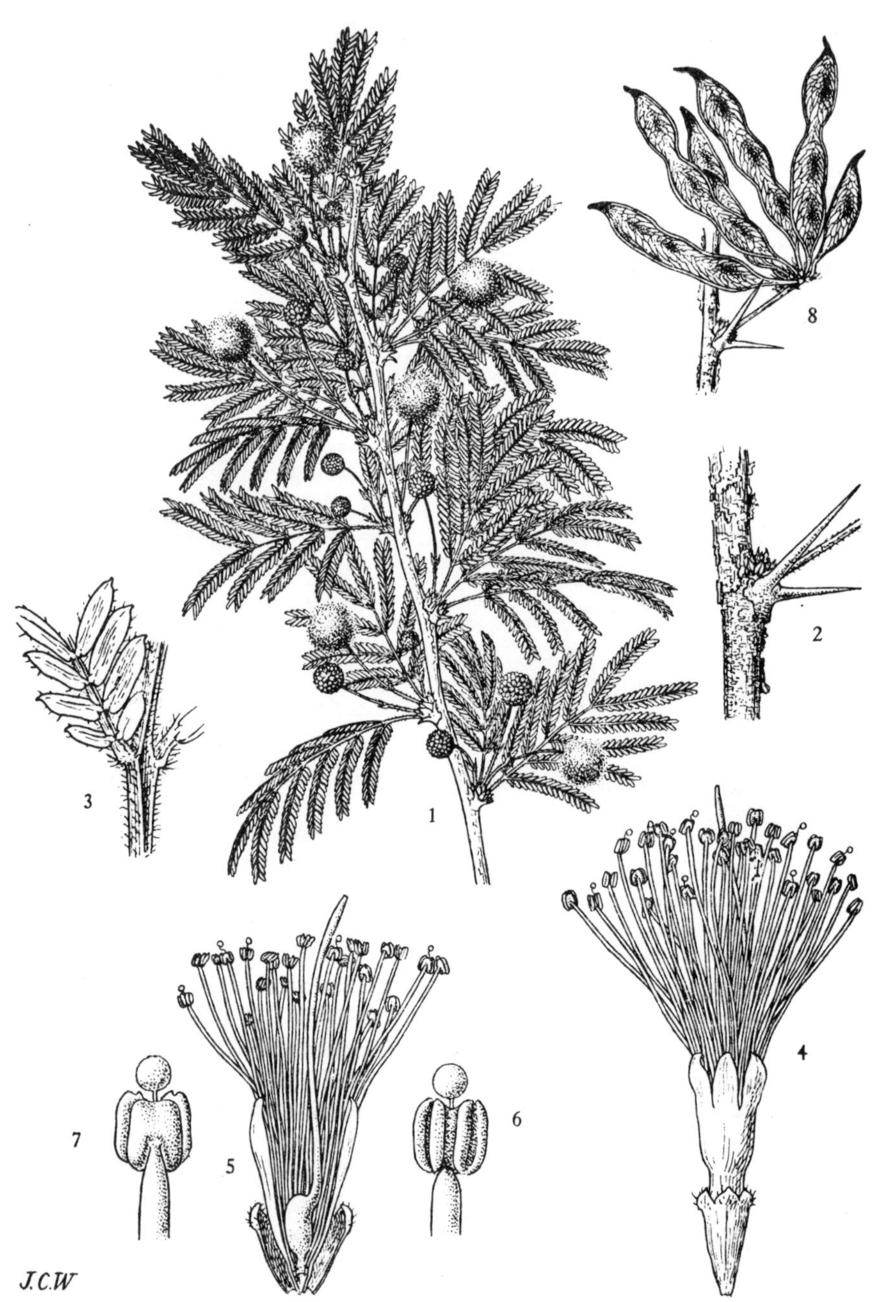

J.C.W

FIG. 10.—**Acacia xanthophloea.** 1, flowering branch, × ⅔, *Gomes e Sousa* 3604; 2, part of branch, showing bark and paired spines, × 1, *Mendonca* 2330; 3, part of leaf-rhachis and pinnae, × 4; 4, flower, × 10; 5, flower, opened out to show ovary and stamens, × 8; 6, anther, front view, × 82; 7, anther, rear view, × 82, all from *Gomes e Sousa* 3604; 8, pods, × ⅔, *Mendonca* 2330. Reproduced by permission of the Editorial Board of Flora Zambesiaca.

below the involucel but sometimes also above; involucel basal to about halfway up the peduncle, conspicuous, 2–4 mm long. *Calyx* glabrous or sometimes lobes very sparingly puberulous, tube 0,8–1,6 mm long, lobes minute, up to 0,2 mm long. *Corolla* often deep pink to purplish apically especially when in bud, glabrous, tube 2,4–3,2 mm long, lobes up to 0,5 mm long. *Stamen-filaments* free, up to 7 mm long; anthers with a deciduous apical gland. *Ovary* up to 1,8 mm long, sessile, with few to many long glandular hairs. *Pods* yellowish-brown to brown, 3,4–13,5 × 0,7–1,4 cm, linear-oblong, straight or slightly curved, ± moniliform or irregularly constricted between some of the seeds, the segments mostly ± as long as wide, indehiscent, breaking up transversely into segments after being shed, valves thin, coriaceous, reticulate-venose, glabrous or almost so, eglandular or sparingly glandular. *Seeds* olive to blackish-olive, 4,5–6,5 × 3,5–5 mm, subcircular to elliptic, compressed; areole 3–4,5 × 2–3 mm.

Found from Kenya southwards to the Transvaal, Swaziland and Zululand. Occurs along river banks, in seasonally flooded areas adjacent to rivers, on the margins of lakes and shallow pans where water collects after rain, and in low lying areas where underground water is available. Often gregarious and at times forming a closed woodland.

TRANSVAAL.—2231 (Pafuri): Kruger National Park, Makuleka, *Lang sub TRV 32250.* 2531 (Komatipoort): Kruger National Park, 4,8 km N.W. of Crocodile Bridge, *Codd 5710*; Komatipoort, *Burtt Davy 365.*

SWAZILAND.—2631 (Mbabane): Stegi-Big Bend road, *Compton 28042.* 2731 (Louwsburg): 22,4 km N. of Gollel on Nsoko road, *Ross 1551.*

NATAL.—2632 (Bela Vista): Ndumu Game Reserve, *Moll & Strey 3733.* 2731 (Louwsburg): 4,8 km upstream from Pongola Poort Dam, *Ross 1415* (NU). 2732 (Ubombo): 19 km N. of Hluhluwe on Mkuze road, *Ross 1373* (K, NH, NU). 2832 (Mtubatuba): 4,8 km W. of Hluhluwe station, *Acocks 13091.*

A. xanthophloea, with its greenish-yellow bark, is a most distinctive species. On account of its preference for moist or swampy situations, which also form the breeding places of the malaria carrying mosquito, the species has always been associated with fever, whence the common name "Fever Tree".

Brenan in F.T.E.A. Legum.–Mimos. : 108 (1959) records that *A. xanthophloea* is "apparently unique among the East African acacias in having flowers either white to pinkish or purplish, or else yellow to golden." White to purplish flowers appear to be confined to Kenya and Tanzania. Although the corollas of many of our specimens are pinkish or even purple, the stamens are bright yellow and the flowers therefore appear bright yellow. At present there is no evidence to suggest that any flower colour other than bright or golden yellow occurs in our area.

A. xanthophloea achieved fame and is known to generations of children through Rudyard Kipling's "Just So Stories" and the adventures of the Elephant's Child who travelled "till he came at last to the banks of the great, grey-green, greasy Limpopo River, all set about with fever trees".

32. **Acacia kirkii** *Oliv.* in F.T.A. 2 : 350 (1871); Benth. in Trans. Linn. Soc. Lond. 30 : 507 (1875); Brenan in Kew Bull. 12 : 361 (1958); in F.T.E.A. Legum.-Mimos. : 106 (1959). Type: Zambia, Southern Province, Batoka country, *Kirk* (K, holo.!).

subsp. **kirkii.**

Brenan in Kew Bull. 12 : 363 (1958); in F.T.E.A. Legum.—Mimos. : 106 (1959); in F.Z. 3,1 : 94 (1970); Schreiber in Mitt. Bot. Staatssamml. Munchen 11 : 118 (1973).

A. kirkii Oliv. in F.T.A. 2 : 350 (1871); Harms in Warb., Kunene–Samb. Exped. 243 (1903); Bak.f., Leg. Trop. Afr. 3 : 848 (1930); O. B. Miller, Checklist Bech. Prot. 19 (1948); in J. S. Afr. Bot. 18 : 22 (1952); Brenan, Checklist Tang. Terr. 333 (1949); Torre in C.F.A. 2 : 285 (1956); F. White, For. Fl. N. Rhod. 86, fig. 18H (1962); Von Breitenbach, Indig. Trees S. Afr. 2 : 303 (1965); Palmer & Pitman, Trees S. Afr. 2 : 787 (1973). Type: Zambia, Southern Prov., Batoka country, *Kirk* (K, holo.!). *A. kirkii* subsp. *kirkii* var. *kirkii*, Brenan in Kew Bull. 12 : 363 (1958). Type as above. *A. kirkii* subsp. *kirkii* var. *intermedia* Brenan in Kew Bull. 12 : 363 (1958); in F.T.E.A. Legum.—Mimos. : 107, fig. 15/35A (1959); Schreiber in F.S.W.A. 58 : 10 (1967). Type: Kenya, Athi Plains, *Van Someren in C.M.* 2700 (K, holo.!). *A. verrucifera* Harms in Warb., Kunene–Samb. Exped. 17, 244 (1903), nomen nudum. *A. harmsiana* Dinter in Feddes Repert. 15 : 80 (1917); Bak.f., Leg. Trop. Afr. 3 : 848 (1930). Type: South West Africa, Tsumeb Distr., Nagusib, 25 km S.E. of Namutoni, *Dinter 2283* (SAM, iso.!). *A. nilotica* (L.) Del. subsp. *adstringens* (Schumach. & Thonn.) Roberty var. *kirkii* (Oliv.) Roberty in Candollea 11 : 151 (1948) pro parte. Type as for *A. kirkii.*

Tree to 10 m high or a many stemmed shrub branching from near the base, branches slender, ascending, crown ± flattened and spreading somewhat. *Bark* grey or yellowish-brown, often with numerous transversely elongated lenticels, papery, flaking or peeling off to reveal a yellowish-green inner layer; young branchlets grey, brown or reddish- to purplish-brown, subglabrous or puberulous, with numerous sessile reddish glands. *Stipules* spinescent, in pairs, 0,5–4,5(8) cm long, straight or almost so, usually greyish-white; "ant-galls" and other prickles absent. *Leaves:* petiole 0,3–1,3 cm long, subglabrous to densely pubescent, adaxial gland usually absent; rhachis (0,8)3–6(7,2) cm long, subglabrous to densely pubescent, a small rounded or oval gland at the junction of the top 1–3 pinnae pairs only, or absent; pinnae

(3)6–14 pairs; rhachillae 0,8–2,1 cm long, subglabrous to densely pubescent; leaflets 9–18 pairs per pinna, 2–5 × 0,5–1,25 mm, narrowly oblong or linear-oblong, apex acute to minutely mucronate, glabrous below, margins with or without cilia. *Inflorescences* capitate, on axillary peduncles, solitary or fascicled. *Flowers* sessile, pinkish-red in bud, cream or white when open; peduncles 1,8–3,7 cm long, usually densely pubescent and with sessile glands throughout, rarely sparingly pubescent; involucel conspicuous, 2–3 mm long, below, at or slightly above the middle of the peduncle. *Calyx* cream and tinged with pinkish-red, or pinkish-red, sparingly to densely pubescent apically especially on the lobes, tube up to 1,6 mm long, lobes up to 0,8 mm long. *Corolla* pinkish-red, glabrous or subglabrous, tube 1,8–2,4 mm long, lobes up to 1 mm long. *Stamen-filaments* white, free, up to 6 mm long; anthers with a deciduous apical gland. *Ovary* up to 1,2 mm long, sessile, glabrous. *Pods* brown or reddish-brown, (2)3,5–8(10) × (0,8)1,3–2,1 cm, oblong, straight or almost so (sometimes bent in a plane at right angles to the flattened plane of the pod), mostly ± moniliform with the segments as wide as or wider than long, stipe up to 1,5 cm long, apex rounded to mucronate, usually prominently venose, indehiscent, fibrous and tough, each segment with a small or medium irregular wart-like projection up to 5 mm long in the centre of each of its flat sides, glabrous or subglabrous. *Seeds* olive or olive-brown, 5–8 × 4–5,5 mm, subcircular to elliptic, compressed; areole 3,5–5 × 2,5–3,5 mm.

Found in Uganda, Kenya, Tanzania, Zambia, Rhodesia, Angola, northern South West Africa and Botswana. Occurs in woodland, wooded grassland, mixed bush and scrub, often in seasonally flooded alluvium by rivers and pans. Often grows in association with *Colophospermum mopane.*

S.W.A.—1713 (Swartbooisdrift): Handungo, 14,4 km N.W. of Ombazu on road to Swartbooisdrift, *De Winter & Leistner 5896.* 1725 (Livingstone): Mpilila Island, confluence of Zambezi and Chobe rivers, *Killick & Leistner 3382.* 1814 (Otjitundua): Otjitoko, ± 56 km S. of Ohopoho, *De Winter & Leistner 5912.* 1817 (Tsintsabis): 89 km N.W. of Tsumeb on road to Namutoni, *De Winter 2957.*

33. **Acacia tortilis** (*Forsk.*) *Hayne,*

Arzneyk. Gebr. Gewächse 10 : t.31 (1827); Oliv. in F.T.A. 2 : 352 (1871); Benth. in Trans. Linn. Soc. Lond. 30 : 506 (1875); Bak.f., Leg. Trop. Afr. 3 : 841 (1930); Torre in C.F.A. 2 : 284 (1956); Brenan in Kew Bull. 12 : 86 (1957); in F.T.E.A. Legum-Mimos. : 117 (1959); F. White, For. Fl.N. Rhod. 84, fig 17J (1962); Brenan in F.Z. 3,1 : 102 (1970). Type: Arabia, "Mons Soudân prope Hás", *Forskal* (C, holo.).

Mimosa tortilis Forsk., Fl. Aegypt. Arab. CXXIII, 176 (1775). Type as above.

Tree to 15 m high or a small shrub or bush, crown typically flattened and spreading, sometimes rounded; trunk to 1 m in diam. *Bark* grey to grey-brown or blackish, at times reddish-brown, rough, fissured, seldom ± smooth; young branchlets greyish- to reddish-brown or purplish-black, glabrous to densely pubescent, lenticellate, often flaking minutely. *Stipules* spinescent, some short, hooked and up to 5 mm long, mixed with other long straight slender whitish spines 1,5–10 cm long; "ant-galls" and other prickles absent. *Leaves:* petiole 0,2–0,8(1,4) cm long, sparingly to densely pubescent, adaxial gland usually present and just below lowest pinna pair; rhachis 0,2–2(4,6) cm long, sparingly to densely pubescent, eglandular or sparingly glandular, a small gland often present at the junction of the top 1–3 and lowest 1–2 pinnae pairs or absent from some, variable; pinnae 2–10(14) pairs; rhachillae 0,3–1,7 cm long, sparingly to densely pubescent; leaflets 6–22 pairs per pinna, 1–4 × 0,6–1 mm, linear to linear-oblong, apex rounded to acute, glabrous or sparingly to densely appressed-pubescent beneath and sometimes also above, indumentum often concentrated along midrib beneath, margins with or without cilia. *Inflorescences* capitate, on axillary peduncles, fascicled or solitary. *Flowers* white to pale yellowish-white, sessile; peduncles 0,4–2,4 cm long, sparingly to densely pubescent; involucel in lower third of peduncle. *Calyx* glabrous except for the lobes which are sparingly to densely pubescent, tube 0,9–1,6 mm long, lobes up to 0,8 mm long. *Corolla* glabrous throughout or apices of lobes sparingly to densely pubescent, tube 1,2–2 mm long, lobes up to 1 mm long. *Stamen-filaments* free, up to 4,5 mm long; anthers with a deciduous apical gland. *Ovary* glabrous, up to 1,5 mm long, shortly stipitate. *Pods* variously contorted or spirally twisted, 0,6–1,2 cm wide, greenish-yellow to olive-brown, longitudinally veined, indehiscent or very tardily dehiscent on the ground, usually glabrous but at times sparingly to densely pubescent, eglandular or with few to many minute scattered reddish glands. *Seeds* olive- to reddish-brown, 4–7 × 3–6 mm, elliptic to subcircular, smooth, compressed; areole 3–6 × 2–4 mm.

A. tortilis is widespread throughout Africa and Arabia. Four subspecies are recognized, two of which occur in our area.

Young branchlets shortly pubescent with hairs usually less than 0,25 mm long, a few to 0,5 mm; petiole and leaf-rhachis similarly shortly pubescent; pods glabrous or almost so, eglandular..............(a) subsp. *heteracantha*

Young branchlets with longer denser pubescence than in subsp. *heteracantha*, hairs 0,25–0,75 mm long; petiole and leaf-rhachis with hairs mostly more than 0,25 mm long; pods tomentellous or pubescent with spreading or curved hairs among which numerous dark reddish glands are clearly visible through a hand lens...............(b) subsp. *spirocarpa*

A. tortilis with its mixture of short recurved spines and long straight spines and its spirally twisted or variously contorted pods is a very distinctive species. Typical subsp. *tortilis*, with narrow (3–5 mm wide) pubescent but eglandular pods, occurs in Egypt, the Sudan, Arabia, Aden and perhaps Israel.

(a) subsp. **heteracantha** (*Burch.*) *Brenan*

in Kew Bull. 12 : 88 (1957); in Kew Bull. 13 : 409 (1959); Palmer & Pitman, Trees S. Afr. 165, t.xi, 46, 47 (1961); Von Breitenbach, Indig. Trees S. Afr. 2 : 287 (1965); De Winter et al, 66 Transv. Trees 58 (1966); Schreiber in F.S.W.A. 58 : 12 (1967); Brenan in F.Z. 3,1 : 102, t. 16/15 (1970); Ross, Acacia Spp. Natal 42, fig. 2/23 (1971); in Bothalia 10 : 549 (1972); Van Wyk, Trees Kruger Nat. Park 1 : 165 (1972); Ross, Fl. Natal 193 (1973); Palmer & Pitman, Trees S. Afr. 2 : 783 (1973). Type: Cape Province, Hay Distr., Spuigslangfontein, between Griquatown and the Orange River, *Burchell* 1710 (K, holo.!)

A. heteracantha Burch., Trav. 1 : 389 (1822); DC., Prodr. 2 : 473 (1825); Harv. in F.C. 2 : 280 (1862); Benth. in Trans. Linn. Soc. Lond. 30 : 505 (1875); Glover in Ann. Bolus Herb. 1 : 151 (1915); Burtt Davy in Kew Bull. 1922 : 327 (1922); Fl. Transv. 2 : 344 (1932); Bak.f., Leg. Trop. Afr. 3 : 843 (1930); Henkel, Woody Pl. Natal 229 (1934); Hutch., Botanist in S. Afr. 398, 428, 664 (1946); Codd, Trees & Shrubs Kruger Nat. Park 46, figs. 38 f, g, 39 (1951); Torre in C.F.A. 2 : 284 (1956); Schreiber in Mitt. Bot. Staatssamml. Munchen 2 : 283 (1957); Story, Mem. Bot. Surv. S. Afr. 30 : 23, t.17 (1958): non Dinter, Deutsch-Sudwest-Afrika Flora Forst-und-land wirtschaftliche Fragmente 76 (1909); in Feddes Repert. 15 : 80 (1917); Ponnighaus in J.S.W. Afr. Sci. Soc. 6 : 13 (1933); Walter & Volk, Grundlagen der Weiderwirtschaft in Sudwestafrika 211, t.68B (1954). Type as above. *A. litakunensis* Burch., Trav. 2 : 452,

t.6 (1824); Harv. in F.C. 2 : 283 (1862); Benth. in Trans. Linn. Soc. Lond. 30 : 505 (1875); N.E. Br. in Kew Bull. 1909 : 107 (1909); Glover in Ann. Bolus Herb. 1 : 151 (1915); Burtt Davy in Kew Bull. 1922 : 329 (1922); Fl. Transv. 2 : 345 (1932); Marloth, Fl. S. Afr. 2 : 53, t.18B, fig. 33 (1925); Hutch., Botanist in S. Afr. 297, 299, 367, 370, 411 (1946); O. B. Miller in J. S. Afr. Bot. 18 : 23 (1952). Type: Cape Province, Kuruman Distr., Takoon [Litakun], *Burchell* 2205 (K, holo.!). *A. spirocarpoides* Engl. in Bot. Jahrb. 10 : 23 (1888); Marloth in Trans. S. Afr. Phil. Soc. 5 : 270 (1889); Sim, For. Fl. Cape Col. 211 (1907); Burtt Davy in Kew Bull. 1908 : 160 (1908); Glover in Ann. Bolus Herb. 1 : 148, t.19/16 (1915); Pole Evans in S. Afr. J. Sci. 17 : figs. 16, 22 (1920); Bews, Fl. Natal 114 (1921). Syntypes: Cape Province, Barkly West Distr., Barkly West, *Marloth* 809 (not traced); Kimberley Distr., near Kimberley, *Marloth* 839 (GRA!). *A. maras* Engl. in Bot. Jahrb. 10 : 24 (1888); Marloth in Trans. S. Afr. Phil. Soc. 5 : 270 (1889); Schinz in Mém. Herb. Boiss. 1 : 115 (1900); Dinter in Feddes Repert. 15 : 81 (1917). Type: South West Africa, Otjimbingwe, *Marloth* 1260 (B, holo. †; GRA!).

Found in South West Africa, Botswana, Rhodesia, Mozambique, the Transvaal, Orange Free State, Swaziland, Natal and the northern Cape Province. Occurs in woodland, wooded grassland, dry thornveld, river valley scrub and bushveld; common on the Kalahari sands in the northern Cape and South West Afrca.

S.W.A.—1821 (Andara): road from Andara to Bagani, *Merxmuller & Giess 1993*. 1913 (Sesfontein): flats just E. of Sesfontein, *De Winter & Leistner 5881*. 1920 (Tsumkwe): Nama Pan, *Story 5341*. 2115 (Karibib): Karibib, *Dinter 6740*. 2115 (Okahandja): Okahandja, *Dinter 310*. 2216 (Otjimbingwe): farm Otjiseva, *Wiss 948*. 2316 (Nauchas): bed of Tsondab River above Abbabis, *Pearson 9127* (K). 2416 (Maltahöhe): farm Bullsport, *Strey 2288*.

TRANSVAAL.—2229 (Waterpoort): Dongola area, farm Neanderthole, *Codd 4860*. 2230 (Messina): P.O. Schroda, *Native collector sub PRE 1954*. 2326 (Mahalapye): Buffelsdrif, *Vahrmeijer 1283*. 2329 (Pietersburg): farm Davel, *Strey 7999*. 2330 (Tzaneen): Elim, *Obermeyer 532*. 2331 (Phalaborwa): Kruger National Park, Letaba, *Lang sub TRV 30873*. 2425 (Gaberones): Lekkerlach, *Louw 601*. 2428 (Nylstroom): farm Krantzberg, near Nylstroom, *Prosser 1699*. 2429 (Zebediela): Potgietersrust, *Thode A 1696*. 2431 (Acornhoek): Kruger National Park, 4,8 km E. of Skukuza on Lower Sabie road, *Codd & De Winter 5018*. 2526 (Zeerust): Swartruggens, *Sutton 1134A*. 2527 (Rustenburg): 16,8 km E. of Rustenburg on road to Pretoria, *De Winter 7801*. 2528 (Pretoria): Hammanskraal, *De Winter 7759*. 2529 (Witbank): Loskop Irrigation Dam, *Mogg 17283*. 2531 (Komatipoort): Crocodile River drift between Komati River drift and Barberton, *Bolus 7754* (K). 2725 (Bloemhof): Bloemhof district, Cawood's Hope, *Burtt Davy 12959*. Grid ref. unknown: Lydenburg district, Sekukuniland, *Barnard 228*.

O.F.S.—2825 (Boshof): N. of Boshof, near Windsorton road, *Van Zinderen Bakker 91*. 2925 (Jagersfontein): in poort on path to Fauresmith, *Verdoorn sub Henrici 2405*.

SWAZILAND.—2631 (Mbabane): near Mhlatusi River, *H. Hutchinson sub PRE 30305.* 2731 (Louwsburg): Maloma, *Compton 29480.*

NATAL.—2632 (Bela Vista): Ndumu Game Reserve, Bunguzane area, *Tinley 997* (NH, NU). 2730 (Vryheid): Lancaster Hill, Vryheid, *Galpin 9762.* 2731 (Louwsburg): 6,4 km from Mkuze on Nongoma road, *Ross 1639.* 2732 (Ubombo): 3,2 km N. of Mkuze, turn-off to Ubombo, *Ross 1692.* 2829 (Harrismith): bank of Tugela River near Colenso, *West 1801.* 2830 (Dundee): Keates Drift, *Ross 824* (NH, NU). 2831 (Nkandla): Umfolozi Game Reserve, Tobothi, *Ross 2007.* 2832 (Mtubatuba): Hluhluwe Game Reserve, *Ward 3005.* 2930 (Pietermaritzburg): Umgeni Dam, *Moll 1709.* 2931 (Stanger): 1,6 km S. of Mandini, *Ross 882* (NU).

CAPE.—2525 (Mafeking): 1,6 km from Mafeking on Zeerust road, *Morris 1120.* 2723 (Kuruman): between Kuruman and Takoon [Litakun], *Gerstner 6276.* 2823 (Griekwastad): Baviaans Kloof, Kaap Plateau, *Hutchinson 3003.* 2824 (Kimberley): Kimberley Commonage, *Schweickerdt 1117.* 2923 (Douglas): Spuigslangfontein, between Griquatown and the Orange River, *Burchell 1710* (K). 2924 (Hopetown): Hopetown district, *Bryant 1167.*

The mixture of short and long spines accounts for the vernacular name "Haak en Steek". The typically flattened spreading crown, on the other hand, accounts for subsp. *heteracantha* also being referred to as the Umbrella Thorn.

De Winter 2719 from 69 km N. of Okahandja on road to Otjiwarongo in South West Africa is difficult to place with certainty. The young branchlets, petioles and leaf-rhachides have the short indumentum of subsp. *heteracantha* but the pods are sparsely clothed with short hairs among which some dark reddish glands are visible. *Tölken & Hardy* 687 from 20 km from Bullsport on road to Ababis is similar. Both specimens seem to be best referred to subsp. *heteracantha*.

The leaves and pods make an excellent fodder. The wood, however, is of little use except for firewood.

(b) subsp. **spirocarpa** (*Hochst. ex A. Rich.*) *Brenan* in Kew Bull. 12 : 88 (1957); in F.T.E.A. Legum.—Mimos. : 117, fig. 16/44 (1959); in F.Z. 3,1 : 102 (1970); Schreiber in Mitt. Bot. Staatssamml. Munchen 11 : 124 (1973). Syntypes: Ethiopia, near Djeladjeranne [Dscheladscheranne], *Schimper 502* (BM!, FI!, K!, M!, P!, Z!); *Schimper 612* (BM!, FI!, K!, M!, P!); *Schimper 658* (BM!, FI!, K!, M!, OXF!, P!, Z!).

A. spirocarpa Hochst. ex A. Rich., Tent. Fl. Abyss. 1 : 239 (1847); Schweinf. in Linnaea 35 : 322 (1867–8); Oliv. in F.T.A. 2 : 352 (1871); Benth. in Trans. Linn. Soc. Lond. 30 : 505 (1875); Bak.f., Leg. Trop. Afr. 3 : 842 (1930); O. B. Miller in J. S. Afr. Bot. 18 : 25 (1952). Syntypes as above.

Found in Ethiopia and the Sudan southwards to Angola, South West Africa, Botswana, Rhodesia, and Mozambique. Occurs in deciduous woodland and dry scrub.

S.W.A.—1917 (Tsumeb): Tsumeb (—BA), *Giess, Volk & Bleissner 6442* (M). 2017 (Waterberg): Waterberg, below Police station (—CA), *Giess 12349* (K, M).

The pods of *Giess* 12349 are densely clothed with whitish spreading hairs 1–2 mm long and show a very close approach to var. *crinita* Chiov. in Res. Sci. Miss. Stefanini-Paoli 1 : 71 (1916). Var. *crinita* is recorded from Somalia, Kenya and Tanzania.

34. **Acacia luederitzii** *Engl.* in Bot. Jahrb. 10 : 23, t.3B (July 1888) pro parte quoad specim. *Marloth* 1328; Schinz in Mém. Herb. Boiss. 1 : 115 (1900) pro parte quoad specim. *Marloth* 1328; Dinter in Feddes Repert. 15 : 80 (1917) pro parte quoad specim. *Marloth* 1328; Bak.f., Leg. Trop. Afr. 3 : 840 (1930); Brenan in F.Z. 3,1 : 101 (1970); Schreiber in Mitt. Bot. Staatssamml. Munchen 11 : 119 (1973); Ross in Bothalia 11 : 233 (1974). Type: South West Africa, Otjimbingwe, *Marloth* 1328 (PRE lecto!; GRA!, M!, OXF!).

Shrub branching from or near the base or a tree to 15 m high with a flattened and spreading or somewhat rounded crown. *Bark* greyish- or reddish-brown to black, rough, longitudinally fissured on older trunks, often ± smooth on younger stems; young branchlets greyish- or reddish-brown to purplish or blackish, flaking minutely, densely clothed with grey or whitish spreading hairs, older branchlets glabrescent. *Stipules* spinescent, in pairs, some short and strongly hooked, 3–10 mm long, usually intermixed with some elongate, slender and straight or slightly curved spines 1–7 cm long or some inflated spines ("ant-galls") up to 1,8 cm in diam. basally, ±bent towards the apex, greyish-white to purplish, sometimes all spines short and hooked; other prickles absent. *Leaves*: petiole 0,3–1,4 cm long, sparingly to densely spreading-pubescent, adaxial gland often absent, usually just below the lowest pinna pair when present, 0,4–0,9×0,2–0,5 mm; rhachis 0,7–4,8 cm long, sparingly to densely clothed with spreading hairs, eglandular or with a small gland at the junction of the top 1–3(5) pinnae pairs; pinnae 3–9(13) pairs; rhachillae 0,7–2,8 cm long, sparingly to densely clothed with spreading hairs; leaflets 11–26 pairs per pinna, 2–5 × 0,5–1,5 mm, linear-oblong or the terminal ones slightly obovate-oblong, apex rounded or obtuse, margins with conspicuous spreading cilia, especially apically, seldom cilia ±

inconspicuous, otherwise glabrous or sometimes ± pubescent beneath. *Inflorescences* capitate, on axillary peduncles, usually fascicled, seldom solitary. *Flowers* yellowish-white, sessile; peduncles 1,2–4,3 cm long, sparingly to densely pubescent, hairs spreading but shorter than the diameter of the peduncle; involucel ⅓–⅔-way up the peduncle. *Calyx* shortly pubescent or puberulous apically, tube 0,8–1,3 mm long, lobes up to 0,4 mm long. *Corolla* glabrous, tube 2,2–3,2 mm long, lobes up to 0,5 mm long. *Stamen-filaments* free, up to 6,5 mm long; anthers with a deciduous apical gland. *Ovary* up to 1,2 mm long, very shortly stipitate. *Pods* brown or reddish-brown to purplish, 3,2–13 × (0,9)1–1,9 cm, straight or sometimes slightly curved, linear-oblong, longitudinally dehiscent, valves rather thin, brittle, longitudinally or obliquely veined, attenuate basally, rounded to acuminate apically, finely puberulous especially on the margins and near the base, sometimes subglabrous. *Seeds* olive-brown, 5,5–11,5 × 5–8 mm, elliptic to subcircular, usually longitudinal in the pod, smooth, compressed; areole 3–7 × 2,75–5 mm.

Found in South West Africa, Botswana, western Zambia, Rhodesia, Mozambique, the Transvaal, Swaziland, Natal and the northern Cape Province.

Larger spines elongate, 1–7 cm long, 1,5–5 mm thick, straight or almost so, not inflated; found in South West Africa and the northern Cape ....(a) var. *luederitzii*

Larger spines 3–5,5 cm long, 0,8–1,8 cm in diam. basally, inflated, usually ± uncinate-deflexed near the apex; straight elongate non-inflated spines absent; found in the Transvaal, Swaziland and Natal....(b) var. *retinens*

(a) var. **luederitzii.**

Ross & Brenan in Kew Bull. 21 : 72 (1967); Brenan in F.Z. 3,1 : 101, t.16/14 (1970); Palmer & Pitman, Trees S. Afr. 2 : 811 (1973).

A. luederitzii Engl. in Bot. Jahrb. 10 : 23, t.3B (1888) pro parte quoad specim. *Marloth 1328. A. goeringii* Schinz in Verh. Bot. Ver. Prov. Brandenb. 30 : 239 (Sept. 1888); Dinter in Feddes Repert. 15 : 79 (1917); Bak.f., Leg. Trop. Afr. 3 : 841(1930); O. B. Miller, Checklist Bech. Prot. 18 (1948); in J. S. Afr. Bot. 18 : 21 (1952). Type: Botswana, Ghanzi [Chansis], *Schinz 251* (Z, holo.!). *A. retinens* sensu O. B. Miller, Checklist Bech. Prot. 20 (1948); in J. S. Afr. Bot. 18 : 24 (1952); Tinley, Moremi Wildlife Reserve 115 (1966), non Sim. *A. uncinata* sensu O. B. Miller in J. S. Afr. Bot. 18 : 25 (1952); Story in Mem. Bot. Surv. S. Afr. 30 : 23 (1958); Boughey in J. S. Afr. Bot. 30 : 158 (1964), non Engl. *A. reficiens* sensu

Schreiber in F.S.W.A. 58 : 11 (1967) pro parte; Leistner in Mem. Bot. Surv. S. Afr. 38 : 123 (1967), non Wawra.

Found in South West Africa, Botswana, western Zambia, Rhodesia and the northern Cape Province. Occurs in savanna, bush and thornveld, particularly on Kalahari Sand.

S.W.A.—1715 (Ondangua): 24 km W. of Ndola Store on road to Ombalantu, *De Winter 3631.* 1718 (Kuring-Kuru): course of Omuramba Mpungu on Tsinsabis–Kuring–Kuru road, *De Winter 3921.* 1719 (Runtu): 8 km E. of Runtu, *Barnard 51.* 1816 (Namutoni): N. of Namutoni, *Giess & Smook 10570* (M). 1917 (Tsumeb): Otarifontein, *Dinter 5365.* 1920 (Tsumkwe): 16 km S. of Tsumkwe, *Giess 9908* (M, W). 2017 (Waterberg): Okakarara, *Liebenberg 4697.* 2117 (Otjosondu): Quickborn, *Bradfield 19.* 2118 (Steinhausen): Sturmfeld, *Walter 4084* (M). 2216 (Otjimbingwe): farm Otjiseva, *Wiss 946.* 2219 (Sandfontein): Babi Babi, *Liebenberg 4677.* 2319 (Aminuis): 25,6 km N.E. of Vogelweide, *Codd 5858.* 2419 (Aranos): farm Bethel, E. of Aranos, 9,6 km from Botswana border, *Van Vuuren & Giess 1129.*

CAPE.—2520 (Mata-Mata): Kalahari Gemsbok National Park, 14,4 km N.W. of Kwang Pan near Nossob River, *Leistner 1890.* 2524 (Vergeleë): 134,4 km W. of Mafeking, *Acocks 18767.* 2622 (Tsabong): 32 km N. of Aansluit on road to Tsabong, *Leistner 1574.* 2624 (Vryburg): near Mosito, *Breuckner 260.*

(b) var. **retinens** (*Sim*) *Ross & Brenan* in Kew Bull. 21 : 72 (1967); Brenan in F.Z. 3,1 : 101 (1970); Ross, Acacia Spp. Natal 34, fig. 2/12 (1971); Van Wyk, Trees Kruger Nat. Park 1 : 145 (1972); Ross, Fl. Natal 193 (1973); Palmer & Pitman, Trees S. Afr. 2 : 813 (1973). Type: Mozambique, "Umbeluzi and Lebombo", *Sim 6391* (whereabouts unknown, presumed lost). In the absence of a specimen, Sim l.c.: t.40 fig. A will suffice as the type.

A. retinens Sim, For. Fl. P.E. Afr. 157, t.40 fig. A (1909); Henkel, Woody Pl. Natal 230 (1934); O. B. Miller, Checklist Bech. Prot. 20 (1948); Ross in J. S. Afr. Bot. 31 : 219 (1965); Von Breitenbach, Indig. Trees S. Afr. 2 : 307 (1965). Type as above. *A. gillettiae* Burtt Davy, Fl. Transv. 2 : xvii, 343 (1932); O. B. Miller, Checklist Bech. Prot. 18 (1948); in J.S. Afr. Bot. 18 : 21 (1952); Von Breitenbach, Indig. Trees S. Afr. 2 : 306 (1965). Type: Transvaal, Mosdene, Naboomspruit, *Galpin M114* (K, holo.!; FHO!, P!, PRE!, UPS!).

Found in Mozambique, the Transvaal, Swaziland and Natal (Zululand). Occurs in dry thornveld and bushveld; often forms dense impenetrable thickets.

TRANSVAAL.—2428 (Nylstroom): Mosdene, Naboomspruit, *Galpin M114.* 2429 (Zebediela): 19 km S. of Olifants River on Chuniespoort to Burgersfort road, *Codd 1705.* 2528 (Pretoria): Rust der Winter, *Gerstner 5526.* 2530 (Lydenburg): Schoemanskloof, *Pole Evans s.n.* 2731 (Louwsburg): 1,6 km N. of Pongola River on road to Gollel, *Ross 1706* (NH, NU).

SWAZILAND.—2631 (Mbabane): Mtindekwa, *Compton 32059*. 2731 (Louwsburg): Ingwavuma Poort, *Compton 29793*.

NATAL.—2632 (Bela Vista): Ndumu Game Reserve, *Ward 2020*. 2731 (Louwsburg): near entrance to Pongola Poort, *Edwards 3260*. 2732 (Ubombo): Mkuzi Game Reserve, *Ross 1712* (NH, NU). 2831 (Nkandla): Umfolozi Game Reserve, junction of the Gqoyini jeep track and main loop road, *Downing 686* (NU). 2832 (Mtubatuba): Hluhluwe, *Ward 3933*.

Some of the short recurved spines of *A. luederitzii* var. *retinens* are invariably greatly enlarged and characteristically swollen. These distinctive swollen spines are often occupied by ants (or sometimes by other insects or small spiders) which gain access through a small aperture near the apex of the spine. There is no evidence that the enlarged spines are caused by the ants and it seems more likely that the occupation by the ants is secondary and that they merely take advantage of the presence of these suitable domatia. The swollen spines develop rapidly and soon become hard, persisting for several years.

Although the distinction between the two varieties of *A. luederitzii* rests mainly on the enlarged spines (which are not always present), there are other inconstant differential tendencies. Var. *luederitzii* is often rather taller than var. *retinens*, usually 4,5–15 m high as opposed to 1–4,5 m high in var. *retinens*. Var. *luederitzii* usually has 5–8 pairs of pinnae per leaf, while var. *retinens* has usually about 2–4 pairs per leaf in the Transvaal, and 5–13 pairs in Swaziland and Natal. The pods in var. *luederitzii* are 10–19 mm wide, while those of var. *retinens* do not seem to exceed 15 mm.

A. luederitzii var. *luederitzii* is very closely related to *A. reficiens* Wawra, but, although a few specimens from South West Africa are difficult to place, it seems preferable to maintain the two as distinct species. *A. reficiens* is distinguished by its puberulous or pulverulent indumentum, fewer pinnae pairs and few pairs of leaflets (up to 11(13) pairs per pinna) which are glabrous or almost so on the margins.

A. luederitzii var. *retinens* is usually readily distinguished from *A. reficiens* and, as it occupies a different geographical range to *A. reficiens*, is unlikely to be confused with the latter. However, the main difficulty in this complex is that while var. *retinens* and *A. reficiens* are themselves distinct, they are almost linked through typical var. *luederitzii*.

35. **Acacia reficiens** *Wawra* in Sber. Akad. Wiss. Wein 38 : 555 (1859); Oliv. in F.T.A. 2 : 348 (1871) pro parte quoad specim. *Wawra*, excl. specim. *Welwitsch*; Benth. in Trans. Linn. Soc. Lond. 30 : 505 (1875) pro parte quoad specim. *Wawra*, excl. specim. *Welwitsch*; Bak.f., Leg. Trop. Afr. 3 : 841 (1930); Torre in C.F.A. 2 : 283 (1956) pro parte; Brenan in Kew Bull. 12 : 89 (1957); in F.T.E.A. Legum.—Mimos. : 116 (1959); Schreiber in Mitt. Bot. Staats-samml. Munchen 11 : 121 (1973); Ross in Bothalia 11 : 233 (1974). Type: Angola, between Benguela and Catumbela, *Wawra* 248 (W, holo.!, K, fragm.!).

subsp. **reficiens.**

Brenan in Kew Bull. 12 : 90 (1957); Ross & Brenan in Kew Bull. 21 : 72 (1967); Schreiber in F.S.W.A. 58 : 11 (1967) pro parte. Type as above.

A. uncinata Engl. in Bot. Jahrb. 10 : 21, t. 3/A (July 1888) nom. illegit,. non *A. uncinata* Lindl. in Bot. Reg. 16 : t.1332 (1830); Schinz in Mém. Herb. Boiss. 1 : 116 (1900); Dinter in Feddes Repert. 15 : 81 (1917); Bak. f., Leg. Trop. Afr. 3 : 840 (1930). Type: South West Africa, Karibib Distr., Usakos, *Marloth* 1215 (B, holo. †; BOL!, GRA!). *A. luederitzii* Engl. in Bot. Jahrb. 10 : 23 (July 1888) pro parte quoad specim. *Marloth* 1270, excl. lectotypum; Dinter in Feddes Repert. 15 : 80 (1917) pro parte quoad specim. *Marloth* 1270, excl. lectotypum. *A.* cf. *uncinata* sensu Torre in C.F.A. 2 : 283 (1956), non Engl. *A. etbaica* sensu Torre in C.F.A. 2 : 283 (1956) saltem quoad specim. *Gossweiler* 9732.

Obconical shrub branching from or near the base or a tree to 6 m high with a flattened or somewhat rounded spreading crown. *Bark* greyish- or reddish-brown to black, rough, fissured; young branchlets greyish- or reddish-brown to purplish, sometimes as though whitewashed over a reddish or purplish background, flaking minutely, shortly puberulous to pulverulent, older branchlets glabrescent. *Stipules* spinescent, in pairs, all short, strongly hooked, 0,2–0,6 cm long, very occasionally with a few long straight spines up to 7,2 cm long intermixed, coloured like the twigs; "ant-galls" and other prickles absent. *Leaves*: petiole 0,2–1 cm long, puberulous, adaxial gland often absent; rhachis (0)0,8–2(2,4) cm long puberulous, eglandular or with a small gland at the junction of the top pinna pair; pinnae 1–4 pairs; rhachillae 0,4–1,2 cm long, puberulous; leaflets (5)7–11(13) pairs per pinna, 2–4,5 × 0,5–1,25 mm, linear to linear–oblong, apex rounded to acute, margins glabrous or almost so, sometimes a few appressed marginal cilia present, usually glabrous beneath but occasionally sparingly appressed-pubescent. *Inflorescences* capitate, on axillary peduncles, usually fascicled, seldom solitary. *Flowers* yellowish-white, sessile; peduncles 0,6–2,6 cm long, sparingly to densely puberulous; involucel basal or in lower third of peduncle. *Calyx* puberulous throughout or only apically, tube 1,1–1,6 mm long, lobes up to 0,6 mm long. *Corolla* glabrous, tube 1,8–2,8 mm long, lobes up to

1,2 mm long. *Stamen-filaments* free, up to 6 mm long; anthers with a deciduous apical gland. *Ovary* up to 1,3 mm long, very shortly stipitate. *Pods* brown or reddish-brown to purplish, 2,6–8,8 × 0,6–1,1 cm, straight, linear-oblong, longitudinally dehiscent, valves rather thin, brittle, finely longitudinally veined, attenuate basally, obtuse to acuminate apically, ± pulverulent to glabrous or almost so. *Seeds* olive-brown, 5–7,5 × 3,5–5,5 mm, elliptic to subcircular, usually longitudinal in the pod, smooth, compressed; areole 1,5–4,5 × 1–3 mm.

Found in Angola and South West Africa. Occurs in dry scrub.

S.W.A.—1712 (Posto Velho): Ombepera, *De Winter & Leistner 5483*. 1813 (Ohopoho): Kaoko–Otavi, *Abner 55*. 1814 (Otjitundua): Otjitoko, ± 56 km S. of Ohopoho, *De Winter & Leistner 5911*. 1914 (Kamanjab): Kamanjab, *De Winter & Leistner 5091*. 1915 (Okaukuejo): Otjitambi *Walter 2/92* (M). 2015 (Otjihorongo): Fransfontein, *Liebenberg 4920*. 2016 (Otjiwarongo): Omatjenne, *Keet 1683*. 2114 (Uis): Welwitschia plain, south west Brandberg, *Giess, Volk & Bleissner 6251*. 2115 (Karibib): Karibib, *Kinges 3617* (M). 2214 (Swakopmund): 59 km E. of Swakopmund on road to Usakos, *De Winter 3199*. 2215 (Trekkopje): farm Nudis, *Seydel 1679b* (K, M); farm Okongava, *Seydel 3133* (K). 2216 (Otjimbingwe): farm Auchabis 31, S. of Otjimbingwe, *De Winter 2639*. 2217 (Windhoek): hills around Windhoek, *Keet 1686*. Grid ref. unknown: Etosha District (West), S.W. of Etosha Pan, *Giess 2048*.

Subsp. *misera* (Vatke) Brenan is recorded from the Sudan, Somali Republic, Uganda and Kenya.

A. reficiens is very closely related to *A. luederitzii*. For the differences between these two species see the notes under *A. luederitzii*. *A. reficiens* subsp. *reficiens* tends to occupy a somewhat different geographical range to that occupied by *A. luederitzii* var. *luederitzii* in South West Africa; the former occurring in the western areas of the territory and the latter in the eastern, although the ranges of the two do show some overlap.

36. Acacia gerrardii *Benth.* in Trans. Linn. Soc. Lond. 30 : 508 (1875); Bak.f., Leg. Trop. Afr. 3 : 846 (1930); Brenan in Kew Bull. 12 : 369 (1958); in F.T.E.A. Legum.-Mimos. : 119 (1959). Type: Natal, locality unknown, *Gerrard 1702* (K, holo.!; BM!, TCD!)

var. gerrardii.

Brenan in Kew Bull. 12 : 369 (1958); in F.T.E.A. Legum.-Mimos. : 119, fig. 16/46 (1959); in F.Z. 3,1 : 105, t.16/17 (1970); Ross, Acacia Spp. Natal 28, fig. 2/22 (1971); Van Wyk, Trees Kruger Nat. Park 1 : 137 (1972); Ross, Fl. Natal 193 (1973); Palmer & Pitman, Trees S. Afr. 2 : 803 (1973).

A. gerrardii Benth. in Trans. Linn. Soc. Lond. 30 : 508 (1875); Burtt Davy in Kew Bull. 1908 : 157 (1908); Glover in Ann. Bolus Herb. 1 : 149, t.19/17 (1915); Bews, Fl. Natal 115 (1921); Burtt Davy, Fl. Transv. 2 : 343 (1932); Henkel, Woody Pl. Natal 230 (1934); Codd, Trees & Shrubs Kruger Nat. Park 44, figs. 36, 38a, b (1951); O. B. Miller in J.S. Afr. Bot. 18 : 21 (1952); F. White, For. Fl. N. Rhod. 85, fig. 18E (1962); Von Breitenbach, Indig. Trees S. Afr. 2 : 300 (1965). *A. hebecladoides* Harms in [Notizbl. Bot. Gart. Berl. 3 : 195 (1902) nomen nudum] Bot. Jahrb. 36 : 208 (1905); R.E. Fr., Schwed. Rhod.-Kongo-Exped. 1 : 63, t.2 fig. 4 (1914); Bak. f., Leg. Trop. Afr. 3 : 846 (1930). Type: Tanzania, "Masai Steppe" in the Kilimanjaro region, *Merker* (B, holo.†).

Tree to 12 m high or less frequently a shrub to 3 m high; trunk to 0,3 m in diam.; crown often flattened, usually irregularly so, branches ascending and spreading somewhat. *Bark* grey- or reddish-brown to blackish, rough, fissured; young branchlets sparingly to densely grey-pubescent, rarely subglabrous, epidermis usually splitting or flaking to reveal a rusty-red inner layer. *Stipules* spinescent, in pairs, usually 0,4–1,5 cm long, rarely to 6(12,5) cm long, usually straight or almost so, sometimes slightly recurved, densely pubescent when young but becoming glabrous with age; other prickles absent. *Leaves* usually borne on distinct "cushions": petiole (0,2)0,7–2(2,5) cm long, adaxial gland usually present on primary leaves but absent from secondary leaves, sometimes absent from both, variable in position, 1–1,7 × 0,7–1 mm; rhachis (0,2)2–7(10) cm long, densely pubescent, usually with a gland at the junction of the top 1–2(3) pinnae pairs; pinnae (2)5–10(12) pairs; rhachillae (0,7) 1,5–2,5(3,3) cm long, densely pubescent; leaflets (9)12–23(26) pairs per pinna, (2)3–6(7) × 0,8–1,3(1,7) mm, linear to linear-oblong, apex rounded to acute, margins with or without spreading cilia, usually glabrous or nearly so beneath, rarely pubescent. *Inflorescences* capitate, fascicled on axillary peduncles. *Flowers* white or cream, sessile; peduncles (1,5)2,2–4,5(5,2) cm long, densely grey-pubescent, eglandular or inconspicuously glandular, seldom densely glandular; involucel at or shortly above the base or sometimes to ⅓-way up the peduncle. *Calyx* glabrous apart from pubescence towards the apices of the lobes, tube 0,8,-1,9 mm long, lobes up to 0,5 mm long. *Corolla* glabrous or very slightly and inconspicuously pubescent, tube 2–3,8 mm long, lobes up to 1 mm long. *Stamen-filaments* free, up to 9,5 mm long; anthers with a deciduous apical gland. *Ovary* up to 1,5 mm long, sessile, glabrous at first but soon becoming pubescent. *Pods* mostly dark brown, densely grey-puberulous to -tomentellous, rarely subglabrous, (4,5)6,5–15,5 × (0,6)0,8–1,2 cm, usually falcate, linear or linear-oblong, longitudinally dehiscent, apex acute to acuminate, valves rather thin, brittle. *Seeds* olive-brown, 9–12 × 5–7 mm, ± quadrate, compressed; areole 5–7,5 × 3,5–5 mm.

Widespread in tropical Africa from Nigeria in the west to the Sudan in the north-east and southwards to Botswana, the Transvaal, Swaziland and Natal. Occurs in dry river valley scrub, dry thornveld, bushveld and woodland.

TRANSVAAL.—2231 (Pafuri): Kruger National Park, 4 km E. of Punda Milia, *Codd 4241.* 2329 (Pietersburg): near Mara, *Schlieben & Strey 8288.* 2330 (Tzaneen): 13 km N.W. of Duiwelskloof on road to Soekmekaar, *De Winter 7729.* 2428 (Nylstroom): Mosdene, Naboomspruit, *Galpin 474M.* 2429 (Zebediela): 3,2 km beyond Chuniespoort Hotel, *Obermeyer & Verdoorn 9.* 2431 (Acornhoek): near Klaserie, *Strey 7936.* 2529 (Witbank): hills on N. side of Loskop Dam, *Mogg 17309.* 2530 (Lydenburg): Lowveld Botanic Garden, Nelspruit, *Buitendag 717.* 2531 (Komatipoort): Barberton, *Pott 5303.*

SWAZILAND.—2631 (Mbabane): Umtintegwa, Stegi–Sipofaneni road, *Compton 28028.* 2731 (Louwsburg): near Gollel, *Compton 29095.*

NATAL.—2632 (Bela Vista): Ndumu, *Ross 1265* (NH, NU). 2731 (Louwsburg): 17,6 km from Nongoma on Magudu road, *Ross 1079* (NH, NU). 2732 (Ubombo): eastern foothills of Lebombo Mountains, *Strey 6576.* 2831 (Nkandla): Umfolozi Game Reserve, road from Tobothi to Ngoloti, *Ross 2016.* 2832 (Mtubatuba): Hluhluwe Game Reserve, *Ward 1908.* 2930 (Pietermaritzburg): Bisley, near Pietermaritzburg, *Ross 1684* (K, NH, NU). 2931 (Stanger): 4,8 km S. of Mandini on old main road, *Ross 868* (NH, NU). 3030 (Port Shepstone): Campbellton, *Rudatis 2053* (NH).

A. gerrardii is a variable species but within our area only one variant occurs, namely, var. *gerrardii.* To be precise, our plants should be referred to as subsp. *gerrardii* var. *gerrardii* as Zohary in Israel J. Bot. 13 : 39 (1964) described subsp. *negevensis* from the Negev Desert in Israel.

Although fairly widespread in our area, var. *gerrardii* is seldom dominant to the exclusion of all other *Acacia* species except very locally.

In Sekukuniland in the eastern Transvaal, specimens of *A. karroo* with densely pubescent young branchlets, leaf-rhachides, leaflets and pods occur which, superficially, bear a strong resemblance to specimens of *A. gerrardii.* However, these specimens are readily distinguished from *A. gerrardii* as they have bright yellow flowers, elliptic seeds, and lack the large cushion-like abbreviated shoots from which the leaves arise.

37. Acacia robusta *Burch.,* Trav. 2 : 442 (1824); Harv. in F.C. 2 : 281 (1862); Oliv. in F.T.A. 2 : 349 (1871) pro parte excl. specim. *Welwitsch;* Benth. in Trans. Linn. Soc. Lond. 30 : 510 (1875) pro parte excl. specim. *Welwitsch;* Burtt Davy in Kew Bull. 1908 : 159 (1908); Glover in Ann. Bolus Herb. 1 : 148, t.19/22 (1915); Bews, Fl. Natal 115 (1921); Bak.f., Leg. Trop. Afr. 3 : 841 (1930) pro parte excl. specim. Angola; Burtt Davy, Fl. Transv. 2 : 342 (1932); Henkel, Woody Pl. Natal 227 (1934); Verdoorn in Flow. Pl. S. Afr. 22 : t.851 (1942); Hutch., Botanist in S. Afr. 297, 302 (1946); Codd, Trees & Shrubs Kruger Nat. Park 48, fig. 43a, b (1951); O.B. Miller in J.S. Afr. Bot. 18 : 24 (1952); Brenan in Kew Bull. 12 : 365 (1958); Palmer & Pitman, Trees S. Afr. 163, t.ix (flower colour incorrect), 44 (1961); Letty, Wild Flow. Transv. 156, t.77/2 (1962); Gordon-Gray in Brittonia 17 : 202 (1965); Von Breitenbach, Indig. Trees S. Afr. 2 : 307 (1965); De Winter et al, 66 Transv. Trees 54 (1966); Brenan in F.Z. 3,1 : 103 (1970); Palmer & Pitman, Trees S. Afr. 2 : 807 (1973). Type: Cape Province, Kuruman Distr, Takoon [Litakun], *Burchell* 2265 (K, holo.!)

Tree to 20 m high, crown irregularly rounded or flattened and spreading, branches usually ascending. *Bark* grey to dark brown or blackish, usually rough and fissured; young branchlets robust, grey to greyish- or reddish-brown to purplish-black, lenticellate, glabrous to pubescent, smooth, not flaking off to reveal a rusty-red inner layer, eglandular. *Stipules* spinescent, mostly short and up to 1,2 cm long, sometimes longer, to 7(12,5) cm long, straight or slightly curved, whitish but becoming greyish with age; "ant-galls" and other prickles absent. *Leaves* usually borne on distinct "cushions": petiole (0,3)0,7–2,1(3,1) cm long, eglandular or with an oval gland up to 1,5 × 1,2 mm a short distance below the lowest pinna pair, glabrous to puberulous; rhachis (0)2,2–5,1(7,4) cm long, glabrous to pubescent, a small gland at the junction of the top 2 or the lowest pinna pair, sometimes between each pair when few pinnae are present; pinnae (1)3–5(7) pairs; rhachillae (1,2)2–5,6(7,4) cm long, glabrous to pubescent; leaflets (6)10–22(27) pairs per pinna, (2,5)3,5–7,5(17) × 1–3,6(8,5) mm, linear- or obovate-oblong, apex obtuse or rounded, glabrous or occasionally sparingly puberulous beneath, margins glabrous or with conspicuous or inconspicuous cilia. *Inflorescences* capitate, on axillary peduncles, usually fascicled, seldom solitary. *Flowers* white, sessile; peduncles 1,2–5,4 cm long, glabrous to shortly pubescent or puberulous, eglandular or inconspicuously glandular; involucel from near the base to just over ⅓-way up the peduncle. *Calyx* glabrous or apices of lobes sometimes sparingly pubescent, tube 1,2–2,8 mm long, lobes 0,3–0,6 mm long. *Corolla* glabrous, tube 2,6–3,4 mm long, lobes 0,4–0,8 mm long. *Stamen-filaments* free, up to 7 mm long; anthers with a deciduous apical gland. *Ovary* 0,7–1,8 mm long, glabrous, sessile. *Pods* brown to dark reddish-brown or sometimes blackish, (2,4)6–15(22) × (0,9)1,2–3,1 cm, linear, straight to falcate, apex rounded to acute, longitudinally dehiscent, valves thinly woody, brittle, smooth, ± longitudinally veined, glabrous, attenuate basally. *Seeds* dark olive- to reddish-brown, 7,5–12 × 5–9 mm, quadrate to subcircular-lenticular, smooth, compressed; areole 5,5–9 × 3,5–6 mm.

Found in Kenya, Tanzania and southwards to South West Africa (Caprivi Strip), Botswana, the Transvaal, Swaziland, Natal and the Cape Province. Three subspecies are recognized within *A. robusta*, two of which occur in our area.

Leaf-rhachis glabrous or almost so; peduncles glabrous or almost so; pods straight or slightly curved, (1,3)1,7–3,1 cm wide ....(a) subsp. *robusta*
Leaf-rhachis sparingly to densely pubescent; peduncles sparingly to densely pubescent; pods usually±falcate, 1,2–1,7 cm wide ....(b) subsp. *clavigera*

(a) subsp. **robusta.**

Brenan in F.Z. 3,1 : 103 (1970); Ross, Acacia Spp. Natal 37, fig. 2/21 (1971); Fl. Natal 193 (1973).

A. robusta Burch., Trav. 2 : 442 (1824).

Leaf-rhachides glabrous or almost so; pinnae mostly 2–5 pairs; leaflets 10–15 pairs per pinna, (2)2,5–7 mm wide. Peduncles and calyx-lobes glabrous or almost so. Pods straight or slightly curved, (1,3)1,7–3,1 cm wide.

Found in Rhodesia, Botswana, the Transvaal, Natal and the Cape. Occurs on wooded slopes, in woodland, thornveld and river valley scrub in the interior regions of southern Africa.

TRANSVAAL.—2229 (Waterpoort): 19,2 km N. of Louis Trichardt, *Gerstner 5982.* 2230 (Messina): Messina, *Rogers 21828* (K). 2329 (Pietersburg): between Vivo and Bochum, *Strey 7984.* 2427 (Thabazimbi): Leeupoort, *Rogers 19078* (K). 2428 (Nylstroom): Mosdene, Naboomspruit, *Galpin M107.*

2429 (Zebediela): 16 km W. of Steelpoort River on road to Jane Furse Hospital, *Morris 1186.* 2526 (Zeerust): near Zeerust, *Marloth 10165.* 2527 (Rustenburg): ± 1 km N. of Nooitgedacht, *J. Phillips 26.* 2528 (Pretoria): Fountains valley, *De Winter 7664.* 2529 (Witbank): Loskop Irrigation Dam, ± ½ km W. of dam wall on banks of Olifants River, *Mogg 17296.* 2530 (Lydenburg): Schoemanskloof, 33 km from Machadadorp, *Marais 903.* 2726 (Odendaalsrus): 8 km from Wolmaransstad on Leeudoringstad road, *Morris 1116.*

NATAL.—2829 (Harrismith): 4,8 km from Ladysmith on Helpmekaar road, *Ross 544* (NU). 2830 (Dundee): 22,4 km from Greytown on Muden road, *Ross 636* (NU). 2929 (Underberg): Moorleigh, *Strey 7808.* 2930 (Pietermaritzburg): Ashburton, *Ross 958* (K, NU).

CAPE.—2624 (Vryburg): Taungs, *Pole Evans sub PRE 15831.* 2723 (Kuruman): Takoon, *Burchell 2265* (K). Grid ref. unknown: Witrand, Barkly West Division, *Acocks 8503.*

The distinction between subsp. *robusta* and subsp. *clavigera* (E. Mey.) Brenan in Natal and, to a lesser extent, in the eastern Cape is sometimes not particularly clear and intermediates and mixed populations, showing various combinations of characters, occur. Consequently, in these areas difficulty is sometimes experienced in referring specimens to either subspecies with certainty.

Subsp. *robusta* typically has glabrous or subglabrous leaf-rhachides and peduncles, while the pods are usually straight or only slightly curved and 1,7–3,1 cm wide. However, some specimens, for example, *Ross 984* from the Umgeni valley near Camperdown in Natal, with large leaflets, wide pods and the general facies of subsp. *robusta*, have quite densely pubescent leaf-rhachides as in subsp. *clavigera*. Other intermediate specimens have pubescent leaf-rhachides and small leaflets as in subsp. *clavigera* but pods up to 2 cm wide as in subsp. *robusta*. Leaflets in subsp. *robusta* tend to be larger than in subsp. *clavigera* but there is complete overlap in leaflet size between the two subspecies. Similarly, there is overlap in pod width. Despite the presence of some intermediates, most specimens in our area can be readily placed either in subsp. *robusta* or in subsp. *clavigera*.

In Natal subsp. *robusta* occurs in the dry interior areas whereas subsp. *clavigera* is mainly coastal in distribution.

(b) subsp. **clavigera** (*E. Mey.*) *Brenan* in F.Z. 3,1 : 104 (1970); Ross, Acacia Spp. Natal 37, fig. 2/20 (1971); Van Wyk, Trees Kruger Nat. Park 1 : 156 (1972); Ross, Fl. Natal 193 (1973). Type: Natal, near Durban [Port Natal], *Drege* (K, iso.!, P, fragm.!)

A. clavigera E. Mey., Comm. 1 : 168 (1836); Benth. in Trans. Linn. Soc. Lond. 30 : 510 (1875); Glover in Ann. Bolus Herb. 1 : 148, t.19/13 (1915); Harms in Engl., Pflanzenw. Afr. 3,1 : 366 (1915); Bews, Fl. Natal 114 (1921); Brenan in Kew Bull. 12 : 365 (1958); in F.T.E.A. Legum.–Mimos. : 118 (1959); F. White, For. Fl. N. Rhod. 86, fig. 18F (1962). Type as above. *A. sambesiaca* Schinz in Denkschr. Math.-Nat. Kl. Acad. Wiss Wein 78 : 50 (1905); Gomes e Sousa, Pl. Menyharth 69 (1936). Type: Mozambique, *Menyharth 1003* (W, holo., Z, iso!). *A. hirtella* sensu Sim, For. Fl. P.E. Afr. 57, t.35 fig. A (1909), non E. Mey. *A. robusta* sensu Codd, Trees & Shrubs Kruger Nat. Park 48, fig. 43a, b (1951). *A. clavigera* E. Mey. subsp. *clavigera*, Brenan in Kew Bull. 12 : 367 (1958); in F.T.E.A. Legum.-Mimos. : 118 (1959); Gordon-Gray in Brittonia 17 : 202 (1965); Von Breitenbach, Indig. Trees S. Afr. 2 : 302 (1965) pro parte excl. syn. *A. grandicornuta*. Type as for *A. clavigera*.

Leaf-rhachides sparingly to densely pubescent; pinnae often 5-7 pairs; leaflets mostly 12–25 pairs per pinna, (1)1,5–3,5 mm wide. Peduncles and calyx-lobes ± densely pubescent. Pods usually ± falcate, mostly 1,2–1,7 cm wide.

Found in South West Africa (Caprivi Strip), Botswana, Zambia, Rhodesia, Mozambique, the Transvaal, Swaziland, Natal and the eastern Cape Province. Occurs in wooded grassland, woodland, thornveld and river valley scrub; often found along river banks or on the margins of pans.

S.W.A.—1725 (Livingstone): Mpilila Island, confluence of Zambesi and Chobe Rivers, *Killick & Leistner 3425*.

TRANSVAAL.—2229 (Waterpoort): Waterpoort, N. side of Soutpansberg range, *Codd 4341*. 2230 (Messina): banks of Nzhele River at Tshipise store, *De Winter 7739*. 2231 (Pafuri): Kruger National Park, Pafuri Camp, bank of Pafuri River, *Codd & Dyer 4629*. 2431 (Acornhoek): Kruger National Park, 2,4 km E. of Skukuza, *Codd & Verdoorn 5488*. 2531 (Komatipoort): Kruger National Park, 13,6 km W. of Crocodile River bridge Camp, *Codd 4403*.

SWAZILAND.—2631 (Mbabane): Big Bend, *Compton 30297*. 2632 (Bela Vista): Mbuluzi Poort, *Compton 30730*.

NATAL.—2632 (Bela Vista): Ndumu Game Reserve, edge of Fantana Pan, *Ward 3705*. 2731 (Louwsburg): 22,4 km from Mkuze on Nongoma road, *Ross 1039* (NU). 2732 (Ubombo): banks of Mkuzi River, Mkuze, *Galpin 133117*. 2831 (Nkandla): Umfolozi Game Reserve, *Feely 57*. 2832 (Mtubatuba): Hluhluwe Game Reserve, edge of Amanzimyama stream, *Ward 1478*. 2930 (Pietermaritzburg): 5,2 km S. of Marianhill, *Moll 1910*. 2931 (Stanger): 3,2 km S. of Mandini on old main road, *Ross 873* (NH, NU). 3030 (Port Shepstone): bank of Izotsha River near Port Shepstone, *Nicholson 155* (NH).

CAPE.—3129 (Port St Johns): 8 km S.W. of Tombo Post Office, *Codd 9289*. 3227 (Stutterheim): Kei River bridge, *Sim 20193*. 3228 (Butterworth): Mpetu, *Flanagan 1311*.

The Zulu name for subsp. *clavigera* is "umNgamanzi" which means "fond of water".

Subsp. *usambarensis* (Taub.) Brenan has not been recorded so far in our area.

38. Acacia grandicornuta *Gerstn.* in J.S. Afr. Bot. 4 : 55, fig. 1 (1938); O.B. Miller, Checklist Bech. Prot. 19 (1948); Codd, Trees & Shrubs Kruger Nat. Park 44, fig.

38c, d, e (1951); O.B. Miller in J.S. Afr. Bot. 18 : 21 (1952); Brenan in F.Z. 3,1 : 104 (1970); Ross, Acacia Spp. Natal 29, fig. 2/19 (1971); Van Wyk, Trees Kruger Nat. Park 1 : 140 (1972); Ross, Fl. Natal 193 (1973); Palmer & Pitman, Trees S. Afr. 2 : 805 (1973). Syntypes: Natal, flowered at Emkunzana and Mkuzi Drift between Nongoma and Magudu, 6 Jan. 1936, *Gerstner 2870* (BOL!); fruits found at same places and at lower Pongola, 13 May 1936, *Gerstner 2870* (BOL!).

A. clavigera E. Mey. subsp. *clavigera* pro parte quoad syn. *A. grandicornuta* sensu Von Breitenbach, Indig. Trees S. Afr. 2 : 302 (1965).

Tree to 12 m high, crown ± rounded, often irregularly so, branches usually ascending. *Bark* grey to brownish-black or black, rough, longitudinally fissured; young branchlets greyish to reddish-brown or purplish, lenticellate, glabrous, smooth, not flaking off to reveal a rusty-red inner layer. *Stipules* spinescent, in pairs, 0,2–10(14) cm long, straight or slightly curved, sometimes deflexed, typically stout and slightly swollen, whitish but becoming greyish with age; other prickles absent. *Leaves*: petiole 0,4–2 cm long, glabrous, eglandular or with a rounded to elliptic gland up to 1,5 × 1 mm, variable in position; rhachis 0–3,6 cm long, glabrous or subglabrous, a small gland at the junction of the top or top 1–3 pinnae pairs; pinnae (1)2–3(5) pairs; rhachillae 0,9–3,6(5,2) cm long, glabrous; leaflets 7–18 pairs per pinna, 3–8 × (1)1,5–2,5(3,2) mm, linear to linear- or obovate-oblong, apex obtuse or rounded, glabrous throughout or rarely margins minutely ciliate, lateral nerves invisible or slightly conspicuous beneath. *Inflorescences* capitate, on axillary peduncles, fascicled. *Flowers* white, sessile; peduncles 1,2–2,5 cm long, glabrous or occasionally very sparingly puberulous, eglandular; involucel ⅓–½-way up the peduncle. *Calyx* glabrous throughout or apices of lobes occasionally with a few hairs, tube 0,8–1,3 mm long, lobes up to 0,3 mm long. *Corolla* glabrous, tube 1,6–2,2 mm long, lobes up to 0,6 mm long. *Stamen-filaments* free, up to 5 mm long; anthers with a deciduous apical gland. *Ovary* up to 1 mm long, glabrous, very shortly stipitate. *Pods* brown to reddish-brown or purplish, (4,8)6–12,8(15,2) × 0,6–1,1 cm, falcate or occasionally straightish, sometimes irregularly constricted between some of the seeds, longitudinally dehiscent, valves rather thin, brittle, finely longitudinally veined, glabrous, attenuate basally. *Seeds* olive-brown, 6–10 × 5–7 mm, ± oblong, smooth, compressed; areole 4–6 × 2,5,4–5 mm.

Found in south-eastern Botswana, Rhodesia, Mozambique, the Transvaal, Swaziland and Zululand. Occurs in dry thornveld and woodland; favours deep soils. Sometimes dominant over fairly large areas and forming thickets.

TRANSVAAL.—2230 (Messina): Gaandrik area, *Van der Schijff 5261*. 2328 (Baltimore): near Villa Nora, *Acocks 8817*. 2426 (Mochudi): 4 km S.W. of Rooibokkraal, *Codd 8654*. 2429 (Zebediela): 19,2 km S. of Olifants River on Chuniespoort to Burgersfort road, *Codd 1706*. 2430 (Pilgrim's Rest): 20,8 km S. of Steelpoort, *Morris 1184*. 2431 (Acornhoek): Kruger National Park, 4,8 km E. of Skukuza, *Codd & Verdoorn 5493*. 2531 (Komatipoort): Kruger National Park, 4,8 km from Skukuza on Crocodile Bridge road, *Story 3940*. 2731 (Louwsburg): 3,2 km N. of Pongola River on road to Gollel, *Ross 1707* (NH, NU).

SWAZILAND.—2631 (Mbabane): Big Bend, *Compton 30293*. 2731 (Louwsburg): Nsoko, *Compton 30320*.

NATAL.—2632 (Bela Vista): Ndumu Game Reserve, *Ward 4519*. 2731 (Louwsburg): Mkuzi Drift, 16 km S. of Magudu on Nongoma road, *Ross 1107* (NH, NU). 2732 (Ubombo): 1 km N. of Mkuzi River on road to Candover, *Ross 1641* (NH, NU). 2831 (Nkandla): Umfolozi Game Reserve, *Ward 3319* (NH).

A. grandicornuta is closely related to *A. robusta* Burch. in its wide sense and a case could perhaps be made out for placing it under the latter species. However, *A. grandicornuta* is a very distinctive taxon and it seems best maintained as a distinct species.

The more slender branches and the narrow falcate pods distinguish *A. grandicornuta* from *A. robusta* subsp. *robusta*, while the glabrous leaf-rhachides, typically fewer pinnae and leaflets, and the narrow pods distinguish *A. grandicornuta* from *A. robusta* subsp. *clavigera*.

39. Acacia arenaria *Schinz* in Mém. Herb. Boiss. 1 : 105 (1900); Harms in Engl., Pflanzenw. Afr. 3,1 : 376 (1915); Dinter in Feddes Repert. 15 : 78 (1917): Bak.f., Leg. Trop. Afr. 3 : 839 (1930); Hutch., Botanist in S. Afr. 523 (1946); O.B. Miller in J. S. Afr. Bot. 18 : 18 (1952); Torre in C.F.A. 2 : 282 (1956); Brenan in F.T.E.A. Legum.-Mimos. : 126, fig. 17/55 (1959); Schreiber in F.S.W.A. 58 : 7 (1967); Brenan in F.Z. 3,1 : 106 (1970); Palmer & Pitman, Trees S. Afr. 2 : 805 (1973). Syntypes: South West Africa, Ovamboland, Olukonda-Oshiheke, *Schinz 2071* (Z!); Amboland, "Omatope", *Schinz 2072* (Z!)

A. hermannii Bak.f. in J. Bot., Lond. 67 : 198 (1929); Bak.f., Leg. Trop. Afr. 3 : 847 (1930); Brenan, Checklist Tang. Terr. 337 (1949); Wild, S. Rhod. Bot.

Dict. 47 (1953). Type: Tanzania, Singida District, near Manyugi [? Manyigi], *B.D. Burtt 1379* (BM, holo.!; FHO!, K!, Z!). *A. rufobrunnea* N.E. Br. in Kew Bull. 1909 : 107 (1909). Type: Botswana, Ngamiland, Botletle valley, *E.J. Lugard 245* (K, holo.!). *A. seyal* Del. var. *multijuga* sensu O.B. Miller in J. S. Afr. Bot. 18 : 24 (1952) quoad specim. *Pole Evans 3251*, non Schweinf.

Shrub or small tree 1,5–9 m high, branching from near the base. *Bark* pale to dark grey- or reddish-brown to black, rough; young branchlets pale to dark grey- or reddish-brown, often as though whitewashed over a purplish background, sparingly to densely puberulous. *Stipules* spinescent, in pairs, up to 6 cm long, straight, slender, usually the same colour as the young branchlets and sparingly to densely puberulous; other prickles absent. *Leaves*: petiole 0,3–1,4 cm long, adaxial gland often absent, when present usually just below the lowest pinna pair, slightly raised, round to oval, 0,6–1,4 × 0,5–1,2 mm; rhachis 5–21,5 cm long, sparingly to densely pubescent, with a gland at the junction of each of the top 1–3 pairs of pinnae; well-developed leaves with 15–36 pairs of pinnae (reduced leaves with as few as 9 pairs of pinnae sometimes also present); rhachillae 1,2–4,1 cm long; leaflets (11)18–26(32) pairs per pinna, 1,5–5 × 0,5–1,1 mm, linear to linear-oblong, apex rounded to subacute, margins glabrous or ciliate, glabrous abaxially. *Inflorescences* capitate, solitary or more usually fascicled, often crowded into an irregular terminal "raceme". *Flowers* white or pale pink, sessile; peduncles (0,6)1,1–2,2(2,9) cm long, sparingly to densely pubescent, usually glandular; involucel at or above the middle or at the apex of the peduncle. *Calyx* cupular, white or pink, sparingly to densely pubescent, especially at apices of lobes, tube 0,3–0,9 mm long, lobes up to 0,6 mm long. *Corolla* tubular, glabrous, tube up to 2,5 mm long, lobes up to 0,7 mm long, usually 2–4 times as long as the calyx. *Stamen-filaments* free, up to 5 mm long; anthers 0,2–0,25 mm across, with a deciduous apical gland. *Ovary* up to 1 mm long, glabrous. *Pods* pale to dark reddish-brown, 8-22 × 0,5–0,8 cm, arcuate, sometimes slightly constricted between the seeds, flat, longitudinally dehiscent; valves thin, glabrous to sparingly pubescent and glandular. *Seeds* olive-grey, 7–9 × 3–4,5 mm, oblong or quadrate, smooth, compressed; areole 3,5–4,5 × 1,5–2,25 mm.

Occurs in Tanzania, Angola, South West Africa, Botswana and Rhodesia. Found usually in drier types of mixed woodland, grassland or scrub, sometimes with *Colophospermum mopane*.

S.W.A.—1715 (Ondangua): Okapsa, road dam near Ondangua, *De Winter & Giess 6898*. 1721 (Mbambi): between Shamvura and Kangongo, *De Winter 4209*. 1813 (Ohopoho): 32 km S. of Kaoko Otavi on road to Ombombo, *De Winter & Leistner 5609*. 1821 (Andara): road from Andara to Bagani, *Merxmuller & Giess 1992*. 1824 (Kachikau): E. Caprivi Strip, 62 km from Katima Mulilo on road to Linyanti, *Killick & Leistner 3127*.

A. davyi bears a superficial resemblance to *A. arenaria*. For the differences between the two species see the notes under *A. davyi*.

40. **Acacia rehmanniana** *Schinz* in Bull. Herb. Boiss. 6 : 525 (1898); Burtt Davy in Kew Bull. 1908 : 159 (1908); Glover in Ann. Bolus Herb. 1 : 151 (1915); Eyles in Trans. Roy. Soc. S. Afr. 5 : 362 (1916); Bak.f., Leg. Trop. Afr. 3 : 838 (1930); Burtt Davy, Fl. Transv. 2 : 343 (1932); O.B. Miller, Checklist Bech. Prot. 20 (1948); in J.S. Afr. Bot. 18 : 24 (1952); Pardy in Rhod. Agric. J. 51 : 376 (1954); F. White, For. Fl. N. Rhod. 86, fig. 18i (1962); Von Breitenbach, Indig. Trees S. Afr. 2 : 304 (1965); Brenan in F.Z. 3,1 : 99 (1970); Palmer & Pitman, Trees S. Afr. 2 : 783 (1973). Type: Transvaal, Streydpoort, Makapansberge, *Rehmann* 5517 (Z, holo.!).

Shrub or tree to 10 m high with a somewhat flattened spreading crown. *Bark* reddish-brown to black, rough; young branchlets clothed with dense spreading hairs, the hairs golden at first but turning greyish-white, sometimes becoming glabrescent with age, the epidermis later peeling or flaking off to reveal a rusty-red inner layer. *Stipules* spinescent, in pairs, 0,5–5,4 cm long, straight or almost so, sparingly to densely pubescent with spreading hairs especially basally, becoming glabrescent with age; other prickles absent. *Leaves* densely clothed with spreading hairs, at first golden but turning greyish-white, sometimes becoming glabrescent: petiole 0,1–0,5 mm long, adaxial gland present, yellowish to reddish-brown, oval, somewhat flattened, up to 1,5(3) × 1(2) mm; rhachis 2,4–12 cm long, a small gland usually at the junction of the top 1–3 and bottom 1–6 pinnae pairs, sometimes present between each pinna pair or absent from some; well-developed leaves of mature shoots with 15–44 pinnae pairs (reduced leaves with as few as 8 pinnae pairs are sometimes

also present); rhachillae 0,6–2,8 cm long; leaflets 24–48 pairs per pinna, 1,2–2,8 × 0,4–0,9 mm, linear to linear-oblong, apex rounded to acute, margins with conspicuous spreading hairs especially apically. *Inflorescences* capitate, on axillary peduncles, solitary or more usually fascicled, aggregated into a terminal "raceme", each fascicle of heads subtended by young to scarcely developed leaves. *Flowers* white or cream, sessile; peduncles 0,5–2 cm long, clothed with dense spreading hairs, the hairs golden at first but soon turning greyish-white, eglandular; involucel below the middle of the peduncle. *Calyx* sparingly to densely pubescent towards the apices of the lobes, sometimes a small basal tuft of hairs also present, tube 1,6–2,2 mm long, lobes up to 0,7 mm long. *Corolla* glabrous apart from the sparingly to densely pubescent lobes, tube 2–2,8 mm long, lobes up to 1,2 mm long. *Stamen-filaments* free, up to 7 mm long; anthers with a deciduous apical gland. *Ovary* up to 1,6 mm long, sessile, sparingly to densely pubescent. *Pods* grey- or reddish-brown to olive, (2,9) 7–10(14) × 1,1–1,8(2,2) cm, linear-oblong, straight or almost so, longitudinally dehiscent, flattened, not constricted between the seeds, slightly venose, glabrous to sparingly pubescent. *Seeds* olive- or reddish-brown, 4,5–8 × 4–7 mm, ellipsoid to suborbicular, scarcely compressed; areole 2,5–4,5 × 1,25–2,8 mm, completely closed.

Found in southern Zambia, northern Botswana, Rhodesia and the Transvaal. Occurs in wooded grassland and bushveld; sometimes near rivers or streams.

TRANSVAAL.—2329 (Pietersburg): Mara, *Rogers 21706*. 2330 (Tzaneen): Duiwelskloof, *Schweickerdt 1556*. 2429 (Zebediela): 32 km S. of Pietersburg on road to Chuniespoort, *Van Vuuren 1603*.

A. rehmanniana has been often confused with *A. sieberana* DC. in the past. However, *A. rehmanniana* is a distinctive species, easily separated from *A. sieberana* by the more numerous pairs of closely arranged pinnae, the much smaller and thinner textured pods, the involucels in the lower half of the peduncle, and the way in which the capitula are clustered in the axils and aggregated into terminal "racemes".

A. rehmanniana was reported to occur in Natal in F.Z. 3,1 : 100 (1970), but this record was probably based on a very poor specimen (*Sidey 3574*) of *A. sieberana* DC. var. *woodii* (Burtt Davy) Keay & Brenan.

41. **Acacia sieberana** *DC.*, Prodr. 2 : 463 (1825); Oliv. in F.T.A. 2 : 347 (1871); Benth. in Trans. Linn. Soc. Lond. 30 : 503 (1875);

Bak.f., Leg. Trop. Afr. 3 : 836 (1930);
Gilbert & Boutique in F.C.B. 3 : 166 (1952);
Brenan in F.T.E.A. Legum.–Mimos. : 127,
fig. 17/57 (1959); F. White, For. Fl. N. Rhod.
84, fig. 17K (1962); Troupin in Bull. Jard.
Bot. Brux. 35 : 449 (1965); Brenan in F.Z.
3,1 : 107 (1970); Ross in Bothalia 11 : 128
(1973). Type: Senegal, *Sieber* 43 (G, holo.;
K!).

var. **woodii** (*Burtt Davy*) *Keay & Brenan*
in Kew Bull. 5 : 364 (1951); Pardy in Rhod.
Agric. J. 48 : 406 (1951); Palgrave, Trees
Cent. Afr. 254 (1956); Torre in C.F.A.
2 : 281 (1956); Brenan in F.T.E.A. Legum.-
Mimos. : 128 (1959); Palmer & Pitman,
Trees S. Afr. 163, t.x, 45 (1961); Von Breiten-
bach, Indig. Trees S. Afr. 2 : 292 (1965);
De Winter et al, 66 Transv. Trees 56 (1966);
Brenan in F.Z. 3,1 : 108, t.16/18 (1970);
Ross, Acacia Spp. Natal 40, fig. 2/8 (1971);
Flow. Pl. Afr. 40 : t.1653 (1972); Van Wyk,
Trees Kruger Nat. Park 1: 162 (1972); Ross,
Fl. Natal 193 (1973); in Bothalia 11 : 131
(1973): Palmer & Pitman, Trees S. Afr. 2 :
779 (1973); Schreiber in Mitt. Bot. Staats-
samml. Munchen 11 : 122 (1973). Type:
Natal, Estcourt Distr., between Estcourt and
Colenso, *Wood* 3528 (K, holo.!, NH!).

A. amboensis Schinz in Mém. Herb. Boiss. 1 : 105
(1900); Dinter in Feddes Repert. 15 : 78 (1917);
Bak.f., Leg. Trop. Afr. 3 : 838 (1930); O.B. Miller,
Checklist Bech. Prot. 16 (1948); in J.S. Afr. Bot.
18 : 18 (1952). Syntypes: Angola, "Omupanda in
Uukuanjama", *Wulfhorst* 2 (Z!); precise locality
unknown, "Kilevi am Kunene", *Schinz* 763 (Z!).
A. lasiopetala sensu Burtt Davy in Kew Bull. 1908 :
158 (1908); Glover in Ann. Bolus Herb. 1 : 149,
t.19/23 (1915); Bews, Fl. Natal 115 (1921); Henkel,
Woody Pl. Natal 226 (1934); Stapleton, Common
Transv. Trees 6 (1937); Suesseng. & Merxm. in
Proc. & Trans. Rhod. Sci. Assoc. 43 : 16 (1951), non
Oliv. *A.* cf. *hebeclada* sensu Wood in Trans. S. Afr.
Phil. Soc. 18 : 152 (1908), non DC. *A. hebeclada*
sensu Bews, Fl. Natal 114 (1921), non DC. *A. woodii*
Burtt Davy in Kew Bull. 1922 : 332 (1922); Fl.
Transv. 2 : 344 (1932); Steedman, Trees etc. S.
Rhod. 15 (1933); Hutch., Botanist in S. Afr. 394
(1946); West in Rhod. Agric. J. 47 : 208 (1950);
Codd, Trees & Shrubs Kruger Nat. Park 51, figs. 44a,
b, 45 (1951); O.B. Miller in J. S. Afr. Bot. 18 : 26
(1952). Type as for *A. sieberana* var. *woodii*. *A.
vermoesenii* De Wild., Pl. Bequaert 3 : 68 (1925);
Brenan, Checklist Tang. Terr. 335 (1949). Type:
Congo–Brazzaville, Boma, *Vermoesen* 1378 (BR,
holo.!). *A. sieberana* var. *vermoesenii* (De Wild.)
Keay & Brenan in Kew Bull. 5 : 364 (1951); Pardy in
Rhod. Agric. J. 48 : 406 (1951); Brenan in F.T.E.A.
Legum.-Mimos. : 128 (1959); in F.Z. 3,1 : 108 (1970).
Type as for *A. vermoesenii*. *A.* cf. *stolonifera* sensu
Torre in C.F.A. 2 : 282 (1956)) quoad specim.

Gossweiler 11035 (K) pro parte, non Burch. *A.
sieberana* subsp. *vermoesenii* (De Wild.) Troupin var.
vermoesenii, Troupin in Bull. Jard. Bot. Brux. 35 :
455 (1965); Schreiber in F.S.W.A. 58 : 12 (1967).
Type as for *A.* ` *vermoesenii*. *A. sieberana* subsp.
vermoesenii var. *woodii* (Burtt Davy) Keay & Brenan,
Troupin in Bull. Jard. Bot. Brux. 35 : 457 (1965).
Type as for *A. woodii*.

Tree to 18 m high, branches usually
spreading, crown typically flattened and
spreading or umbrella-shaped, sometimes
branches ascending and crown conical,
particularly in young plants. *Bark* yellowish-
or greyish-brown, rough, typically papery,
the outer layers flaking off to expose a
yellowish inner layer, sometimes $\pm$ smooth
and not flaking; young branchlets yellowish-
or greyish-brown, smooth or the outer bark
flaking away to reveal a yellowish inner layer,
glabrous to densely tomentose, indumentum
golden to greyish-white. *Stipules* spinescent,
in pairs, 0,3–7,8 cm long, straight or almost
so, whitish, glabrous to densely tomentose;
other prickles absent. *Leaves* glabrous to
densely tomentose, indumentum golden to
greyish-white: petiole 0,2–1 cm long, adaxial
gland present, often just below the lowest
pinna pair, yellowish to reddish-brown, up
to 2 × 1,5 mm; rhachis 2,5–12,8 cm long,
a small gland present at the junction of the
top 1–6 pinnae pairs only, between each
pair, or absent from some; pinnae (4)8–30
pairs; rhachillae 0,8–4(6,5) cm long; leaflets 13–45 pairs per pinna, 1,2–6,5 × 0,5–
1,6 mm, linear to linear-oblong, apex
rounded to acute, margins with or without
cilia, usually glabrous beneath, midrib and
sometimes a few lateral nerves often fairly
prominent beneath. *Inflorescences* capitate,
on axillary peduncles, solitary or fascicled.
Flowers pale yellowish-white, sessile or very
shortly pedicelled; peduncles 1,4–5,8 cm
long, glabrous to densely tomentose, indumentum golden to greyish-white, eglandular; involucel apical or in the upper half of
the peduncle. *Calyx* sparingly pubescent
throughout or glabrous except for pubescence towards the apices of the lobes, tube
1,6–2,4 mm long, lobes up to 0,8 mm long.
Corolla glabrous throughout or apices of
lobes sparingly pubescent, tube 2,4–3,6 mm
long, lobes up to 2 mm long. *Stamen-filaments* free, up to 9 mm long; anthers with a
deciduous apical gland. *Ovary* sessile, up to
2,1 mm long, glabrous. *Pods* pale to dark
yellowish- or reddish-brown, 5–19,5 ×

1,3–3,3 cm, straight or slightly curved but sometimes ± falcate, somewhat flattened or at times oval in cross-section, thick and almost woody in texture when dry, tardily dehiscent, often only dehiscing on the ground, smooth, glossy and glabrous or subglabrous to densely pubescent especially on margins and stipe, indumentum frequently golden. *Seeds* olive to grey-brown, 7–12 × 5–8 mm, elliptic to subcircular, slightly compressed, smooth; areole 5–9,5 × 4–6 mm.

Found from the Sudan and Ethiopia southwards to South West Africa, Botswana, the Transvaal, Swaziland and Natal. Occurs in grassland, savanna, woodland and thornveld; favours deep soils or shallow soil overlying shale.

S.W.A.—1714 (Ruacana Falls): 20,8 km W. of Ruacana at the Kunene River, *Giess & Wiss 3346*. 1718 (Kuring-Kuru): 9,6 km E. of Tondoro mission on road to Lupala, *De Winter 3965*.

TRANSVAAL.—2229 (Waterpoort): Soutpan 193, *Obermeyer, Schweickerdt & Verdoorn 12*. 2230 (Messina): Palmaryville, *Van den Berg 23*. 2329 (Pietersburg): Blouberg Police Station kloof, *Strey & Schlieben 8595*. 2330 (Tzaneen): Elim, *Obermeyer 853*. 2430 (Pilgrim's Rest): Strydom Tunnel, *Strey 7901*. 2527 (Rustenburg): between Rustenburg Kloof and Rustenburg town, *De Winter 8637*. 2528 (Pretoria): Great North road, near Onderstepoort, *Verdoorn 2359*. 2529 (Witbank): Loskop Dam Nature Reserve, *Mogg 17513*. 2530 (Lydenburg): Schoemans-kloof, *Smuts 306*. 2531 (Komatipoort): Kruger National Park, 3,2 km N. of Pretorius Kop, *Codd & De Winter 5167*. 2730 (Vryheid): Piet Retief, *Pole Evans 10*.

SWAZILAND.—2531 (Komatipoort): Ngonini bushveld, *Compton 28210*.

NATAL.—2729 (Volksrust): S. of Newcastle, *Dyer & Verdoorn 2377*. 2730 (Vryheid): Vryheid, *Hovy sub NH 54544*. 2731 (Louwsburg): 20,8 km from Louwsburg on Magudu road, *Ross 1233* (NH, NU). 2828 (Bethlehem): Royal Natal National Park, bank of Tugela River, *J. D. Ross 1* (NU). 2829 (Harrismith): base of Van Reenen's Pass, *Schweickerdt 945*. 2830 (Dundee): 8 km from Dundee on Wasbank road, *Ross 1256* (NH, NU). 2831 (Nkandla): Nkwaleni valley, *McClean 975*. 2832 (Mtubatuba): Hluhluwe Game Reserve, *Ward 2701*. 2929 (Underberg): 4,8 km S. of Estcourt, *Ross 739* (NH, NU). 2930 (Pietermaritzburg): Golf road, Pietermaritzburg, *Ross 458* (K, NH, NU). 2931 (Stanger): New Guelderland, *Stewart 84*. 3030 (Port Shepstone): Umkomaas, *Ross 793* (NU).

The indumentum on the young branchlets, leaves, peduncles and pods in typical var. *woodii* is villous and distinctly golden, especially when young. However, there is considerable variation in the degree of pubescence and in the colour of the indumentum. The young branchlets vary from glabrous or subglabrous to densely pubescent, while the indumentum varies in colour from golden to greyish-white. Often in amongst the greyish-white indumentum a faint tinge of gold is visible, especially at the base of the petioles, spines and peduncles. Occasionally the indumentum on the old shoots is greyish-white while on young branchlets from the same plant the indumentum is golden.

Var. *woodii* typically has yellowish-brown papery bark that flakes off irregularly, whence the common names "Paperbark Thorn" and "Papierbas-doring". Often, however, the bark does not flake or peel off at all. Both forms frequently occur in the same population. There seems to be a tendency for specimens with non-peeling bark to have glabrous or glabrescent branchlets and in Natal these plants are often, but by no means always, confined to the coastal areas. These plants sometimes seem to occupy slightly different ecological conditions to those occupied by the densely pubescent specimens with papery peeling bark. However, these ± glabrous specimens seem best regarded as ± glabrous forms of var. *woodii* since they are otherwise indistinguishable from specimens of var. *woodii*.

The typical forms of var. *woodii* and var. *vermoesenii* appear to be different but as they are linked by so many and varied intermediates, and as difficulty is frequently experienced in attempting to refer specimens to one variety or the other with certainty, it is no longer considered desirable to maintain both of these varieties. Consequently, var. *vermoesenii* has been relegated to synonymy under var. *woodii*.

The pods of *A. sieberana*, which dehisce very slowly and usually only after they have been shed from the plant, are eaten by livestock and by game and are reputed to cause tainting of dairy produce when eaten in quantity. The leaves contain large amounts of prussic acid particularly when wilted and have caused some stock losses.

In common with some other *Acacia* species, a few flowers sometimes develop in the involucel on the peduncle, giving the appearance of a smaller secondary capitulum below the main one.

A. sieberana has in the past been confused with *A. rehmanniana* Schinz. The differences between the two species are given under *A. rehmanniana*.

42. **Acacia hebeclada** *DC.*, Cat. Hort. Monsp. 73 (1813); Prodr. 2 : 461 (1825); Brenan in F.Z. 3,1 : 109 (1970); Palmer & Pitman, Trees S. Afr. 2 : 775 (1973). Type: Cape Province, Kuruman Distr., between Kuruman [New Litakun] and the Matlowing [Moshowing] River, *Burchell 2267* (G, holo.; K!, PRE!).

Low spreading shrub or tree 0,4–9 m high, shrubs branching near ground-level or with aerial stems arising from a subterranean stolon, often forming large dense thickets; crown in arborescent forms rounded with the branches sometimes drooping to the ground or flattened and spreading somewhat. *Bark* dark grey- or reddish-brown to blackish, fissured; young branchlets pale to dark grey- or reddish-brown to purplish, sometimes appearing as though whitewashed over a purplish background, sparingly to densely grey-pubescent or tomentose, sometimes becoming glabrescent with age, epidermis sometimes splitting and flaking minutely. *Stipules* spinescent, in pairs, straight to arcuate or hooked, either short and 0,4–± 1,5 cm long or up to 6 cm long and then straight or very slightly recurved apically, greyish-white to reddish, pubescent when young but becoming subglabrous or glabrous; "ant-galls" and other prickles absent. *Leaves:* petiole 0,3–0,9(1,9) cm long, densely spreading-pubescent, adaxial gland present or absent, rounded or elongated along the petiole, up to 2 × 1,6 mm, often just below lowest pinna pair; rhachis (0)1–4,5(6) cm long, sparingly to densely spreading-pubescent, often with minute scattered glands, a small gland often at the junction of the top 1–3 pinnae pairs and sometimes the lowest pair; pinnae (1)4–9(12) pairs; rhachillae (0,4)1,2–2(3,4) cm long, sparingly to densely spreading-pubescent; leaflets 7–18 pairs per pinna, (1,5)2,5–5(7) × (0,75)0,9–1,5(2) mm, linear to linear- or obovate-oblong, apex rounded to subacute, margins usually with spreading cilia, sometimes cilia inconspicuous or absent, usually glabrous beneath but sometimes sparingly appressed pubescent. *Inflorescences* capitate, on axillary peduncles, usually fascicled and scattered along the shoots, seldom solitary. *Flowers* yellowish-white or cream, sessile; peduncles 0,5–2(4) cm long, sparingly to densely spreading-pubescent, hairs on peduncle often equalling or longer than its diameter, usually eglandular, becoming glabrescent and thick and ± woody with age; involucel at or shortly above the base or occasionally to almost halfway up the peduncle. *Calyx* glabrous except for the apices of the lobes which are sparingly to densely pubescent, tube 1,2–1,8 mm long, lobes 0,5–0,8 mm long. *Corolla* glabrous throughout or apices of lobes sparingly to densely pubescent, tube 2,5–3,2 mm long, lobes 0,2–0,6 mm long. *Stamen-filaments* free, up to 6,5 mm long; anthers with a deciduous apical gland. *Ovary* up to 1,6 mm long, shortly stipitate, glabrous. *Pods* yellowish- to greyish-brown, 4–21 × 1–4,5 cm, up to 1,5 cm thick, straight or nearly so, seldom ± falcate, turgid, oblong-ellipsoid, cylindric or fusiform, apex rounded to acute or distinctly

pointed, erect or pendulous, finally longitudinally dehiscent; valves thick, hard, densely tomentellous outside, longitudinally nerved, sometimes very conspicuously so, sparingly to densely glandular. *Seeds* olive- to reddish-brown, on a long funicle, 6–15 × 4–11 mm, subcircular-lenticular, sometimes scarcely compressed; areole 5–12 × 2–7 mm.

Found in Angola, South West Africa, Botswana, western Zambia, western Rhodesia, the Transvaal, western Orange Free State and northern Cape Province.

Pods erect, 4–12(14,5) × 1,5–4,5 cm, straight or nearly so:

    Pods (1)1,5–2,3(2,5) cm wide, seed 7–9 × 5,5–7 mm; flowering peduncles mostly 0,5–1,5 cm long; low spreading shrub or tree, often gregarious; widespread in arid areas in South West Africa, the Transvaal, Orange Free State and the northern Cape........(a) subsp. *hebeclada*

    Pods 2,5–3,5(4,5) cm wide; seed 10–15 × 7–10,5 mm; flowering peduncles mostly 1,5–3 cm long, large riverine shrub or a tree occuring in north-east South West Africa (Caprivi Strip)..(b) subsp. *chobiensis*

Pods pendulous (at least when mature), 10–21 × 1–1,5 cm, often somewhat falcate; tree or large shrub occurring in northern South West Africa......................(c) subsp. *tristis*

(a) subsp. **hebeclada.**

Schreiber in Mitt. Bot. Staatssamml. Munchen 6 : 251 (1966); in F.S.W.A. 58 : 8 (1967).

A. hebeclada DC., Cat. Hort. Monsp. 73 (1813); Prodr. 2 : 461 (1825); Benth. in Hook., Lond. J. Bot. 1 : 499 (1842); in Hook., Lond. J. Bot. 5 : 95 (1846); Harv. in F.C. 2 : 280 (1862); Benth. in Trans. Linn. Soc. Lond. 30 : 504 (1875); Engl. in Bot. Jahrb. 10 : 22 (1888); Marloth in Trans. S. Afr. Phil. Soc. 5 : 270 (1889); Schinz in Mém. Herb. Boiss. 1 : 112 (1900); Burtt Davy in Kew Bull. 1908 : 158 (1908); N.E. Br. in Kew Bull. 1909 : 107 (1909); Glover in Ann. Bolus Herb. 1 : 149, t.19/19, 19/20 (1915); Harms in Engl., Pflanzenw. Afr. 3,1 : 355 (1915); Marloth, Fl. S. Afr. 2 : 53, t.18c (1925); Milne-Redhead in Kew Bull. 1937 : 416 (1937); Hutch., Botanist in S. Afr. 398, 399 (1946); Von Breitenbach, Indig. Trees S. Afr. 2 : 294 (1965); Leistner, Mem. Bot. Surv. S. Afr. 38 : 123, t.25 (1967); Brenan in F.Z. 3,1 : 109 (1970) pro parte. Type: Cape, Kuruman Distr., between Kuruman [New Litakun] and the Matlowing [Moshowing] River, *Burchell* 2267 (G, holo; K!, PRE!). *A. stolonifera* Burch., Trav. 2 : 241 (1824); Harv. in F.C. 2 : 284 (1862); Engl. in Bot. Jahrb. 10 : 22 (1888); Marloth in Trans. S. Afr. Phil. Soc. 5 : 268 (1889); Schinz in Mém. Herb. Boiss. 1 : 115 (1900); Sim, For. Fl. Cape Col. 211 (1907); Burtt Davy in Kew Bull. 1908 : 160 (1908); in Kew Bull. 1922 : 331 (1922); Fl. Transv. 2 : 340, fig. 58 (1932); Bak.f., Leg. Trop. Afr. 3 : 836 (1930); Hutch., Botanist in S. Afr. 398, 418, 433, 632 (1946); O.B. Miller in J. S. Afr. Bot. 18 : 25 (1952).

Type: Cape Province, Hay Distr., Ongeluk's-fontein, between Griquatown and Kuruman, *Burchell* 2138 (K, holo.!). *A. hebeclada* var. *stolonifera* (Burch.) Dinter in Feddes Repert. 15 : 80 (1917). Type as for *A. stolonifera* Burch.

Found in South West Africa, Botswana, the Transvaal, western Orange Free State and northern Cape Province. Occurs in dry thornveld, bushveld, savanna, open grassland and along dry watercourses; often on sand or alluvium.

S.W.A.—1824 (Kachikau): Caprivi Strip, near Linyanti, *Killick & Leistner 3164*. 1917 (Tsumeb): between Grootfontein and Otavi, *Werdermann & Oberdieck 2392*. 1920 (Tsumkwe): Tsumkwe, 251 km E. of Grootfontein, *Story 6207*. 2115 (Karibib): Usakos, *Marloth 1261*. 2116 (Okahandja); Okahandja, *Dinter 266*. 2117 (Otjosondu): Quickborn, *Bradfield 17*. 2215 (Trekkopje): Okongava, *Seydel 3158* (K). 2216 (Otjimbingwe): farm Friedenau, in Khomas Hochland, 25,6 km S.W. of Windhoek, *De Winter 2591*. 2217 (Windhoek): Windhoek, *Rogers 29513* (GRA). 2317 (Rehoboth): Uhlenhorst, farm Sib, *De Hoogh s.n.* (K). 2519 (Koes): near Eindpaal, near Auob River, *Kinges 2001*. 2616 (Aus): Kuibis ravine, *Pearson 8009* (K). 2718 (Grunau): Klein Karas, *Dinter 4988*.

TRANSVAAL.—2329 (Pietersburg): near Bandelierkop, *Strey 7938*. 2428 (Nylstroom): 3,2 km S. of Warmbad on road to Pretoria, *Codd 908*. 2526 (Zeerust): Marico, *J. van der Merwe 24*. 2528 (Pretoria): Grosvenor Square, Pretoria, *Repton 3520*. 2626 (Klerksdorp): 16 km from Lichtenburg on Mafeking road, *Morris 1167*. 2627 (Potchefstroom): 12,8 km E. of Potchefstroom, *Prosser 1865*. 2725 (Bloemhof): Boskuil, *Sutton 121*.

O.F.S.—2726 (Odendaalsrus): Bothaville, *Goossens 1229*. 2825 (Boshof): Smitskraal, *Burtt Davy 9893*. 2925 (Jagersfontein): on path to Fauresmith in poort, *Verdoorn sub Henrici 2406*. 2926 (Bloemfontein): 19 km from Bloemfontein on road to Dealesville, *Potts 2939*.

CAPE.—2620 (Twee Rivieren): Kalahari Gemsbok National Park, 14,4 km N.E. of Twee Rivieren in bed of Nossob River, *Leistner 1493*. 2623 (Morokweng): Lolwanen, *Burtt Davy 13850* (K). 2624 (Vryburg): Taungs, *Pole Evans sub PRE 15830*. 2722 (Olifantshoek): 3,2 km S. of Olifantshoek, *Leistner & Joynt 2735*. 2723 (Kuruman): between Kuruman and the Matlowing River, *Burchell 2267*. 2823 (Griekwastad): Ongeluk's-fontein, between Griquatown and Kuruman, *Burchell 2138* (K). 2824 (Kimberley): Barkly West, *Marloth 956*. Grid ref. unknown: Vryburg district, along Molopo River, *De Winter 7832*.

A. hebeclada subsp. *hebeclada* occurs either as a low spreading shrub branching near ground level or with stems arising from a subterranean stolon, or as a tree. Plants often form large dense impenetrable thickets several metres in diameter. In the Kalahari these thickets are frequently utilized by the large carnivores for shelter during the heat of the day.

Specimens of subsp. *hebeclada* are usually readily distinguished from subsp. *chobiensis* which just reaches our area. However, some specimens, but not from our area, are difficult to place. In addition to the smaller pods, subsp. *hebeclada* tends to have shorter internodes $\pm$ 1–1,5 cm long, smaller leaves and shorter peduncles. Subsp. *hebeclada* has a different distributional range and different ecological preferences to subsp. *chobiensis*.

A. hebeclada subsp. *hebeclada* when in pod cannot be mistaken. Flowering specimens, however, particularly those with inadequate habit notes, can be confused with *A. luederitzii* Engl. The pods in subsp. *hebeclada* often persist on the plant for more than one season so that it is possible to find flowers and the previous season's pods on a plant together. The flowers are sometimes produced when the plants are leafless.

(b) subsp. **chobiensis** (*O.B. Miller*) *Schreiber* in Mitt. Bot. Staatssamml. Munchen 6 : 251 (1966); in F.S.W.A. 58 : 9 (1967). Type: Botswana, Serondella, *O.B. Miller B/1069* (K, holo.!).

A. stolonifera Burch. var. *chobiensis* O.B. Miller in J. S. Afr. Bot. 18 : 25 (1952). Type as above. *A. hebeclada* sensu F. White, For. Fl. N. Rhod. 85, fig. 18A (1962); Brenan in F.Z. 3,1 : 109 (1970) pro parte.

Found in north-eastern South West Africa (Caprivi Strip), northern Botswana, western Rhodesia and Zambia. Occurs on alluvial soils along river banks, often partially submerged.

S.W.A.—1719 (Runtu): Okavango River at Runtu, *De Winter & Marias 4915*. 1724 (Katima Mulilo): Lisikili, 24 km E. of Katima Mulilo, *Codd 7091*. 1821 (Andara): Andara, *Merxmuller & Giess 1946*.

Subspecies *chobiensis* occurs as a tree or large shrub. The growth form is often hemi-spherical, the lower branches touching the ground.

(c) subsp. **tristis** *Schreiber* in Mitt. Bot. Staatssamml. Munchen 6 : 251 (1966); in F.S.W.A. 58 : 9 (1967); Ross in Bothalia 11 : 131 (1973); Schreiber in Mitt. Bot. Staatssamml. Munchen 11 : 117 (1973). Type: Angola, Huila district, between Lopolo e Ferrão da Sola, *Welwitsch 1829* (LISU, holo.; BM!, K!).

A. tristis Welw. ex Oliv. in F.T.A. 2 : 349 (1871) nom. illegit., non *A. tristis* R. Graham in Bot. Mag. 62 : t.3420 (1835); Benth. in Trans. Linn. Soc. Lond. 30 : 510 (1875); Bak.f., Leg. Trop. Afr. 3 : 838 (1930) pro parte excl. specim. *Munro 453*; Torre in C.F.A. 2 : 282 (1956). Type as above. *A. hebeclada* sensu Harms in Warb., Kunene Samb. Exped. 243 (1903).

Found in southern Angola and in northern South West Africa. Occurs in open bush, mixed woodland and thornveld; usually on sandy soils.

S.W.A.—1715 (Ondangua): 24 km W. of Ndola Store on road to Ombalantu, *De Winter 3632*; Oshikango, *De Winter & Giess 7055*. 1718 (Kuring-Kuru); Makambu Camp, 32,8 km W. of Kuring-Kuru on road to Katwitwi, *De Winter & Marais 5015*. 1813 (Ohopoho): near Ohopoho, *De Winter & Leistner 5305*; 8 km N. of Ombombo, *De Winter & Leistner 5887*.

Subspecies *tristis* grows as a tree with a rounded or flattened and somewhat spreading crown, or as a shrub. The narrow pendulous pods readily distinguish subsp. *tristis* from subsp. *hebeclada* and subsp. *chobiensis*.

FIG. 11.—**Acacia stuhlmannii.** 1, part of leafy branch, × 1; 2, gland on petiole, × 6; 3, leaflet × 6; 4, leaflet from lowest part of pinna, × 6, all from *Burtt* 5503; 5, part of flowering branch, × 1; 6, flower-head, × 2; 7, bract subtending flower, × 6; 8, flower-bud, × 6; 9, flower, × 6; 10, calyx, opened out, × 6; 11, corolla, opened out, × 6; 12, stamens, × 6; 13, anther with part of filament, × 12; 14, ovary, × 6; 15, pod, × 1; 16, seed, × 2, all from *Burtt* 3400. Reproduced by permission of the Editor of Flora of Tropical East Africa.

43. Acacia stuhlmannii *Taub.* in Engl., Pflanzenw. Ost-Afr. C : 194, t.21, E,F (1895); Harms in Notizbl. Bot. Gart. Berl. 4 : 196, fig.1 (1906); Harms in Engl., Pflanzenw. Afr. 3,1 : 355, fig. 213 (1915); Bak.f., Leg. Trop. Afr. 3 : 836 (1930); Brenan, Checklist Tang. Terr. 334 (1949); in F.T.E.A. Legum.-Mimos. : 131, figs. 17/61, 19 (1959); in F.Z. 3,1 : 110, t.16/20 (1970). Syntypes: Tanzania: Dar es Salaam, *Stuhlmann* 6755 (B, †); Pangani, *Stuhlmann* 282 (B, †); Tanga, *Volkens* 189 (B, †); Amboni, *Holst* 2202 (K!, Z!); Tanzania/Kenya, Lake Jipe, *Volkens* 2383 (B, †).

Obconical shrub to 2,5 m high, branching from the base. *Bark* olive- to dark reddish-brown; young shoots with spreading golden villous hairs 1,5–3 mm long, hairs later going greyish-white; branchlets olive- to reddish-brown, longitudinally wrinkled, with large yellowish somewhat transversely elongated lenticels, becoming glabrescent with age. *Stipules* spinescent, in pairs, 0,7–6,5 cm long, straight or slightly deflexed, sparingly to densely pubescent especially basally, becoming glabrescent with age; "ant-galls" and other prickles absent. *Leaves* sparingly to densely clothed with spreading whitish hairs: petiole 0,3–1 cm long, adaxial gland usually present on primary leaves but absent from secondary leaves; rhachis 2–5,5(7,5) cm long, a gland often at the junction of the top 1–3 pinnae pairs or absent; pinnae 4–12 pairs (occasionally leaves on juvenile non-flowering shoots may have up to 17 pairs of pinnae and rhachides up to 9 cm long); rhachillae 1,1–1,9 cm long; leaflets 6–15 pairs per pinna, 2–5,5 × 0,6–1,5 mm, linear-oblong, apex subacute to acute, glabrous beneath, margins usually with appressed or spreading cilia. *Inflorescences* capitate, on axillary peduncles, fascicled or solitary, usually produced before the leaves. *Flowers* white, sessile; peduncles 0,6–1,6 cm long, densely hairy or tomentose, eglandular; involucel basal or in lower half of the peduncle, up to 3 mm long. *Calyx* sparingly to densely pubescent apically, tube 1,6–2,2 mm long, lobes up to 0,4 mm long. *Corolla* sparingly to densely pubescent apically, tube 2,2–3,2 mm long, lobes up to 0,5 mm long. *Stamen-filaments* free, up to 6 mm long; anthers with a deciduous apical gland. *Ovary* up to 1,4 mm long, sessile or very shortly stipitate. *Pods* dark grey-brown to black, densely clothed with long spreading greyish-white hairs, 2,2–6,5 × (0,9)1,1–2,2 cm, straight or somewhat curved, fibrous, indehiscent, usually much attenuate basally, the stipe up to 2 cm long. *Seeds* olive, 4,5–9 mm in diam., ellipsoid to subglobose, minutely punctate; areole 4–6 × 2,5–4,5 mm, often indistinct.

Found in Somalia, Kenya, Tanzania, south-eastern Botswana, southern Rhodesia and the northern Transvaal. Occurs in low lying areas or flats, often in heavy alluvial soils; sometimes in *Colophospermum mopane* veld.

TRANSVAAL.—2229 (Waterpoort): Dongola Reserve, Farm Breslau 619, *Codd & Dyer 3787*; *Codd 4323*. 2328 (Baltimore): N. of Villa Nora, *Strey & Schlieben 8674*; 12,8 km from Villa Nora on road to Marnitz, *Van der Schijff 5331A*. Grid ref. unknown: Messina Distr., Limpopo River, *Smuts & Gillett 4035*.

A. stuhlmannii has a very disjunct distribution in Africa occurring in Somalia, Kenya and Tanzania in the north and in Botswana, Rhodesia and the Transvaal in the south. Although rather a variable species in East Africa, *A. stuhlmannii* is relatively uniform in our area being always shrubby and with short peduncles 0,6–1,6 cm long, and small pods 2,2–6,5 cm long and up to 2,2 cm wide.

44. Acacia redacta *J. H. Ross* in Bothalia 11 : 231 (1974). Type: Cape, Namaqualand, 22,4 km N. of Stinkfontein on way to Jenkinskop, *Werger* 1518 (PRE, holo.!; K!).

Much branched shrub 0,3–0,6 m high. *Bark* dark grey-brown, flaking minutely; young branchlets reddish-brown, densely and persistently appressed pubescent, with numerous small conspicuous dark purplish glands scattered in amongst the hairs. *Stipules* spinescent, in pairs, 0,8–1,4 cm long, reddish-brown, straight or often deflexed, slender. *Leaves*: petiole short, mostly 2–6 mm long, grey-puberulous, adaxial gland absent; pinna 1 pair; rhachillae 0,4–1,8 cm long, subglabrous or puberulous; leaflets 2–4 pairs per pinna, 2–5,5 × 1,2–3,5 mm, oblique, oblong or elliptic or ± subrotund, apex rounded or obtuse, sparingly to densely appressed pubescent on both surfaces or on the lower surface only, midrib and lateral nerves not visible or inconspicuous beneath, with minute reddish glands at the point of attachment of the leaflets. *Inflorescences* apparently capitate, reduced, on axillary peduncles; flowers 2–4 per inflorescence or rarely apparently solitary. *Flowers* apparently pinkish; peduncles 2–6 mm long, conspicuously glandular, usually densely appressed pubescent. *Calyx* cupular, 0,8–1,2 mm long, densely pubescent. *Corolla* 4–6 mm long, lobes tinged with red or purple apically, densely appressed pubescent or tomentellous. *Stamen-filaments* 14–17 mm long, pinkish, shortly connate basally and tubular for ± 2 mm. *Ovary* ± 1,5 mm long, shortly stipitate. *Pods* pinkish-brown, 2,6–3,2 × 0,9–1,1 cm, linear-oblong, straight, 1–2-seeded, densely appressed grey-puberulous, with small conspicuous dark purplish sessile glands among the hairs, dehiscent, acute or acuminate apically. *Seeds* not seen.

Restricted to the north western Cape Province. Occurs most frequently on schistoid granite ridges.

CAPE.—2817 (Vioolsdrif): 20,8 km N. of Stinkfontein (—CB), *Leistner 3401*; 22,4 km N. of Stinkfontein on way to Rosyntjieberg (—CB), *Werger 428*; 22,4 km N. of Stinkfontein on way to Jenkinskop (—CB), *Werger 1518*; near Rosyntjieberg (±—CB), *Van der Merwe 1828*.

A. redacta differs from all other species in our area in having reduced inflorescences which contain only 2–4 flowers or, rarely, the flowers appear to be solitary. This reduction in the number of flowers per "head" is coupled with an increase in the size of the individual flowers, the flowers being larger than in most other species of African *Acacia. A. redacta* is also unusual in that the stamen-filaments are shortly united basally. The pods dehisce longitudinally from the apex downwards and the two valves diverge. More material is required and the flower colour needs confirmation.

A. redacta appears to have a very restricted distribution. The plants are of very small stature but this is thought to be the result of the extreme and inhospitable environment they occupy, an environment which is conducive to shrubbiness.

Insufficiently known species

45. **Acacia** sp.

Shrub; young branchlets glabrous, yellowish- to reddish-brown, epidermis splitting and peeling away to reveal a rusty-red inner layer. *Stipules* spinescent, in pairs, up to 3,2 cm long, straight. *Leaves* glabrous throughout: petiole 0,5–2 cm long, primary leaves with a large gland up to 4 × 2 mm situated immediately above the pulvinus, secondary leaves without a gland; rhachis 0–2 cm long, projecting at the end in a short rigid persistent deflexed hook or claw, with a slightly raised gland at the point of attachment of each pinna pair; pinnae 1–2 pairs, primary leaves with 2 pairs, secondary leaves with 1 pair; leaflets 5–8 pairs per pinna, 8–18 × 3–9 mm, oblong to obovate-oblong, acute or rounded apically, sometimes slightly emarginate, midrib and usually several other basal and lateral nerves ± raised and conspicuous on the lower surface, glabrous throughout. *Inflorescences* capitate, solitary or fascicled on axillary peduncles, terminal. *Flowers* sessile; peduncles 3,2–4,2 cm long, glabrous but with few to many scattered sessile glands; involucel at or near the middle of the peduncle. *Calyx* very shortly lobed, ± 2,25 mm long, glabrous except for apices of lobes. *Corolla* ± 3 mm long, glabrous except for apices of lobes. *Stamens* free, up to 6 mm long; anthers with a deciduous apical gland. *Ovary* shortly stipitate, ± 1,25 mm long; style ± 4 mm long. *Pods* immature, 3,5–7 × 0,9–1 cm, slightly curved to falcate, apex obtuse or acute, probably longitudinally dehiscent, valves distinctly venose, glabrous, very sparingly glandular. *Seeds* immature.

Known from a single gathering from the eastern Transvaal.

TRANSVAAL.—2430 (Pilgrim's Rest): 10 km N. of Burgersfort (–CB), *W.F. Stuurman W 34*.

There are 9 sheets of *Stuurman* W 34 and they exhibit quite a range of variation in leaflet shape and size. Unfortunately there is no information on flower colour, the ecological preferences of the plant, or the species with which it was associated. It is not known whether this was a single isolated plant or whether it is locally common.

Stuurman W 34 does not match material of any other *Acacia* species. Superficially the specimens look like extremely robust specimens of *A. karroo*. However, they differ from *A. karroo* in having much larger leaflets with ± raised and conspicuous venation on the lower surface, longer peduncles and slightly wider pods than in "typical" *A. karroo*. Although there is no information on flower colour, it appears as though the flowers were bright golden yellow as in *A. karroo*.

The robust leaves and large leaflets of *Stuurman* W 34 are reminiscent of *A. robusta* subsp. *robusta* but the largest leaflets are larger than those recorded in *A. robusta* and, in addition, differ in having some ± raised and conspicuous nerves on the lower surface. In *A. robusta* the flowers are pale yellowish-white and the involucel is near the base to just over ⅓ way up the peduncle, while in *Stuurman* W 34 the involucels are at or near the middle of the peduncle. The pods are quite different and much smaller than those of *A. robusta*.

From the limited material and information available it appears that *Stuurman* W 34 is most closely allied to *A. karroo* or to a related species. More material and detailed field observations and notes are required to enable the identity and affinities of *Stuurman* W 34 to be established with certainty. The possibility that *Stuurman* W 34 is of hybrid origin cannot be excluded at this stage.

46. Acacia schlechteri *Harms* in Bot. Jahrb. 51 : 367 (1914); Ross in Bothalia 11 : 234 (1974). Type: Mozambique, Ressano Garcia, *Schlechter* 11901 (B, holo. † ; Z, iso!).

? Tree. *Branchlets* dark grey or blackish-brown, young extremities very sparingly pubescent. *Prickles* paired, recurved, short. *Leaves*: petioles, rhachides and rhachillae sparingly pubescent: petiole and rhachis together ± 4–6 cm long; pinnae 2–5 pairs; rhachillae 1–4 cm long; leaflets 3–6 pairs per pinna, very variable in size and shape, 7–14 × 4–8 mm, obliquely obovate-oblong or oblong, asymmetric basally, discolorous, glabrous throughout or with few marginal cilia, sometimes with a small basal tuft of hairs to one side of the midrib on the lower surface. *Inflorescences* spicate; spikes 5–9 cm long, axes sparingly pubescent. *Calyx* glabrous or with occasional scattered hairs, sometimes sparingly pubescent. *Pods* unknown.

Young, in Candollea 15 : 123 (1955), recorded *A. schlechteri* from the eastern Transvaal but the specimen cited, and on which he based his description of the pod, namely *Rogers* 18537 (PRE), is in fact *Albizia anthelmintica* (A. Rich.) Brongn. The precise identity of *A. schlechteri* is not absolutely certain but no specimen matching the type or fitting the description has been recorded from our area.

A. schlechteri is possibly a local variant of *A. goetzei* Harms subsp. *goetzei*. More material from the type locality and field studies are required in an attempt to establish the identity of *A. schlechteri*.

47. Acacia callicoma *Meisn.* in Hook., Lond. J. Bot. 2 : 104 (1843). Type unknown.

? Tree. *Branches* unarmed, glabrous, lenticellate. *Stipules* not evident. *Leaves*: petiole and rhachis together 10 cm long, terete, glabrous, petiole with an oblong sessile adaxial gland; pinnae 8 pairs; rhachillae 2,5–3,75 cm long; leaflets 12–14 pairs per pinna, 14–16 × 6–8 mm, semihastate-oblong, acute apically, rounded-truncate basally, margins ciliate. *Inflorescence* a terminal panicle, branchlets patent; peduncles 4–8 mm long, solitary or rarely paired. *Flowers* in round heads, sessile, hermaphrodite. *Calyx* and corolla infundibuliform, green, puberulous outside. *Calyx* half as long as the corolla, shortly 5–lobed. *Corolla* 5 mm long, 5–lobed. *Stamens* ± 20; filaments united below, reddish, exceeding the corolla. *Pods* unknown.

Meisner based his description of *A. callicoma* on a specimen without fruits and of unknown provenance, "loco natali incerto, aut Port Natal, aut Ins. S.

Yago Promontorii viridis", seen in Krauss's herbarium. No type specimen was cited and no specimen bearing this name or any further reference to the species in literature has been traced. Neither Harvey in Fl. Cap., vol. 2 (1862), nor Bentham in his revision of Mimoseae in Trans. Linn. Soc. Lond., Vol. 30 (1875) mention *A. callicoma*. The description suggests that the specimen Meisner saw was not an *Acacia* but, in the absence of a specimen, the precise identity of *A. callicoma* remains unknown.

Naturalized and cultivated species

The distribution of most of the naturalized and cultivated species in our area is much wider than indicated here but few specimens are available for citation.

48. Acacia farnesiana (*L.*) *Willd.*, Sp. Pl. 4 : 1083 (1806); Benth. in Fl. Austral. 2 : 419 (1864); Oliv. in F.T.A. 2 : 346 (1871); Benth. in Trans. Linn. Soc. Lond. 30 : 502 (1875); Bak.f., Leg. Trop. Afr. 3 : 835 (1930); Brenan, Checklist Tang. Terr. 334 (1949); Gilbert & Boutique in F.C.B. 3 : 164 (1952); Torre in C.F.A. 2 : 278 (1956); Keay in F.W.T.A. ed.2, 1 : 499 (1958); Brenan in F.T.E.A. Legum.-Mimos. : 111, fig. 16/38 (1959); in F.Z. 3,1 : 111 (1970); Ross, Fl. Natal 193 (1973); in Bothalia 11 : 465 (1975). Type: Aldinus, Exact. Descr. Rar. Pl.Romae Hort. Farnesiano 4 (1625) (lecto.!).

Mimosa farnesiana L., Sp. Pl. 1 : 521 (1753); Ross in Bothalia 11:471 (1975). Type as above.

Shrub or small tree to 4 m high; young branchlets grey to reddish-brown or purplish, epidermis not obviously peeling off, with numerous somewhat transversely elongated lenticels, glabrous or almost so. *Stipules* spinescent, in pairs, usually short, up to 1,5(3) cm long, straight, slender, never inflated; "ant-galls" and other prickles absent. *Leaves* bipinnate: petiole sparingly to ± densely pubescent, usually with a small gland; rhachis sparingly to ± densely pubescent, often with a small gland below the junction of the top pinna pair; pinnae 2–7 pairs; leaflets 10–21 pairs, 2–7 × 0,75–1,75 mm, midrib and lateral nerves visible and somewhat raised beneath, glabrous throughout or with few inconspicuous marginal cilia. *Inflorescences* capitate, on axillary peduncles, solitary or in pairs or threes. *Flowers* bright golden-yellow, sweetly scented; peduncles sparingly to ± densely pubescent basally, sparingly glandular; involucel apical. *Calyx* and corolla glabrous except for the apices of the lobes. *Pods* dark brown to blackish, 4–7,5 × 0,9–1,5 cm, straight or curved, subterete and turgid, glabrous, tardily dehiscent, finely longitudinally striate. *Seeds* chestnut-brown, 7–8 × 5,5 mm, elliptic, thick, only slightly compressed; areole 6–7 × 4 mm.

Probably a native of tropical America, doubtfully so in Africa. Widely introduced in the tropics and often becoming wild. Only planted or an escape from cultivation in our area.

TRANSVAAL.—2528 (Pretoria): Wonderboom Poort, at the footpath of stones laid across the Aapies River, *Gerstner 5519.* 2531 (Komatipoort): Komatipoort, *Pole Evans sub PRE 18281.*

NATAL.—2930 (Pietermaritzburg): Durban Botanic Gardens, *Ross 1714.*

CAPE.—2824 (Kimberley): Kimberley, *Wilman sub BOL 15701.*

A. farnesiana is grown for ornament and for its fragrant flowers which are used to make perfume. The pods of *A. farnesiana* are very distinctive and enable the species to be easily recognized. In the absence of pods, it will be helpful to recall that no other African *Acacia* has the following combination of characters: absence of "ant-galls", leaflets with the lateral nerves raised and somewhat prominent beneath, apical involucels, and bright golden-yellow flowers in non-paniculate heads.

A further very significant distinguishing feature of *A. farnesiana* is that the anthers lack, even in bud, the small deciduous apical gland which is present in all of the indigenous capitate-flowered acacias occurring in our area.

49. **Acacia mearnsii** *De Wild.*, Pl. Bequaert. 3 : 61 (1925); Brenan in F.T.E.A. Legum.-Mimos. : 95, fig. 15/21 (1959); Brenan & Melville in Kew Bull. 14 : 37 (1960); Tindale in Beadle, Evans & Carolin, Handb. Vasc. Pl. Sydney Distr. & Blue Mts. 231 (1962); Brenan in F.Z. 3,1 : 111 (1970); Court in Willis, Handb. Pl. Victoria 2 : 243 (1972); Tindale in Beadle, Evans & Carolin, Fl. Sydney Region 275 (1972); Ross, Fl. Natal 193 (1973); in Bothalia 11 : 465 (1975). Type: Kenya, near Thika, *Mearns 1092* (BR, lecto.; BM.!).

A. decurrens sensu Bak.f., Leg. Trop. Afr. 3 : 853 (1930) saltem pro parte, non Willd. sensu stricto. *A. mollissima* sensu auct. mult., Benth. in Hook., Lond. J. Bot. 1 : 385 (1842); Burtt Davy, Fl. Transv. 2 : 345 (1932); Brenan, Checklist Tang. Terr. 333 (1949); Salter in Adamson & Salter, Fl. Cape Penins. 454 (1950); F. White, For. Fl. N. Rhod. 82 (1962), non Willd.

Unarmed tree up to 15 m high with a conical or rounded crown; bark grey-brown to blackish, smooth or rough on very old trunks; young branchlets angular; all parts (except flowers) ± densely pubescent or puberulous, indumentum on young parts often golden. *Leaves* bipinnate: petiole 1–2,5 cm long, often with a gland above; rhachis usually 4–12 cm long, with numerous raised glands all along its upper surface both at and between the junctions of the pinnae pairs; pinnae 8–21 pairs; leaflets 15–70 pairs, 1,5–4 × 0,5–0,75 mm, linear-oblong, appressed-pubescent or glabrous beneath, margins usually with cilia. *Inflorescences* capitate, in terminal panicles. *Flowers* pale yellow, fragrant; peduncles 2–6 mm long. *Calyx* sparingly pubescent especially towards the apices of the lobes. *Corolla* glabrous or almost so. *Pods* (1,6)3–10 × 0,5–0,8 cm, jointed, almost moniliform, ± grey- puberulous, dehiscing longitudinally along one margin only, straight or slightly curved. *Seeds* black, ± 5 ×3,5 mm, elliptic, compressed, smooth; caruncle conspicuous; areole 3,5 × 2 mm.

Introduced from Australia and now widespread in parts of the Transvaal, Swaziland, Natal and the Cape Province.

TRANSVAAL.—2430 (Pilgrim's Rest): Belvedere 26N forest, ± 29 km from Pilgrim's Rest, *Davidson & Mogg 33515.* 2528 (Pretoria): Waterkloof, Pretoria, *Schlieben 10090.* 2530 (Lydenburg): 26,4 km S.S.E. of Lydenburg, *D. Morris 58.*

SWAZILAND.—2631 (Mbabane): 10 km W. of Mbabane on main road to Transvaal, *Brummitt 12423* (K). 2731 (Louwsburg): 3,2 km E. of Goedgegun, *Ross 1767.*

NATAL.—2929 (Underberg): Cathkin Peak area, *Strey 7809.* 2930 (Pietermaritzburg): Winterskloof, *Ross 2129; Ross 2131.* 3030 (Port Shepstone): Mtwalume, *Wood 10589.*

CAPE.—3318 (Cape Town): Ida's Valley, bottom of Hell's Hoogte Pass, Stellenbosch, *Thompson 837.* 3319 (Worcester): Bain's Kloof, *White 5657.* 3225 (Somerset East): Glen Avon Falls area, *P. T. van der Walt 190.* 3326 (Grahamstown): Grahamstown, *Troughton 49.*

A. mearnsii is the well-known Black Wattle which is economically important for the tannin content of the bark. The wood is used for firewood and for building.

By a strange mischance *A. mearnsii*, the earliest valid name for this Australian species, is based on a specimen collected in Kenya and thought by De Wildeman to be endemic there.

50. **Acacia dealbata** *Link*, Enum. Hort. Berol. 2 : 445 (1822), non *A. dealbata* A. Cunn. (1825); Benth. in Fl. Austral. 2 : 415 (1864); in Trans. Linn. Soc. Lond. 30 : 497 (1875); Burtt Davy, Fl. Transv. 2 : 346 (1932); Brenan, Checklist Tang. Terr. 332 (1949); in F.T.E.A. Legum.-Mimos. : 50 (1959); Salter in Adamson & Salter, Fl. Cape Penins. 455 (1950); Tindale in Beadle, Evans & Carolin, Handb. Vasc. Pl.

Sydney Distr. & Blue Mts. 231 (1962); Brenan in F.Z. 3,1 : 112 (1970); Court in Willis, Handb. Pl. Victoria 2 : 245 (1972); Tindale in Beadle, Evans & Carolin, Fl. Sydney Region 273 (1972); Ross, Fl. Natal 193 (1973); in Bothalia 11 : 465 (1975). Type: a plant cultivated at Berlin.

A. decurrens var. *mollis* Lindl. in Bot. Reg. 5 : t.371 (1819).

Unarmed shrub or tree up to 15 m high with a conical or rounded crown; bark grey-brown to blackish, smooth or rough on very old trunks; young branchlets usually densely short-pubescent, rarely subglabrous, ± grey-pruinose, indumentum grey or sometimes yellowish at first and then grey. *Leaves* bipinnate, often glaucous; petiole 0,5–2 cm long, eglandular; rhachis 2,5–10 cm long, with a raised gland on its upper surface at the junction of each pair of pinnae, but without other glands in between the pinnae pairs as in *A. mearnsii*; pinnae (5)10–26 pairs; leaflets in 17–50 pairs, 2–5,5 × 0,4–0,7 mm, linear-oblong, sparingly to ± densely pubescent or glabrous beneath, margins with or without cilia. *Inflorescences* capitate, panicled or racemose. *Flowers* bright yellow; peduncles densely pubescent, up to 6 mm long. *Calyx* and corolla glabrous except for apices of the lobes. *Pods* 3–8 × 0,7–1,3 cm, not or only slightly moniliform, dehiscing longitudinally along one margin only, straight or slightly curved. *Seeds* brown to blackish-brown, 5–6 × 3–3,5 mm, elliptic, compressed, smooth; caruncle conspicuous; areole 3,5–4 × 0,75–1,5 mm.

Introduced from Australia.

TRANSVAAL.—2526 (Zeerust): Swartruggens, *Sutton 1031*. 2528 (Pretoria): E. of Pretoria, *Kinges 1781*. 2529 (Witbank): Loskop Dam Reserve, *Theron 1752*. 2531 (Komatipoort): Kruger National Park, Pretoriuskop–Seekoeigat, *Van der Schijff 3177*. 2628 (Johannesburg): Melville Koppies, Johannesburg, *MacNae 1161* (BOL). 2629 (Bethal): Ermelo, *Burtt Davy 594*.

NATAL.—2730 (Vryheid): near Grootspruit, *Strey 8053*. 2930 (Pietermaritzburg): Hilton Road, *Ross 2105*; farm Mountain Glen, Dargle, *Taat 1025*.

LESOTHO.—2927 (Maseru): Roma, *Ruch 16*; Mamathe's *Jacot-Guillarmod 1426*.

CAPE.—3318 (Cape Town): Cape Town, *Gerstner 6147*. 3326 (Grahamstown): 1820 Settlers Nature Reserve, *Troughton 227*.

A. dealbata, commonly known as the Silver Wattle, is sometimes confused with *A. mearnsii*. It differs from the latter in lacking the glands in between the pinnae pairs along the upper surface of the leaf-rhachis, in being more pruinose, and in having wider usually less moniliform pods.

51. Acacia decurrens *Willd.*, Sp. Pl. 4 : 1072 (1806); Benth. in Fl. Austral. 2 : 414 (1864); in Trans. Linn. Soc. Lond. 30 : 496 (1875); Burtt Davy, Fl. Transv. 2 : 345 (1932); Tindale in Beadle, Evans & Carolin, Handb. Vasc. Pl. Sydney Distr. & Blue Mts. 230 (1962); Court in Willis, Handb. Pl. Victoria 2 : 244 (1972); Tindale in Beadle, Evans & Carolin, Fl. Sydney Region 273 (1972); Ross, Fl. Natal 193 (1973); in Bothalia 11 : 466 (1975). Type from Australia, unknown.

Mimosa decurrens Donn, Hort. Cant. 1 : 114 (1796) nomen nudum.

Unarmed tree up to 12 m high with a conical or rounded crown; young branchlets prominently angled, sometimes with wing-like ridges 1–2 mm high, glabrous or the very young shoots slightly tomentose-pubescent. *Leaves* bipinnate, green, decurrent: petiole angular, 1,5–3,5 cm long, often eglandular; rhachis 3–10 cm long, with a raised gland just below the junction of each pinna pair; pinnae (5)8–15 pairs; leaflets 15–35 pairs, 6–15 × 0,3–0,75 mm, linear, usually glabrous throughout. *Inflorescences* capitate, panicled or racemose. *Flowers* bright golden-yellow; peduncles 2–5 mm long. *Calyx* sparingly pubescent on apices of lobes. *Corolla* glabrous or almost so. *Pods* brown or dark brown, 3,5–10 × 0,4–0,7 cm, not or only slightly moniliform, dehiscing longitudinally along one margin only, straight or slightly curved. *Seeds* brown to blackish-brown, ± 5 × 3,5 mm, elliptic, compressed, smooth; caruncle conspicuous; areole ± 3,5 × 2 mm.

Introduced from Australia.

TRANSVAAL.—2630 (Carolina): 8 km from Carolina on road to Badplaas, *Brummitt 12418* (K).

NATAL.—2930 (Pietermaritzburg): Winterskloof, *Ross 2130*.

CAPE.—3318 (Cape Town): Stellenbosch, *Garside 1246* (K).

A. decurrens, commonly known as the Green Wattle, is readily distinguished by its long narrow leaflets from all of the other introduced *Acacia* species with bipinnate leaves in our area.

A. decurrens is usually attributed to "(Wendl.) Willd." with *Mimosa decurrens* Wendl., Bot. Beob. 57 (1798), being taken as the basionym. However, Willdenow cited only *Mimosa decurrens* Donn, Hort. Cant. 1 : 114 (1796) which is a nomen nudum. As he provided no reference to Wendland, either direct or indirect, Willdenow's binomial must be treated as a new name.

52. Acacia baileyana *F. Muell.* in Trans. & Proc. Roy. Soc. Victoria 24 : 168 (1887); Brenan in F.T.E.A. Legum-Mimos. : 50 (1959); Tindale in Beadle, Evans & Carolin, Handb. Vasc. Pl. Sydney Distr. & Blue Mts. 231 (1962); Brenan in F.Z. 3, 1: 112 (1970); Court in Willis, Handb. Pl. Victoria 2 : 244 (1972); Tindale in Beadle, Evans & Carolin, Fl. Sydney Region 273 (1972); Ross in Bothalia 11 : 466 (1975). Type from Australia.

Unarmed shrub or tree up to 5 m high; young branchlets subglabrous to sparingly pubescent. *Leaves* bipinnate, glaucous: petiole very short, 2–8 mm long; rhachis 0–1,2 cm long, with a gland at the junction of each or only the top few pinnae pairs; pinnae (1)2–4 pairs, crowded; leaflets 12–20 pairs, 3–7 × 0,8–1,5 mm, linear-oblong, often slightly falcate, glabrous throughout or with few marginal cilia. *Inflorescences* capitate, in axillary racemes or panicles longer than the leaves. *Flowers* bright yellow; peduncles 2–5 mm long. *Calyx* and corolla glabrous or almost so. *Pods* brown, 4–10 × 0,8–1,4 cm, straight or slightly curved, margins entire or only slightly and irregularly constricted between some of the seeds, dehiscing longitudinally along one margin. *Seeds* blackish, ± 6 × 3 mm, smooth; caruncle conspicuous; areole ± 5 × 2 mm.

Introduced from Australia.

TRANSVAAL.—2528 (Pretoria): Prince's Park, *Repton 1B*. 2626 (Klerksdorp): near Rooijantjies-fontein, *Kinges 1475*.

SWAZILAND.—2631 (Mbabane): 10 km W. of Mbabane on main road to Transvaal, *Brummitt 12425* (K).

NATAL.—2930 (Pietermaritzburg): Manderston, *Ross 1203*.

CAPE.—3125 (Steynsburg): Grootfontein, *Theron 612*. 3326 (Grahamstown): Grahamstown, *Troughton 65*.

53. Acacia armata *R.Br.* in Ait.f. Hort. Kew, ed. 2, 5 : 463 (? Dec. 1813); DC., Prodr. 2 : 449 (1825); Benth. in Fl. Austral. 2 : 347 (1864); in Trans. Linn. Soc. Lond. 30 : 461 (1875); Salter in Adamson & Salter, Fl. Cape Penins. 453 (1950); Beadle, Evans & Carolin, Handb. Vasc. Pl. Sydney Distr. & Blue Mts. 224 (1962); Court in Willis, Handb. Pl. Victoria 2 : 216 (1972); Beadle, Evans & Carolin, Fl. Sydney Region 265 (1972); Ross in Bothalia 11 : 466 (1975).

Type: South Australia, Bay IX Memory Cove, Kangaroo Island, *R. Brown* (BM, holo.!; E).

Shrub up to 3,5 m high; young branchlets reddish-brown or brown, angular-striate, usually hirsute-pubescent, seldom glabrous. *Stipules* spinescent, in pairs, slender, divaricate, up to 1 cm long. *Leaves* phyllodic, apparently simple, 0,5–1(1,5) × 0,2–0,6 cm, obliquely-ovate to oblong or narrowly lanceolate, undulate, with a single nearly centric midrib, apex obtuse or distinctly mucronate, glabrous throughout or sometimes with hairs on the margins and on the midrib. *Inflorescences* capitate, on axillary peduncles which are about as long as the phyllodes. *Flowers* bright yellow. *Calyx* lobed but not separating into sepals, ± half as long as the corolla. *Petals* narrow, glabrous. *Pods* straight or ± falcate, 2,5–6 × 0,2–0,6 cm, dehiscent, villous, rarely glabrous or hispid. *Seeds* dark brownish-black, ± 7× 2,5 mm, smooth; caruncle conspicuous.

Introduced from Australia.

CAPE.—3318 (Cape Town): Cape Peninsula, Rhodes Estate, *Salter 7619* (BOL); above Rhodes Memorial and Groote Schuur Hospital, *Gerstner 6141*.

It is quite probable that *A. paradoxa* DC., Cat. Hort. Monsp. 74 (Feb.–Mar. 1813), is an earlier name for this species. However, until this has been definitely established, the name *A. armata* is retained.

54. Acacia podalyriifolia *A. Cunn. ex G. Don*, Gen. Syst. 2 : 405 (1832); Benth. in Fl. Austral. 2 : 374 (1864); in Trans. Linn. Soc. Lond. 30 : 474 (1875); Brenan, Check-list Tang. Terr. 332 (1949); in F.T.E.A. Legum.-Mimos. : 51 (1959); Beadle, Evans & Carolin, Handb. Vasc. Fl. Sydney Distr. & Blue Mts. 225 (1962); Brenan in F.Z. 3,1 : 113 (1970); Beadle, Evans & Carolin, Fl. Sydney Region 267 (1972); Ross, Fl. Natal 193 (1973); in Bothalia 11 : 466 (1975). Type: Australia, Queensland, Birnam Range, Brisbane River, *A. Cunningham 157/1828* (K, holo.!).

Unarmed shrub or small tree up to 6 m high; young branchlets densely grey-pubescent. *Leaves* phyllodic, apparently simple, glaucous, mostly 1,5–4 × 1–2 cm, ovate to elliptic or elliptic-oblong, often oblique, with a single main longitudinal nerve and finely but distinctly penninerved, sparingly to densely pubescent, with 1 or 2 marginal glands.

Inflorescences capitate, in axillary racemes which are usually longer than the phyllodes, mostly terminal. *Flowers* bright yellow; peduncles pubescent, up to 7 mm long. *Calyx* less than half as long as the corolla, pubescent apically. *Petals* ± free, hirsute. *Pods* brown, glabrous or pubescent, 4–8,5 × 1,5–2 cm, straight or almost so, flattened, margins often ± undulate, dehiscing longitudinally. *Seeds* dark brownish-black, 6–7 × ± 3,5 mm, smooth, compressed; caruncle conspicuous; areole 3,5–4 × ± 1,5 mm.

Introduced from Australia.

TRANSVAAL.—2528 (Pretoria): Riviera, Pretoria, *Schlieben 10083.*

NATAL.—2930 (Pietermaritzburg): slopes below World's View, *Ross 2104.*

CAPE.—3318 (Cape Town): Stellenbosch, *Louw 5.* 3326 (Grahamstown): Grahamstown, *Troughton 44.*

55. **Acacia saligna** (*Labill.*) *Wendl.*, Comm. Acac. 26 (1820); Benth. in Fl. Austral. 2 : 364 (1864); in Trans. Linn. Soc. Lond. 30 : 469 (1875); Court in Willis, Handb. Pl. Victoria 2 : 229 (1972); Maslin in Nuytsia 1(4) : 334 (1974); Ross in Bothalia 11 : 467 (1975). Type from Western Australia, *Labillardiere* (FI, lecto.).

Mimosa saligna Labill., Pl. Nov. Holl. 2 : 86, t.235 (1806). Type as above.

Acacia cyanophylla Lindl., Bot. Reg. 25 : Misc. 45 (1839); Benth. in Fl. Austr. 2 : 364 (1864); in Trans. Linn. Soc. Lond. 30 : 469 (1875); Salter in Adamson & Salter, Fl. Cape Penins. 454 (1950); F. White, For. Fl. N. Rhod. 82 (1962); Roux & Middlemiss in S. Afr. J. Sci. 59 : 286 (1963); Henderson & Anderson, Mem. Bot. Surv. S. Afr. 37 : 170, fig. 84a, b, c (1966); Brenan in F.Z. 3,1 : 112 (1970); Beadle, Evans & Carolin, Fl. Sydney Region 269 (1972); Ross, Fl. Natal 193 (1973). Type from Australia.

Unarmed shrub or tree up to 10 m high; young branchlets slightly angular, glabrous. *Leaves* phyllodic, apparently simple, glabrous, mostly 8–22 × 0,5–1,4 cm (the lower ones sometimes much longer and 4 cm or more wide), usually narrow, linear-lanceolate to linear-oblong or oblanceolate, straight or slightly falcate, much narrowed basally, with a single main longitudinal nerve and finely but distinctly penninerved, sometimes glaucous, with a basal gland (on young plants and coppice shoots bipinnate leaves are sometimes produced at the apex of the phyllode). *Inflorescences* globose, 6,5–9 mm in diameter, in short axillary racemes. *Flowers* bright yellow; peduncles 0,6–2,2 cm long. *Calyx* slightly pubescent apically.

Corolla glabrous. *Pods* brown, 5,5–15 × 0,5–0,6 cm, straight or slightly falcate, flattened, margins slightly constricted between some of the seeds, dehiscing longitudinally. *Seeds* dark brown, 5–7 × 2,75–3,5 mm, smooth, compressed; caruncle conspicuous; areole 3,5–5 × ± 1,5 mm.

Introduced into the Cape Province from Australia and now fairly widespread from the Cape Peninsula to the eastern Cape; also introduced into Natal more recently.

S.W.A.—2615 (Luderitz): Luderitz, *Kinges 2736.*

TRANSVAAL.—2528 (Pretoria): Sunnyside, *Repton 1861.*

NATAL.—2930 (Pietermaritzburg): Botha's Hill, *Ross 2132.* 2931 (Stanger): Virginia Airport, *Ross 2139.*

CAPE.—3318 (Cape Town): Ida's Valley, bottom of Hell's Hoogte Pass, Stellenbosch, *Thompson 836.* 3325 (Port Elizabeth): Port Elizabeth, *Begg s.n.* (GRA). 3326 (Grahamstown): road from Port Elizabeth to Grahamstown, *Wells 2603.* 3418 (Simonstown): Tokai, *Burtt Davy sub FHO 20021* (K). 3422 (Mossel Bay): Sedgefield, farm Karawater, bank of Karatara River, *Ross 2410.*

A. saligna, commonly called the "Port Jackson Willow" on account of *its pendulous branches and phyllodes*, was introduced on the Cape Flats in the 1870's in an attempt to stabilize the shifting dune sands. It proved highly successful for this purpose and soon started spreading by natural means. *A. saligna* is now found far beyond the area of the Cape Flats and has become a serious menace in many parts of the Cape Peninsula and on the mainland by invading and displacing the indigenous vegetation. *A. saligna* coppices freely when cut down and in many areas occurs in dense stands. The wood is relatively soft and the branches are brittle.

56. **Acacia pycnantha** *Benth.* in Hook., Lond. J. Bot. 1 : 351 (1842); Benth. in Fl. Austral. 2 : 365 (1864); in Trans. Linn. Soc. Lond. 30 : 469 (1875); Salter in Adamson & Salter, Fl. Cape Penins. 455 (1950); Beadle, Evans & Carolin, Handb. Vasc. Pl. Sydney Distr. & Blue Mts. 226 (1962); Court in Willis, Handb. Pl. Victoria 2 : 226 (1972); Beadle, Evans & Carolin, Fl. Sydney Region 269 (1972); Ross in Bothalia 11 : 467 (1975). Type from Australia.

Unarmed shrub or tree up to 10 m high; young branchlets terete or almost so, glabrous. *Leaves* phyllodic, apparently simple, glabrous, 10–20 × (1)1,5–3 cm, obovate-lanceolate, distinctly falcate, mostly obtuse apically, narrowed basally, with a single main longitudinal nerve and finely but distinctly penninerved, margin nerve-like, with a fairly large marginal gland near the base (on young plants and coppice shoots bipinnate leaves are sometimes produced at the apex of the phyllode). *Inflorescences* globose, in axillary racemes or panicles. *Flowers* bright yellow; peduncles stout, up to 7 mm long. *Calyx* about ⅔ as long as the corolla, pubescent apically. *Corolla* ± glabrous. *Pods* brown, 6–12 × 0,4–0,7 cm, straight or slightly curved, flattened, margins slightly constricted between some of the seeds, dehiscing longitudinally. *Seeds* dark brownish-black, 5–7 × 2,75–3,5 mm, smooth, compressed; caruncle conspicuous.

Introduced from Australia.

CAPE.—3318 (Cape Town): Pinelands, *Salter 8767;* Cape Town University, *Leighton sub BOL 25537.* 3418 (Simonstown): Somerset West, *Parker 3517* (K). 3420 (Bredasdorp): Potteberg, *Van Niekerk sub BOL 23359.*

57. Acacia longifolia (*Andr.*) *Willd.,* Sp. Pl. 4 : 1052 (1806), non *A. longifolia* Paxt. (1846); Benth. in Fl. Austral. 2 : 397 (1864); in Trans. Linn. Soc. Lond. 30 : 487 (1875); Salter in Adamson & Salter, Fl. Cape Penins. 454 (1950); Beadle, Evans & Carolin, Handb. Vasc. Pl. Sydney Distr. & Blue Mts. 228 (1962); Henderson & Anderson, Mem. Bot. Surv. S. Afr. 37 : 170, fig. 84d, e, f (1966); Court in Willis, Handb. Pl. Victoria 2 : 241 (1972); Beadle, Evans & Carolin, Fl. Sydney Region 272 (1972); Ross, Fl. Natal 193 (1973); in Bothalia 11 : 467 (1975). Type from Australia.

Mimosa longifolia Andr., Bot. Rep. t.207 (1802). Type as above.

Unarmed shrub or tree to 8 m high; young branchlets angular, glabrous or the young shoots minutely pubescent. *Leaves* phyllodic, apparently simple, glabrous, 6–18 × 0,7–2 cm, linear-lanceolate or narrowly oblong to oblanceolate, straight or almost so, mucronate apically, sometimes obliquely so, narrowed basally, with 2–5 prominent longitudinal nerves and faintly or conspicuously anastomosing almost longitudinal veins between the nerves. *Inflorescences* spicate, axillary, solitary or in pairs; spikes up to 4 cm long. *Flowers* bright yellow, sessile. *Calyx* very short. *Corolla* glabrous. *Pods* brown, 7–14 × 0,4–0,6 cm, cylindrical, straight or slightly curved, margins constricted between the seeds, dehiscing longitudinally along both margins, valves longitudinally wrinkled or striate, acuminate apically, glabrous. *Seeds* dark brownish-black, 4–7 × ± 2,5 mm, more or less oblong, smooth, compressed; areole ± 3,5 × 1,5 mm; funicle not much folded, thickened almost from the base into a small ± cupular aril enclosing the apex of the seed.

Introduced from Australia.

TRANSVAAL.—2627 (Potchefstroom): Randfontein, *Barnard sub PRE 32122.* 2628 (Johannesburg): Johannesburg, *Moss 5258* (BM).

NATAL.—2930 (Pietermaritzburg): Town Bush Valley, 1,6 km W. of Cascade Falls, *Ross 1281* (NU); Hilton Road, *Ross 2106.*

CAPE.—3318 (Cape Town): Rondebosch, lower slopes of Devil's Peak behind University, *White 5002.* 3319 (Worcester): Franschhoek. *Van der Merwe 1209.* 3325 (Port Elizabeth): 24 km up Elands River road, *Acocks 21275.* 3326 (Grahamstown): road from Port Elizabeth to Grahamstown, *Wells 2602.* 3418 (Simonstown): near Wynberg, *Schlechter 1061* (GRA). 3419 (Caledon): Kogelberg Reserve, Paardeberg, *Grobler 17140.* 3422 (Mossel Bay): Mossel Bay, *Hutchinson s.n.* (K).

A. longifolia is commonly known as the Golden Wattle. Like several of the other introduced Australian species, *A. longifolia* is also invading and displacing the indigenous vegetation in some areas.

A. longifolia is a variable species. Although some of the extremes look very different, they are connected by an almost continuous range of intermediates and consequently cannot be separated satisfactorily. Bentham l.c. :397 (1864) enumerated six forms of *A. longifolia.*

Beadle, Evans & Carolin l.c. 228 (1962) recognized two varieties, namely, var. *longifolia* and var. *sophorae* (Labill.) F. Muell. ex Benth. Var. *sophorae* has mostly obovate-oblong, oblanceolate or oblong-elliptic phyllodes 1,2–3,6 cm wide and 5–12 cm long, in contrast to the linear or linear-lanceolate phyllodes 0,3–1 cm wide and 7,5–13 cm long of var. *longifolia.* Although there is no distinct morphological discontinuity between the two, in Australia var. *sophorae* has somewhat different ecological preferences and tends to occur as a low plant along the coastal sand-dunes, while var. *longifolia* grows into a larger plant. Specimens from our area are often difficult to place in one variety or the other with certainty.

58. Acacia cyclops *A. Cunn. ex G. Don,* Gen. Syst. 2 : 404 (1832); Benth. in Fl. Austral. 2 : 388 (1864); in Trans. Linn. Soc. Lond. 30 : 481 (1875); Salter in Adamson & Salter, Fl. Cape Penins. 454 (1950); Roux in S. Afr. J. Sci. 57 : 99 (1961); Roux & Middlemiss in S. Afr. J. Sci. 59 : 286 (1963); Middlemiss in S. Afr. J. Sci. 59 : 419 (1963); Henderson & Anderson, Mem. Bot. Surv. S. Afr. 37 : 172, fig. 85 (1966); Ross in Bothalia 11 : 468 (1975). Syntypes: Western Australia, King George's Sound, *A. Cunningham* 104/1818 (K!), 328/1821 (K!).

A. cyclopis A. Cunn. ex Loudon, Hort. Britt. 407 (1830) nomen nudum.

Unarmed shrub or small tree up to 6 m high; young branchlets usually angular and glabrous. *Leaves* phyllodic, apparently simple, glabrous, 3–9 × 0,6–1,5 cm, narrowly-oblong, usually ± straight, sometimes slightly falcate, obliquely mucronate apically, narrowed basally, with 3–5 prominent longitudinal nerves and anastomosing almost longitudinal veins. *Inflorescences* globose, solitary or two or three in short axillary racemes. *Flowers* bright yellow; peduncles up to 7 mm long. *Calyx* pubescent apically, more than half as long as the corolla. *Petals* free. *Pods* brown, 5–15 × 0,8–1,3 cm, oblong, falcate or variously coiled or spirally twisted, flattened, margins not constricted between the seeds, dehiscing longitudinally along both margins. *Seeds* dark brown, 5–7 × 3–4 mm, smooth, compressed; areole ± 4 × 2 mm; funicle thickened, bright red or orange, encircling the seed in a double fold.

Introduced into the Cape Province from Australia and now widespread in coastal areas from Lambert's Bay in the north-west to Kidd's Beach in the north-east.

S.W.A.—2615 (Luderitz): Luderitz, *Kinges 2732*.

CAPE.—3318 (Cape Town): Hell's Hoogte, Stellenbosch, *Taylor 7298*. 3325 (Port Elizabeth): Port Elizabeth, *Theron 1142*. 3326 (Grahamstown): Kowie River, *Wells 2580*. 3418 (Simonstown): Cape Peninsula, *Rodin 3287A*. 3419 (Caledon): near Caledon, *Gilliland A62* (BM). 3422 (Mossel Bay): Sedgefield, farm Karawater, bank of Karatara River, *Ross 2408*. 3423 (Knysna): bank of Lagoon, road to Knysna Heads, *Bos 935*.

Like *A. saligna*, *A. cyclops* was introduced on the Cape Flats in the 1870's in an attempt to stabilize the shifting dune sands. It proved highly successful for this purpose and soon started spreading by natural means. *A. cyclops* is now found far beyond the area of the Cape Flats and has become a serious menace in many parts of the Cape Peninsula and on the mainland by invading and displacing the indigenous vegetation. In many areas *A. cyclops* occurs in dense almost impenetrable stands.

Unlike *A. saligna*, *A. cyclops* does not usually coppice when cut down. The wood of *A. cyclops* provides a useful firewood.

A. cyclops is commonly known as "Rooikrans" on account of the bright red funicle which encircles the seed. The pods usually remain attached to the plant long after the ripe seeds have been shed.

A number of species of birds feed on the conspicuous funicles and assist in the distribution of *A. cyclops* (see Middlemiss in S. Afr. J. Sci. 59 : 419, 1963).

59. Acacia melanoxylon *R.Br.* in Ait.f. Hort. Kew ed. 2,5 : 462 (1813); Benth. in Fl. Austral. 2 : 388 (1864); in Trans. Linn. Soc. Lond. 30 : 481 (1875); J. Phillips in Mem. Bot. Surv. S. Afr. 14 : 291 (1931); Salter in Adamson & Salter, Fl. Cape Penins. 454 (1950); Beadle, Evans & Carolin, Handb. Vasc. Pl. Sydney Distr. & Blue Mts. 227 (1962); Court in Willis, Handb. Pl. Victoria 2 : 236 (1972); Beadle, Evans & Carolin, Fl. Sydney Region 270 (1972); Ross, Fl. Natal 193 (1973); in Bothalia 11 : 468 (1975). Type: Tasmania, Port Dalrymple, *R. Brown* (BM, holo.!).

Unarmed tree up to 20 m high; young branchlets angular, glabrous or the young shoots minutely pubescent. *Leaves* phyllodic, apparently simple, glabrous, mostly 6–12 × 0,6–1,2(2,5) cm, linear-lanceolate to oblanceolate or narrowly obovate, straight to falcate, narrowed basally, with 3–7 prominent longitudinal nerves and a conspicuous reticulate venation between the longitudinal nerves (on young plants bipinnate leaves are sometimes produced at the apex of the phyllode). *Inflorescences* globose, solitary or in short axillary racemes. *Flowers* pale yellowish-white; peduncles up to 6 mm long. *Calyx* more than half as long as the corolla. *Corolla* glabrous. *Pods* brown, 5–15 × 0,6–0,8 cm, oblong, falcate or variously coiled or spirally twisted, flattened, margins thickened, not constricted between the seeds, dehiscing longitudinally along both margins. *Seeds* dark brownish-black, 4–5 × ± 2,5 mm, smooth, compressed; areole ± 3 × 1 mm; funicle very long, thickened, almost encircling the seed in a double fold.

Introduced from Australia.

TRANSVAAL.—2528 (Pretoria): Wonderboom Reserve, *Repton 1871*. 2627 (Potchefstroom): Krugersdorp, *Webster sub PRE 32118*. 2628 (Johannesburg): around Johannesburg, *Moss 7082* (BM).

SWAZILAND.—2631 (Mbabane): 1,6 km from Hlatikulu on Sitobela road, *Ross 1759*.

NATAL.—2730 (Vryheid): Donkerhoek, *Devenish 1020*. 2929 (Underberg): farm Vergelegen, Umkomaas River near Lesotho border, *Rissik s.n.* 2930 (Pietermaritzburg): slope below World's View, *Ross 2128*.

LESOTHO.—2927 (Maseru): Masoeling, *Jacot-Guillarmod 2605*.

CAPE.—3219 (Wuppertal): Cedar Mts., Algeria forest reserve, *Bos 516*. 3318 (Cape Town): Rondebosch, near University of Cape Town, *White 5066*. 3326 (Grahamstown): Grahamstown, *Roux sub PRE 32121*. 3422 (Mossel Bay): Sedgefield, farm Karawater, banks of Karatara River, *Ross 2409*.

A. melanoxylon, the well-known Blackwood, yields a good timber which is used in the manufacture of furniture. Like several of the other introduced Australian species, *A. melanoxylon* is also invading and displacing the indigenous vegetation in some areas.

In addition to the species dealt with in some detail above, several species are cultivated in our area. At present, however, there is no evidence to suggest that any of them have become naturalized. The species cultivated are:

60. **Acacia elata** *A. Cunn. ex Benth.* in Hook., Lond. J. Bot. 1 : 383 (1842), non *A. elata* Wallich, Cat. 5233 (1832) nomen nudum, non *A. elata* R. Grah.; Benth. in Fl. Austral. 2 : 413 (1864); in Trans. Linn. Soc. Lond. 30 : 495 (1875); Summerh. in Bot. Mag. 154 : t.9214 (1930); Brenan in F.T.E.A. Legum.-Mimos. : 50 (1959); in F.Z. 3,1 : 111 (1970); Tindale in Beadle, Evans & Carolin, Fl. Sydney Region 272 (1972); Ross in Bothalia 11 : 469 (1975). Type from New South Wales, Australia.

A. terminalis sensu Court in Handb. Pl. Victoria 2 : 242 (1972).

Unarmed tree. *Leaves* bipinnate, large, 30–40 cm long; pinnae 3–5 pairs; leaflets 8–15 pairs per pinna, mostly 2–6 × 0,4–1 cm, lanceolate to linear-lanceolate, often somewhat falcate, usually finely pubescent at least on the lower surface. *Flowers* pale yellow, in round heads, arranged in axillary racemes or panicles. *Pods* ± 9–15 × 0,9–1,3 cm, linear-oblong, straight or curved, the margins irregularly constricted between the seeds, compressed, dehiscing along both margins.

A. elata is easily distinguished from all of the other species with bipinnate leaves by its large leaflets.

Recorded from Krugersdorp in the Transvaal, *Gerstner* 6671, and Stellenbosch in the Cape, *Taylor* 7968, but much more widely cultivated.

61. **Acacia visite** *Griseb.* in Abh. K. Ges. Wiss. Göttingen 19 : 135 (1874); Ross in Bothalia 11 : 469 (1975). Type from Argentina.

Unarmed tree. *Leaves* bipinnate; pinnae 2–7 pairs; leaflets 24–38 pairs per pinna, 6–9 × 0,8–1,25(2) mm, linear or linear-oblong, acute apically, midrib almost marginal throughout its length and usually pubescent. *Flowers* in round heads; inflorescences

solitary, paired or fascicled in the axils of the leaves. *Pods* 7–12 × 1,4–1,9 cm, valves thin, dehiscing longitudinally.

Recorded from Capital Park, Pretoria, *Repton* 1880; Grounds of Division of Botany, Pretoria, *Verdoorn sub PRE* 32344, *Schlieben* 10106; Mr Loock's garden in Pretoria, *Gerstner sub PRE* 32346; Bloemfontein, *Potts* 3219.

62. **Acacia cultriformis** *A. Cunn. ex G. Don*, Gen. Syst. 2 : 406 (1832); Benth. in Fl. Austral. 2 : 375 (1864); in Trans. Linn. Soc. Lond. 30 : 474 (1875); Brenan in F.Z. 3,1 : 113 (1970); Ross in Bothalia 11 : 469 (1975). Type from New South Wales, Australia.

Unarmed shrub or small tree; young branchlets angular, glabrous. *Leaves* phyllodic, apparently simple, 0,8–3 × 0,6–1,1 cm, obliquely obovate-lanceolate to ovate-triangular, glaucous, glabrous, with a single main longitudinal nerve and finely penni-nerved, usually with 1 marginal gland, sometimes on a prominent angle. *Flowers* in small round heads, arranged in axillary racemes which are longer than the phyllodes and are often ± aggregated terminally. *Pods* 5–9 × 0,5–0,7 cm, linear-oblong, glabrous, longitudinally dehiscent.

Recorded from Stellenbosch, *Garside* 1248 (K).

A. cultriformis differs from *A. podalyriifolia* in being glabrous and in having narrower pods.

63. **Acacia retinodes** *Schlechtend.* in Linnaea 20 : 664 (1847); Benth. in Fl. Austral. 2 : 362 (1864); in Trans. Linn. Soc. Lond. 30 : 468 (1875); Stapf & Ballard, Bot. Mag. 153 : t.9177 (1929); Brenan in F.T.E.A. Legum.-Mimos. : 51 (1959); Court in Willis, Handb. Pl. Victoria 2 : 227 (1972); Ross in Bothalia 11 : 469 (1975). Type from Australia.

Unarmed glabrous shrub or small tree. *Leaves* phyllodic, apparently simple, linear-lanceolate to -oblong or oblanceolate, straight or slightly curved, 4,5–17 cm long, up to 1,5 cm wide, narrowing gradually towards the base, with a single main longitudinal nerve and finely but distinctly penninerved. *Flowers* in round heads up to 6 mm in diameter; inflorescences on peduncles 3–6 mm long, arranged in short axillary racemes. *Pods* 7–12 × 0,5–0,7 cm, linear-oblong, flattened, longitudinally dehiscent; funicle encircling the seed in a double fold.

Recorded from Roodeplaat near Pretoria, *Du Toit* 105, 151, *Schlieben & Mendelsohn* 12717.

A. retinodes differs from *A. saligna* in having smaller flower-heads, shorter peduncles and funicles which encircle the seeds in a double fold.

64. Acacia fimbriata *A. Cunn. ex G. Don*, Gen. Syst. 2 : 406 (1832); Beadle, Evans & Carolin, Fl. Sydney Region 267 (1972); Ross in Bothalia 11 : 469 (1975). Type from New South Wales, Australia.

Unarmed shrub or small tree. *Leaves* phyllodic, apparently simple, linear to narrowly oblong-elliptic, 2–4,5 cm long, 2–5 mm wide, narrowed basally, with a single main longitudinal nerve, margins typically densely ciliate, usually with a rounded gland near the base. *Flowers* in small round heads, arranged in axillary racemes. *Pods* linear-oblong, straight, flattened, up to 7 cm long and 7 mm wide, dehiscent.

Recorded from the Grounds of the Union Buildings, Pretoria, *Repton* 2640, *Schlieben* 10084, *Schlieben & Mendelsohn* 12881; Grahamstown, *Troughton* 228.

65. Acacia adunca *A. Cunn. ex G. Don*, Gen. Syst. 2 : 406 (1832); Maiden, For. Fl. New South Wales 5, part 46 : 113–118, t.173 (1911); Ross in Bothalia 11 : 469 (1975). Type: Australia, New South Wales, Hunters River, *Cunningham* 79/1827 (K, holo.!).

A. accola Maiden & Betche in Proc. Linn. Soc. New South Wales 31(4) : 734 (1907). Syntypes from Australia.

Unarmed small tree; young branchlets angular, glabrous. *Leaves* phyllodic, apparently simple, 5–12 cm long, 1,5–3 mm wide (in our area), linear, with a single main longitudinal nerve, usually with an oblique slightly recurved point apically, a fairly conspicuous marginal gland situated a short distance above the base. *Flowers* in small round heads, arranged in short axillary racemes which are mostly aggregated terminally. *Pods* reddish-brown when mature, 7–10 × 0,8–1 cm, oblong, margins often irregularly constricted, valves thin, umbonate over the seeds, longitudinally dehiscent.

Recorded from the Groot Drakenstein in the Cape Province, *Voorligtingsbeampte* C4.

66. Acacia maidenii *F. Muell.* in Linn. Soc. New South Wales Macleay Mem. Vol. 222 : t.29 (1893); Court in Willis, Handb. Pl. Victoria 2 : 240 (1972); Beadle, Evans & Carolin, Fl. Sydney Region 271 (1972); Ross in Bothalia 11 : 470 (1975). Type from New South Wales, Australia.

Unarmed small to medium-sized tree. *Leaves* phyllodic, apparently simple, 6–15 × 0,8–1,5 cm, with 3–7 main longitudinal nerves and almost anastomosing longitudinal veins. *Flowers* in elongate spikes up to 4 cm long, spikes axillary, solitary or in twos or threes. *Pods* 4–12 cm long, 3–5 mm wide, variously coiled or twisted, pubescent.

Recorded from the Caledonian Grounds, Pretoria, *Repton* 3766.

Differs from *A. longifolia* in having pubescent coiled pods.

67. Acacia viscidula *A. Cunn. ex Benth.* in Hook., Lond. J. Bot. 1 : 363 (1842); in Fl. Austral. 2 : 387 (1864); in Trans. Linn. Soc. Lond. 30 : 480 (1875); Ross in Bothalia 11 : 470 (1975). Type: Australia, New South Wales, banks of Lachlan River, *Fraser* (K, holo.!).

Unarmed shrub or small tree; young branchlets angular, mostly sparingly pubescent, viscid. *Leaves* phyllodic, apparently simple, 4,5–10 cm long, 1,25–3 mm wide, linear, narrowed basally, with several longitudinal nerves. *Flowers* in small round heads, on axillary peduncles, solitary or paired, rarely fascicled; peduncles up to 5 mm long, pubescent. *Sepals* free or shortly united basally. *Corolla* pubescent. *Pods* 4–7 cm long, 3–3,5 mm wide, linear, sparingly to densely pubescent, longitudinally dehiscent.

Recorded on the Cape Peninsula on the slopes below the ruins of Lady Anne Barnard's cottage, *Salter* 9044.

68. Acacia pendula *A. Cunn. ex G. Don*, Gen. Syst. 2 : 404 (1832); Benth. in Fl. Austral. 2 : 383 (1864); in Trans. Linn. Soc. Lond. 30 : 479 (1875); Court in Willis, Handb. Pl. Victoria 2 : 238 (1972); Ross in Bothalia 11 : 470 (1975). Type from New South Wales, Australia.

Unarmed tree or shrub. *Leaves* phyllodic, apparently simple, linear to linear-oblong or lanceolate, 4,5–8 cm long, 3–9 mm wide, narrowed towards the base, coriaceous, with several inconspicuous longitudinal nerves, often greyish or glaucous. *Flowers* in small round heads, usually arranged in very short axillary racemes. *Pods* oblong, flattened, 4–8 × 0,8–1,8 cm, the margins bordered by a narrow wing 0,5–2 mm wide.

Recorded from a Johannesburg park, *Hobson sub PRE* 32341; Middelburg, Cape, *Loock sub PRE* 32340.

3447 3. LEUCAENA

Leucaena *Benth*. in Hook., J. Bot. 4 : 416 (1842); Benth. & Hook. f., Gen. Pl. 1 : 594 (1865); Oliv. in F.T.A. 2 : 337 (1871); Benth. in Trans. Linn. Soc. Lond. 30 : 442 (1875); Taub. in Pflanzenfam. 3, 3 : 115 (1892); Bak. f., Leg. Trop. Afr. 3 : 814 (1930); Gilbert & Boutique in F.C.B. 3 : 231 (1952); Keay in F.W.T.A. ed. 2, 1 : 495 (1958); Brenan in F.T.E.A. Legum.-Mimos.: 48 (1959); Hutch., Gen. Fl. Pl. 1 : 281 (1964); Brenan & Brummitt in F.Z. 3, 1 : 53 (1970). Type species: *L. diversifolia* (Schlechtd.) Benth., fide Williams in Taxon 13 : 300 (1964).

Trees or shrubs, unarmed. *Leaves* bipinnate, a gland often present at the junction of the lowest pair of pinnae, petiole and rhachis otherwise eglandular or rarely with glands between other pairs of pinnae; pinnae each with one to many pairs of leaflets. *Inflorescences* capitate, pedunculate, axillary, 1–3 together, often racemosely aggregated. *Flowers* hermaphrodite, sessile, 5-merous. *Calyx* gamosepalous, 5-toothed. *Petals*, 5, free, pubescent or glabrous outside. *Stamens* 10, free, fertile; anthers mostly eglandular apically. *Ovary* pubescent or sometimes glabrous. *Pods* oblong or linear–oblong, usually thinly subcoriaceous, compressed, dehiscing into 2 non-recurving valves. *Seeds* lying±transversely in the pod, compressed, brown, smooth, unwinged, with endosperm.

A genus of±50 species, one widespread in the tropics and subtropics (*L. leucocephala*), one in the Pacific islands, the rest in tropical America.

The generic name is derived from *leukos*, the Greek word for white; in allusion to the flowers of these plants.

Leucaena leucocephala (*Lam.*) *De Wit* in Taxon 10 : 54 (1961); Brenan & Brummitt in F.Z. 3,1 : 53, t.14 (1970); Ross, Fl. Natal 193 (1973). Type an American plant cultivated in France.

Mimosa leucocephala Lam., Encycl. Méth. Bot. 1 : 12 (1783). Type as above. *M. glauca* sensu L., Sp. Pl. ed. 2,2 : 1504 (1763) pro parte, non L., Sp. Pl. 1 : 520 (1753).

Leucaena glauca sensu auct. mult. : Benth. in Hook., J. Bot. 4 : 416 (1842); Oliv. in F.T.A. 2 : 337 (1871); Benth. in Trans. Linn. Soc. Lond. 30 . 443 (1875); Bak. f., Leg. Trop. Afr. 3 : 814 (1930); Gilbert & Boutique in F.C.B. 3 : 231 (1952); Torre in C.F.A. 2 : 268 (1956); Keay in F.W.T.A. ed. 2,1 : 495 (1958); Brenan in F.T.E.A. Legum.–Mimos. : 48 (1959).

Unarmed shrub or small tree to 4 m high; young branchlets densely grey-puberulous. *Leaves* grey-puberulous: petiole 2–4,5 cm long, often with a gland at the junction of the lowest pair of pinna, glands otherwise absent; rhachis (2,5)7–15 cm long; pinnae (2)3–8 opposite pairs; rhachillae 4–8,5 cm long; leaflets 7–20 pairs, 7–15 × 1,5–4 mm, obliquely oblong-lanceolate, acute apically, puberulous on the margins and sometimes also on the midrib beneath, sometimes glabrous throughout. *Flowers* white to pale yellowish-white, in heads up to 1,8 cm in diameter; peduncles up to 3,5 cm long, grey-puberulous. *Calyx* 2–3,5 mm long, densely pubescent apically. *Petals* 3,5–5 mm long, puberulous apically. *Stamens* 10, free, filaments 6–7,5 mm long; anthers with scattered hairs. *Ovary* up to 2 mm long, densely pubescent apically. *Pods* light to dark brown, 10–18 × 1,4–1,8 cm, with a stipe up to 2,5 cm long, oblong, compressed, thinly subcoriaceous, raised over the seeds, dehiscing into 2 non-recurving valves. *Seeds* 7–9 × 3,5–5 mm, elliptic to obovate, glossy.

Widespread in the tropics and subtropics, possibly native only in the New World. Introduced into our area, sometimes escaping and becoming naturalized in Natal.

NATAL.—2831 (Nkandla): Empangeni, *Lawn 1599* (NH). 2930 (Pietermaritzburg): Durban Berea *Lansdell sub NH 15939.* 2931 (Stanger): Nyoni, *Gerstner sub NH 22631*; 6,4 km N. of Stanger, *Edwards 3307*; Stanger, *Ross 859*; New Guelderland, *Stewart 130* (NH).

The hairs on the anthers (visible with a hand lens) are a most useful diagnostic character of *L. leucocephala*, and distinguish it from all other Mimosoideae in our area.

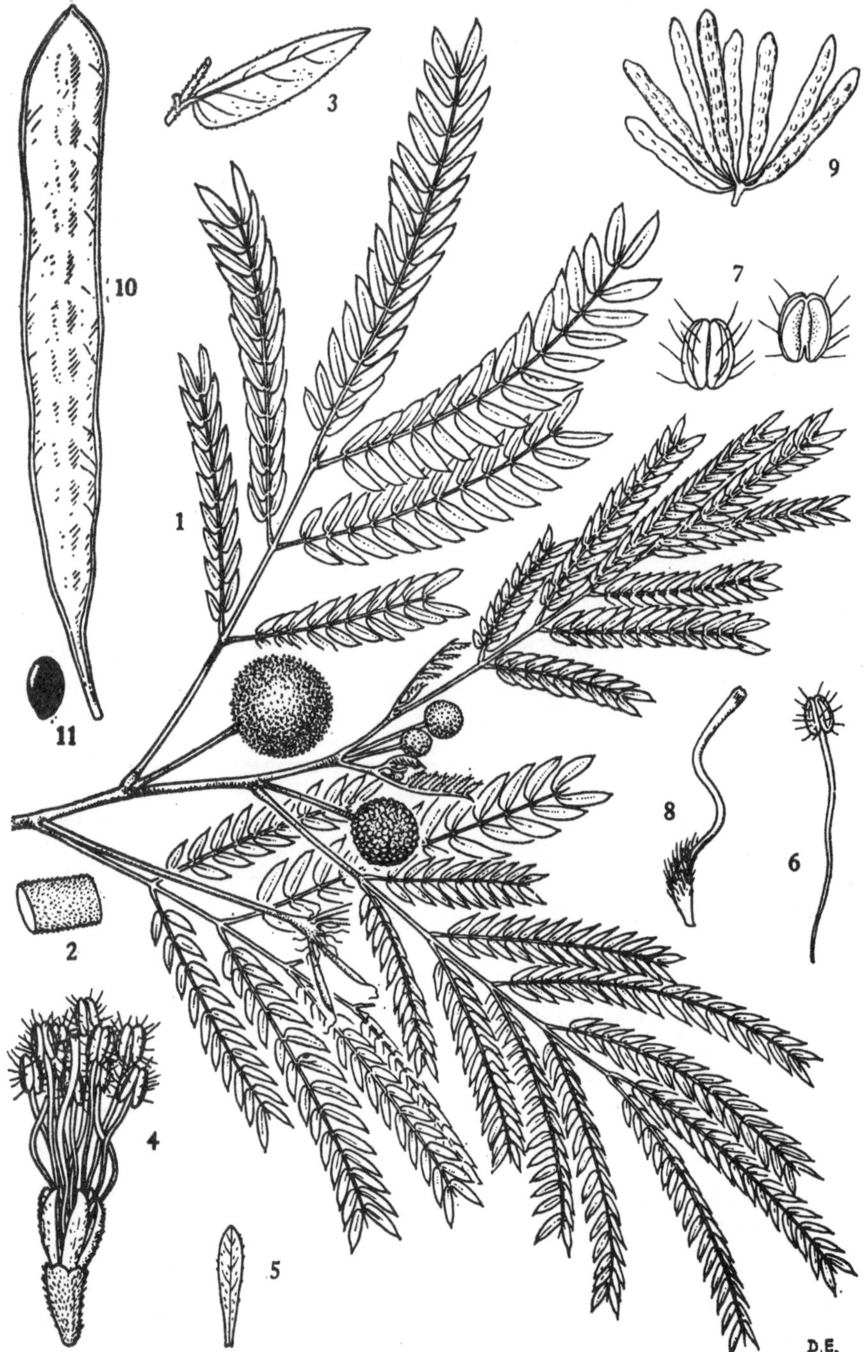

FIG. 12.—**Leucaena leucocephala.** 1, flowering branch, × ⅔, *Lemos & Balsinhas* 22; 2, part of stem to show indumentum, × 4; 3, leaflet, × 3; both from *Faulkner* 576; 4, flower, × 4, 5, petal, × 4; 6, stamen, × 4; 7, two views of anther, × 8; 8, ovary, × 4, all from *Lemos & Balsinhas* 22; 9, cluster of pods, × ⅙; 10, pod, × ⅔; 11, seed, × 1, all from *Faulkner* 576. Reproduced by permission of the Editorial Board of Flora Zambesiaca.

3449   4. MIMOSA

Mimosa *L.*, Sp. Pl. 1 : 516 (1753); Gen. Pl. ed. 5 : 233 (1754); DC., Prodr. 2 : 425 (1825); Benth. in Hook., J. Bot. 4: 358 (1841); Benth. & Hook. f., Gen. Pl. 1 : 593 (1865); Oliv. in F.T.A. 2 : 335 (1871); Benth. in Trans. Linn. Soc. Lond. 30 : 388 (1875); Taub. in Pflanzenfam. 3, 3 : 115 (1892); Harms in Engl., Pflanzenw. Afr. 3,1 : 390 (1915); Fawc. & Rendle, Fl. Jam. 4 : 132 (1920); Bak. f., Leg. Trop. Afr. 3 : 811 (1930); Phill., Gen. 391 (1951); Gilbert & Boutique in F.C.B. 3 : 228 (1952); Brenan in F.T.E.A. Legum.–Mimos. : 42 (1959); Hutch., Gen. Fl. Pl. 1 : 282 (1964); Schreiber in F.S.W.A. 58 : 18 (1967); Brenan & Brummitt in F.Z. 3, 1 : 47 (1970). Type species: *M. pudica* L.

Herbs or shrubs, rarely trees (not in our area), sometimes scrambling or climbing, usually armed with prickles. *Leaves* sensitive, bipinnate or the pinnae seeming almost digitate on account of the very short rhachis, rarely (not in our species) absent or modified to phyllodes; pinnae in 1–21 pairs; pinnae each with few to many pairs of leaflets; stipules often persistent. *Inflorescences* of ovoid or subglobose heads or (not in our species) spikes, which are pedunculate, axillary, solitary or more usually clustered and often aggregated. *Flowers* hermaphrodite or male, small, sessile. *Calyx* often very small and almost inconspicuous. *Corolla* gamopetalous, 4- or sometimes 3-, 5- or 6-lobed. *Stamens* free, as many as or twice as many as the corolla-lobes, fertile; anthers eglandular apically. *Ovary* usually sessile; stigma terminal. *Pods* straight or circinate, flat, in our species densely bristly or prickly; at maturity the valves between the margins splitting $\pm$transversely into 1-seeded segments or rarely (not in our species) remaining entire, the margins persisting as an empty frame. *Seeds*$\pm$compressed, smooth.

A genus of $\pm$ 500 species, widely distributed through the tropics but the vast majority of species found in South America. One species is indigenous in Southern Africa and one species has become naturalized.

The name *Mimosa* is derived from the Greek word *mimos*, a mimic. This is in reference to the sensitive collapse of the leaves of some species when touched.

Pinnae in (2)4–14 pairs, pinnately arranged along the rhachis which is longer than the petiole; leaves with
    a straight, erect or forward-pointing, slender prickle at the junction of each pinna pair; sta-
    mens 8. . . . . .1. *M. pigra*
Pinnae in 1–2 pairs, subdigitately arranged on the very short rhachis which is much exceeded by the petiole;
    leaves without prickles on the petiole or rhachis (sometimes bristly hairs may be present); stamens 4
. . . . .2. *M. pudica*

1. **Mimosa pigra** *L.*, Cent. Pl. 1 : 13 (1755); Fawc. & Rendle, Fl. Jam. 4 : 135 (1920); Brenan, Checklist Tang. Terr. 346 (1949); Gilbert & Boutique in F.C.B. 3 : 230 (1952); O. B. Miller in J.S. Afr. Bot. 18 : 34 (1952); Brenan in Mem. N.Y. Bot. Gard. 8 : 429 (1954); Torre in C.F.A. 2 : 268 (1956); Keay in F.W.T.A. ed. 2, 1 : 495 (1958); Brenan in F.T.E.A. Legum-Mimos. : 43, fig. 13 (1959); F. White, For. Fl. N. Rhod. 93 (1962); Schreiber in F.S.W.A. 58 : 18 (1967); Brenan & Brummitt in F.Z. 3, 1 : 49, t. 13 (1970); Ross, Fl. Natal 193 (1973). Type: *Aeschynomene spinosa quinta* Commelin, Rar. Pl. Amst. 59, t. 30 (1697) (lecto!).

Mimosa asperata L., Syst. Nat. ed. 10, 2 : 1312 (1759); Willd., Sp. Pl. 4 : 1035 (1806); DC., Prodr. 2 : 428 (1825); DC., Mém. Leg. t.63 (1827); Benth. in Hook., J. Bot. 4 : 400 (1842); Oliv. in F.T.A. 2 : 335 (1871); Benth. in Trans. Linn. Soc. Lond. 30 : 437 (1875); Burtt Davy in Kew Bull. 1908 : 162 (1908); Eyles in Trans. Roy. Soc. S. Afr. 5 : 363 (1916); Bak. f., Leg. Trop. Afr. 3 : 812 (1930); Burtt Davy, Fl. Transv. 2 : 333 (1932). Type: Origin unknown, *Herb. Linnaeus* No. 1228·32 (LINN, holo.!).

Shrub to 3 m high, sometimes scandent or rambling; stems armed with broad-based prickles up to 7 mm long, also usually appressed- or sometimes spreading-setose. *Leaves* sensitive; petioles, rhachides and rhachillae usually setulose: petiole 0,3–1,4 cm long; rhachis 2–12 cm long, with a straight, erect or forward-pointing, slender prickle up to 1 cm long at the junction of each of the (2)4–14 pairs of pinnae, often with other stouter, spreading or deflexed prickles between the pinnae pairs; rhachillae 1,8–3,6 cm long; leaflets in 18–33 pairs, 3–9 $\times$ 0,5–

1,25 mm, linear-oblong, ± appressed-pubescent, particularly on the lower surface, margins often setulose, venation nearly parallel to the midrib. *Stipules* up to 5 mm long, often persistent. *Flowers* pink or mauve, in subglobose pedunculate heads ± 1 cm in diameter, 1–2 in the axils of the upper leaves; peduncles 1–3 cm long, setulose. *Calyx* minute, 0,75–1 mm long, laciniate. *Corolla* tubular-campanulate, 2–3 mm long, lobes usually densely pubescent or minutely setulose apically. *Stamens* 8, free, up to 6 mm long; anthers eglandular apically. *Ovary* sessile, up to 2 mm long, densely villous. *Pods* clustered, brown, compressed, straight or slightly falcate, 2,5–7 × 0,8–1,2 cm, bristly all over, breaking up transversely into segments 3–5 mm long, the margins persisting as an empty frame. *Seeds* olive-brown, narrowly elliptic, up to 7 × 3,5 mm, smooth.

Widespread in tropical Africa and America, also in Madagascar and Mauritius. Found on sand or alluvium by rivers and pans and in swamps.

S.W.A.—1712 (Posto Velho): bank of Kunene River at Otjinungua, *De Winter & Leistner 5779*. 1714 (Ruacana Falls): banks of Kunene River, near Ruacana Falls, *De Winter 3657*. 1719 (Runtu): river bank at Runtu, behind Native Commissioner's hut, *De Winter 3721*. 1821 (Andara): Andara Mission Station, *De Winter 4148*.

NATAL.—2632 (Bela Vista): Ndumu Game Reserve (—CC), *Moll & Strey 3764*; Ndumu Game Reserve, N.E. shore of Nyamiti Pan (—CD), *Ross 1933*.

There is some doubt whether *M. pigra* occurs indigenously in the Transvaal. Burtt Davy, Fl. Transv. 2 : 333 (1932), recorded it from the Transvaal stating: "collected by the writer in 1904 (Davy "1535") in subtropical Transvaal (probably nr. Barberton),.....; the precise locality is uncertain, however, as the labels of *M. asperata* and *Acacia karroo* (*Davy* 1535) appear to have been transposed." The specimen referred to, namely, *Burtt Davy* "1535" is *M. pigra* but, as indicated, the locality of collection is not certain.

Repton 401 labelled "probably from northern Transvaal" and *Grobbelaar* 302 from the garden of a house in Groblersdal are the only other specimens examined from the Transvaal. The origin of all three specimens is therefore uncertain. It is possible that *M. pigra* occurs indigenously in the Transvaal, but its occurrence needs confirmation.

2. **Mimosa pudica** *L.*, Sp. Pl. 1 : 518 (1753); Willd. in L., Sp. Pl. ed. 4, 4 : 1031 (1806); DC., Prodr. 2 : 426 (1825); Benth. in Hook., J. Bot. 4 : 367 (1841); Benth. in Trans. Linn. Soc. Lond. 30 : 397 (1875);

Fawc. & Rendle, Fl. Jam. 4 : 133 (1920); Bak. f., Leg. Trop. Afr. 3 : 812 (1930); Brenan, Checklist Tang. Terr. 346 (1949); Gilbert & Boutique in F.C.B. 3 : 229 (1952); Brenan in Kew Bull. 10 : 184 (1955); Keay in F.W.T.A. ed. 2, 1 : 495 (1958); Brenan in F.T.E.A. Legum.-Mimos. : 46 (1959); Brenan & Brummitt in F.Z. 3, 1 : 51 (1970). Type: A specimen of a cultivated plant in Hort. Cliffort., *Linnaeus* (BM, lecto.!).

var. **hispida** *Brenan* in Kew Bull. 10 : 186 (1955); in F.T.E.A. Legum.-Mimos. : 46 (1959). Type: Java, *Junghuhn 719* (K, holo.!).

Annual or perennial herb, sometimes woody below, up to 1 m high, often prostrate or straggling; stems ± sparsely armed with prickles 2–5 mm long, in addition varying from densely hispid (in our variety) to almost subglabrous. *Leaves* sensitive, unarmed; petioles and rhachillae usually setulose: petiole 1–5,5 cm long; rhachis very short so that the 2 (rarely only 1) pairs of pinnae are subdigitate; rhachillae 2,5–8,8 cm long; leaflets 10–26 pairs, 6–12,5(15) × 1,2–2,75 (3) mm, linear-oblong, margins setulose, venation diverging from and not nearly parallel to the midrib. *Stipules* 8–14 mm long, persistent. *Flowers* pink or lilac, in shortly ovoid pedunculate heads ± 1–1,3 cm long and 0,6–1 cm wide, 1–5 together from the axils; peduncles 1–4 cm long, setulose. *Calyx* minute, ± 0,2 mm long. *Corolla* tubular-campanulate, 2–2,25 mm long, lobes densely grey-puberulous apically. *Bracteoles* 1,8–2,2 mm long, longer than the corollas in bud, their margins ciliate with setiform hairs which project from 1–1,5 mm beyond the corolla when in bud. *Stamens* 4, free, up to 5,5 mm long; anthers eglandular apically. *Pods* clustered, brown, compressed, straight or slightly falcate, 1–1,8 × 0,3–0,5 cm (excluding the prickles), densely setose-prickly on the margins, breaking up transversely into segments, the margins persisting as an empty frame. *Seeds* olive-brown, ± 3,5 × 3 mm.

Pantropical. Found in disturbed areas. Introduced into our area but showing signs of becoming established in some localities.

NATAL.—2930 (Pietermaritzburg): Natal Herbarium Grounds, *H. M. Forbes 1242* (NH); Durban Botanic Gardens, *Ross 1993*. 3030 (Port Shepstone): Umbogintwini valley, *Ward 6212* (E, NH).

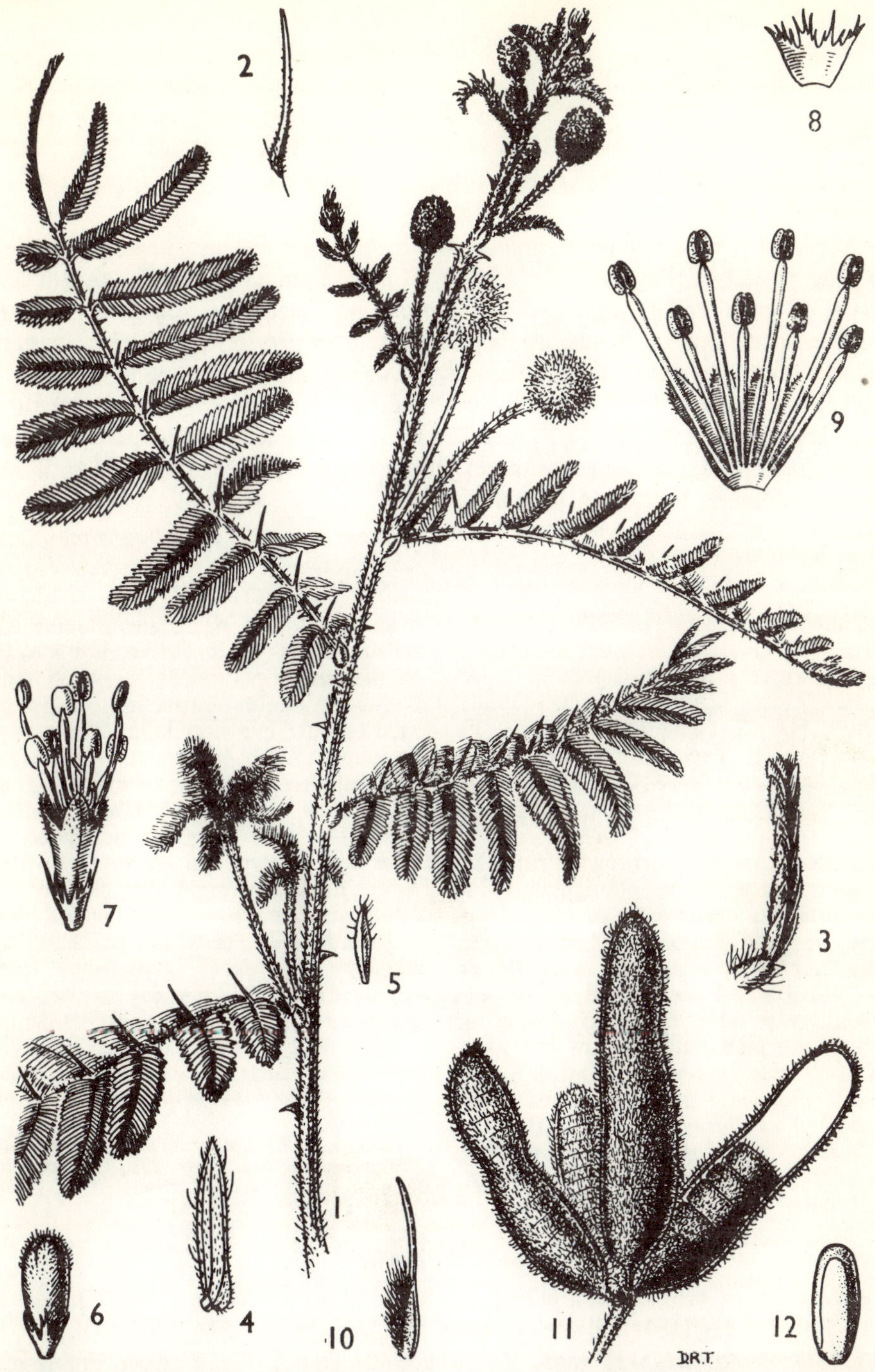

FIG. 13.—**Mimosa pigra.** 1, part of flowering stem, × 1; 2, setiform hair from peduncle, × 6; 3, part of pinna showing leaflets closed up in "sleeping" condition, × 4; 4, leaflet, × 4; 5, bract subtending flower, × 6; 6, flower-bud, × 6; 7, flower, × 6; 8, calyx, opened out, × 6; 9, corolla and stamens, opened out, × 6; 10, ovary, × 6; 11, pods, × 1; 12, seed, × 3, all from *Harris* 45. Reproduced by permission of the Editor of Flora of Tropical East Africa.

3450   5. DESMANTHUS

Desmanthus *Willd.*, Sp. Pl. 4 : 1044 (1806) nom. conserv.; Benth. & Hook. f., Gen. Pl. 1 : 592 (1865); Oliv. in F.T.A. 2 : 334 (1871); Benth. in Trans. Linn. Soc. Lond. 30 : 385 (1875); Taub. in Pflanzenfam. 3, 3 : 117 (1892); Harms in Engl., Pflanzenw. Afr. 3, 1 : 392 (1915); Bak. f., Leg. Trop. Afr. 3 : 811 (1930); Hutch., Gen. Fl. Pl. 1 : 281 (1964). Type species: *D. virgatus* (L.) Willd.

Trees, shrubs or perennial herbs, unarmed. *Stipules* setaceous, persistent. *Leaves* bipinnate; a gland often present at the junction of the lowest pair of pinnae; pinnae each with several to many pairs of small leaflets. *Inflorescences* capitate, pedunculate, axillary, solitary. *Flowers* 5-merous, in ovate-globose heads, all hermaphrodite or the lower neuter and sometimes without petals but with short staminodes. *Calyx* campanulate, shortly dentate. *Petals* free or cohering slightly. *Stamens* 10 or 5, free, exserted; anthers eglandular apically. *Ovary* subsessile; style subulate or thickened above, stigma terminal. *Pods* linear, straight or±falcate, compressed, dehiscing into 2 valves, continuous within or subseptate between seeds. *Seeds* lying lengthwise or obliquely in the pod, compressed.

A genus of ± 22 species occurring mainly in the tropics and subtropics of the New World. One species has become naturalized in our area.

Desmanthus is derived from the Greek words *dĕsmĕ*, a bundle and *anthos*, a flower.

Desmanthus virgatus (*L.*) *Willd.*, Sp. Pl. 4 : 1047 (1806); Oliv. in F.T.A. 2 : 334 (1871); Benth. in Trans. Linn. Soc. Lond. 30 : 385 (1875); Bak. f., Leg. Trop. Afr. 3 : 811 (1930); Torre in C.F.A. 2 : 267 (1956); Ross, Fl. Natal 194 (1973). Type from India, *Herb. Linnaeus No.* 1228.13 (LINN., holo.!).

Mimosa virgata L., Sp. Pl. 1 : 519 (1753). Type as above.

Unarmed perennial herb or suffrutex up to 1 m high; young stems ± angular owing to the prominent decurrent striations from the base of each petiole, subglabrous to sparingly puberulous. *Leaves* sparingly to densely puberulous: petiole 0,2–1,8 cm long, usually with a large flattened discoid or ellipsoid gland immediately below the junction of the lowest pinna pair; rhachis 1–5,5 cm long; pinnae 3–5(7) opposite pairs, rhachillae 1–4 cm long; leaflets 9–20 pairs, 3–8 × 0,8–1,5 mm, linear to linear-oblong, acute apically, glabrous throughout or with marginal cilia. *Stipules* setaceous, up to 5 mm long, linear, persistent. *Flowers* white to pale yellowish-white, in heads up to 0,8 cm in diameter; peduncles up to 5 cm long. *Calyx* campanulate, tube up to 2,5 mm long, lobes up to 1,5 mm long, glabrous. *Petals* free, up to 4 mm long, glabrous. *Stamens* 10, filaments free, linear, 5,5–7 mm long; anthers eglandular apically. *Ovary* subsessile, glabrous; style slightly thickened above. *Pods* dark brown, 4–8 cm long, up to 4 mm wide, linear, straight or slightly curved, thinly subcoriaceous, compressed, dehiscing longitudinally. *Seeds* up to 3 × 2 mm, slightly compressed, sometimes somewhat angular, lying ± obliquely in the pod, dark brown; areole small, up to 1,5 × 1 mm.

Found mainly in the tropics and subtropics of the New World, naturalized elsewhere. Introduced into our area and now established in a few localities in Natal.

NATAL.—2832 (Mtubatuba): Charters Creek Rest Camp, *Ross & Moll 5087*. 2930 (Pietermaritzburg): Jesmond Road, Pietermaritzburg, *K. D. Huntley 820* (NH, NU); Bisley, *Ross 953*.

3451   6. NEPTUNIA

Neptunia *Lour.*, Fl. Cochinch. 653 (1790); Guill. & Perr., Fl. Seneg. 238 (1832); Benth. in Hook., J. Bot. 4 : 354 (1841); Benth. & Hook. f., Gen. Pl. 1 : 592 (1865); Oliv. in F.T.A. 2 : 333 (1871); Benth. in Trans. Linn. Soc. Lond. 30 : 383 (1875); Taub. in Pflanzenfam. 3, 3 : 118 (1892); Harms in Engl., Pflanzenw. Afr. 3, 1 : 403 (1915); Bak. f., Leg. Trop. Afr. 3 : 808 (1930); Gilbert & Boutique in F.C.B. 3 : 198 (1952); Torre in C.F.A. 2 : 267 (1956); Keay in F.W.T.A. ed. 2, 1 : 496 (1958); Brenan in F.T.E.A. Legum.–Mimos. : 40 (1959);

Hutch., Gen. Fl. Pl. 1 : 291 (1964); Windler in Austr. J. Bot. 14 : 379 (1966); Schreiber in F.S.W.A. 58 : 18 (1967); Brenan & Brummitt in F.Z. 3, 1 : 45 (1970). Type species: *N. oleracea* Lour.

Herbs, aquatic or terrestrial, unarmed. *Leaves* bipinnate, frequently sensitive; pinnae each with several to numerous pairs of leaflets. *Stipules* persistent or deciduous. *Inflorescence* a globose to ellipsoid head, pedunculate, usually solitary and axillary. *Flowers* dimorphic, in the upper part of the head hermaphrodite, in the lower part male or neuter with elongate staminodes. *Calyx* 5-toothed. *Petals* 5, free or ±united. *Stamens* 5 or 10, free, exserted, all fertile in hermaphrodite flowers; anthers glandular or eglandular apically. *Pods* clustered, membranous to subcoriaceous, oblong to subcircular, compressed, dehiscent. *Seeds* ±compressed, oblong-ellipsoid to obovoid, smooth.

A genus of 11 species, widely distributed and mostly tropical. Only one species in Africa.

The generic name *Neptunia* is derived from *Neptunus*, latin for Neptune, the god of the sea, rivers and fountains; in allusion to its watery habitat.

Neptunia oleracea *Lour.*, Fl. Cochinch. 654 (1790); Benth. in Hook., J. Bot. 4 : 354 (1841); Oliv. in F.T.A. 2 : 334 (1871); Benth. in Trans. Linn. Soc. Lond. 30 : 383 (1875); Torre in C.F.A. 2 : 267 (1956); Keay in F.W.T.A. ed. 2,1 : 496 (1958); Brenan in F.T.E.A. Legum.-Mimos. : 40, fig. 12 (1959); Windler in Austr. J. Bot. 14 : 401, fig. 10 (1966); Schreiber in F.S.W.A. 58 : 18 (1967); Brenan & Brummitt in F.Z. 3,1 : 45, t.12 (1970); Ross, Fl. Natal 194 (1973). Type: Cochinchina, *Loureiro* (BM, ? holo.!).

Mimosa natans L.f., Suppl. 439 (1781) nomen confusum. Type: India (Tranquebar), Koenig 1777, *Herb. Linnaeus* No. 1228.4 (LINN, holo.!). *M. prostrata* Lam., Encycl. 1 : 10 (1783), excl. β *M. natans* L.f., nom. illegit. Types: *Niti—todda—vaddi* Rheede Hort. Malabar. 9 : 35, t. 20 (1689) (syn.!); *Mimosa orientalis non spinosa* . . . . Pluk., Almagest. Bot. 252, t. 307 fig. 4 (1696) (syn.!).

Desmanthus stolonifer DC., Prodr. 2 : 444 (1825). Type: Senegal, *Perrottet* (G—DC).

Neptunia stolonifera (DC.) Guill. & Perr., Fl. Seneg. 239 (1832). Type as for *Desmanthus stolonifer*. *N. prostrata* (Lam.) Baill. in Bull. Soc. Linn. Par. 1 : 356 (1883); Bak. f., Leg. Trop. Afr. 3 : 809 (1930); Gilbert & Boutique in F.C.B. 3 : 198 (1952). Syntypes as for *Mimosa prostrata* Lam. *N. natans* (L.f.) Druce in Rep. Bot. Soc. Exch. Club Br. Isl. 1916 : 637 (1917); Schreiber in Mitt. Bot. Staatssamml. München 2 : 285 (1957). Type as for *Mimosa natans* L.f.

Aquatic herb with swollen, creeping stems, floating or prostrate near the water's edge, with fibrous roots especially at the nodes, glabrous or rarely puberulous when young. *Leaves* very sensitive: petiole 2–6,5 cm long, eglandular; rhachis 1–4,5 cm long, eglandular; pinnae 2–4 pairs; rhachillae 1–6,5 cm long; leaflets 7–22 pairs, 5–16 × 1,5–4 mm, oblong, basal ones smaller, glabrous or sparsely ciliate on the margins. *Stipules* obliquely-ovate, 5–9 × 3–5 mm, thin, membranous, faintly nerved. *Inflorescences* pedunculate, solitary in the axils of the leaves. *Flowers* yellow, in heads 1–2 cm long; peduncles 6,5–20 cm long, glabrous. *Calyx* 1–3 mm long, glabrous. *Petals* 2,5–4 mm long, free or margins ± cohering. *Stamens* 10, free, 6–9 mm long; anthers eglandular apically even in bud; staminodes petal-like, 7–20 × 0,5–1 mm. *Ovary* 1,2–2 mm long, stipitate, glabrous; style slender, elongate. *Pods* clustered, bent almost at right angles to the short basal stipe, 1,3–2,8 × 0,8–1,2 cm, broadly oblong, compressed, membranous-coriaceous, dehiscent. *Seeds* 4–5,5 × 2,7–3,5 mm, ± compressed, brown.

Found in the tropics of the Old and New Worlds. Occurs in and near fresh water rivers, pools, lakes and swamps; sometimes in stagnant water.

S.W.A.—1719 (Runtu): W. of Runtu, *Volk 1918* (M). 1816 (Namutoni) : 64 km N. of Namutoni on road to Ondangua, *De Winter & Giess 6810.* 1820 (Tarikora): Omuramba 59, 2 km W. of Andara, *Merxmuller & Giess 2075.* Grid ref. unknown : Omuramba Omatako, *Dinter 7189*; Amboland, *Rautanen 313* (K).

TRANSVAAL.—2231 (Pafuri): Kruger National Park, Machayi Pan, near Punda Milia, *Stephen 339*; Pan in Nwambiya sandveld, *Brynard & Pienaar 4248.*

NATAL.—2632 (Bela Vista): Namannini Pan, Pongola Flood Plain, *Tinley 579.* 2732 (Ubombo): Mkuzi Game Reserve, *Ward 3069*; Pongola River near Otobotini, *Vahrmeijer & Tölken 991.*

FIG. 14.—**Neptunia oleracea.** 1, part of flowering stem, × 2; 2, leaflet, × 2; 3, stipule, × 2; 4, fertile flower, × 2; 5, anther, × 10; 6, ovary, × 5; 7, neuter flower, × 2; 8, staminode from neuter flower, × 2, all from *Bally* 6133; 9, part of fruiting stem, × 2; 10, seed, × 2, both from *Peter* 44973. Reproduced by permission of the Editor of Flora of Tropical East Africa.

3452   7. DICHROSTACHYS

Dichrostachys (*DC.*) *Wight & Arn.*, Prodr. Fl. Ind. Or. : 271 (1834) nom. conserv.; Harv. in F.C. 2 : 278 (1862); Benth. & Hook. f., Gen. Pl. 1 : 592 (1865); Harv., Gen. Pl. ed. 2 : 92 (1868); Oliv. in F.T.A. 2 : 332 (1871); Benth. in Trans. Linn. Soc. Lond. 30 : 381 (1875); Taub. in Pflanzenfam. 3, 3 : 118 (1892); Sim, For. Fl. P.E. Afr. 53 (1909); Harms in Engl., Pflanzenw. Afr. 3, 1 : 396 (1915); Bak. f., Leg. Trop. Afr. 3 : 807 (1930); Burtt Davy, Fl. Transv. 2 : 349 (1932); Phill., Gen. 392 (1951); Gilbert & Boutique in F.C.B. 3 : 198 (1952); Torre in C.F.A. 2 : 265 (1956); Brenan in F.T.E.A. Legum.–Mimos. : 36 (1959); Hutch., Gen. Fl. Pl. 1 : 291 (1964); Schreiber in F.S.W.A. 58 : 14 (1967); Brenan & Brummitt in F.Z. 3, 1 : 37 (1970). Type species : *D. cinerea* (L.) Wight & Arn.

Desmanthus Willd. sect. *Dichrostachys* DC., Mém. Leg. 12 : 428 (1826).

Cailliea Guill. & Perr. in Guill., Perr. & A. Rich., Fl. Sen. : 239 (1832).

Shrubs or small trees; unarmed or (in our species) some abbreviated lateral shoots terminating in spines; prickles absent. *Leaves* bipinnate; rhachis glandular at the insertion of some at least of the pinnae; each pinnae with several to many pairs of leaflets. *Inflorescences* of axillary spikes, solitary or appearing fascicled; upper part of spike cylindric, of hermaphrodite flowers, lower part broader, of differently coloured neuter flowers. *Calyx* shortly 5-toothed. *Petals* 5, ± united below. *Stamens* 10, all fertile in hermaphrodite flowers; anthers (in our species) with a stalked apical gland which is deciduous. *Staminodes* of neuter flowers elongate, without anthers. *Pods* clustered, coriaceous, narrowly oblong or linear, compressed, usually irregularly contorted or spiral, indehiscent or opening irregularly or (not in our species) dehiscent. *Seeds* (in the African species at least) ± compressed, ovoid to ellipsoid, smooth.

A genus of ± 20 species in the tropics of the Old World from Africa to Australia, most species in Madagascar. One species occurs in our area. The generic limits are in need of revision.

The generic name *Dichrostachys* is derived from the Greek words *dis* meaning twice, *chroa* meaning colour, and *stachys* meaning a spike.

Dichrostachys cinerea (*L.*) *Wight & Arn.*, Prodr. Fl. Ind. Or. : 271 (1834); Benth. in Hook., J. Bot. 4 : 353 (1841); in Trans. Linn. Soc. Lond. 30 : 382 (1875); Brenan in Kew Bull. 12 : 357 (1958); in F.T.E.A. Legum.-Mimos. : 36, fig. 11 (1959); Palmer & Pitman, Trees S. Afr. 169 (1961); F. White, For. Fl. N. Rhod. 432 (1962); Brenan & Brummitt in Bol. Soc. Brot., Ser. 2, 39 : 61 (1965); Gomes e Sousa, Dendrol. Mocamb. Estudo Geral. 1 : 225, t. 30 (1966); Volk in J.S.W. Afr. Wiss. Ges. 20 : 47, fig. 9 (1966); Brenan & Brummitt in F.Z. 3, 1 : 37 (1970), Van Wyk, Trees Kruger Nat. Park 1 : 170 (1972); Palmer & Pitman, Trees S. Afr. 2 : 817 (1973); Ross, Fl. Natal : 194 (1973); in Bothalia 11 : 265 (1974). Type: Sri Lanka [Ceylon], *Hermann Mus. Zeyl.* No. 215 (BM, syn.!).

Mimosa cinerea L., Sp. Pl. 1 : 520 (1753); Syst. Nat. ed. 10, 2 : 1312 (1759) non *M. cinerea* L., Sp. Pl. 1 : 517 (1753). Type as above. [See Brenan in Kew Bull. 12 : 357 (1958) for an explanation]. *M. glomerata* Forsk., Fl. Aegypt. Arab. 177 (1775). Type from Arabia.

Dichrostachys glomerata (Forsk.) Chiov. in Ann. Bot., Roma 13 : 409 (1915); Hutch. & Dalz. ex Greenway in Kew Bull. 1928 : 204 (1928); Bak. f., Leg. Trop. Afr. 3 : 807 (1930); Burtt Davy, Fl. Transv. 2 : 349 (1932); Henkel, Woody Pl. Natal 226 (1934); Flow. Pl. Afr. 23 : t. 894 (1943); Hutch., Botanist in S. Afr. 298, 299, 334, 343, 664 (1946); Brenan, Checklist Tang. Terr. 344 (1949); Codd, Trees & Shrubs Kruger Nat. Park 58, fig. 54 (1951); Gilbert & Boutique in F.C.B. 3 : 202 (1952); O. B. Miller in J.S. Afr. Bot. 18 : 31 (1952); Brenan in Mem. N.Y. Bot. Gard. 8 : 429 (1954); Torre in C.F.A. 2 : 265 (1956); Keay in F.W.T.A. ed. 2, 1 : 494, fig. 158 (1958); F. White, For. Fl. N. Rhod. 91 (1962). Type as for *M. glomerata*.

Shrub or small tree up to 7 m high, sometimes suckering and forming thickets; armed with spine-tipped abbreviated lateral shoots which often bear leaves and inflorescences, other prickles absent. *Bark* yellowish- to dark greyish-brown or blackish, usually rough, sometimes fissured; young branchlets usually ± pubescent, sometimes puberulous or glabrous. *Leaves* extremely variable in size, usually pubescent but sometimes puberulous or glabrous: petiole 0,1–5 cm long; rhachis 1–16 cm long, with a stalked or less frequently a sessile gland at the junction of each pinna pair, or at least the basal and apical pairs; pinnae (2) 4–19 pairs; rhachillae 0,6–7,5 cm long; leaflets 9–41 pairs, 1–12 × 0,3–3 mm (in our area), linear to oblong, glabrous to densely pubescent, margins with appressed or spreading cilia, sometimes glabrous, venation obscure to prominent beneath. *Inflorescences* of axillary spikes, solitary or apparently fascicled, spikes 2,5–12 cm long (including peduncle), pendulous; yellow in the upper hermaphrodite part, mauve, pink or sometimes white in the lower neuter part. *Calyx* 0,6–1,25 mm long. *Corolla* 1,5–3 mm long. *Stamens* of hermaphrodite flowers 3–3,5 mm long; staminodes 4–17 mm long. *Pods* usually dark brown, 2–10 × 0,5–1,5 cm (in our area), clustered, variously contorted or spiral, indehiscent. *Seeds* 4–6 × 3–4,5 mm, deep brown, glossy, ± compressed.

An extremely variable and taxonomically complex species, widespread in Africa and Asia and reaching Australia. Within our area, *D. cinerea* occurs commonly in South West Africa, the Transvaal, Swaziland, Natal and the northern Cape Province. It occupies a diverse range of habitats and is a conspicuous component of many communities.

An analysis of the variation within *D. cinerea*, which resulted in the recognition of a number of infraspecific taxa, was the subject of a very detailed paper by Brenan & Brummitt in Bol. Soc. Brot., Sér. 2, 39 : 61–115 (1965). During the preparation of the present account it was found that Brenan & Brummitt's treatment of the species was not altogether acceptable in our area. Consequently, certain modifications have been made to it. The difficulties encountered while attempting to name specimens in our area and the decisions arrived at are discussed in Bothalia 11 : 265 (1974). Until the species has been thoroughly investigated in the field, the present treatment can only be regarded as a provisional one. The taxonomic significance of the differential characters employed by Brenan & Brummitt to distinguish some taxa must be evaluated. Field studies may also yield valuable information about the ecological preferences of some of the taxa.

D. cinerea, commonly known as the Sicklebush or Sekelbos, has the ability to encroach rapidly into disturbed areas, particularly where the grass cover has been depleted by overgrazing. In some areas fairly large tracts of formerly open woodland have been transformed into dense thickets within relatively few years. Once established in thickets, *D. cinerea* is difficult to eradicate by mechanical means because, even when the main stems are removed, many young plants usually regenerate from the rootlets remaining in the ground.

The wood of *D. cinerea* is very hard and durable and, being termite-resistant, is considered one of the best for use as fencing posts, if sufficiently straight enough lengths can be found. It is also excellent for fuel and for making charcoal.

When using the following key, the width of the largest leaflets must be used; if any leaflets are 2 mm or more wide the specimen should be referred to subsp. *nyassana*. It is likely that most specimens can be correctly placed, but intermediates occur between most of the taxa, and these may cause difficulty. In particular, it may be difficult to decide whether some specimens should be assigned to subsp. *nyassana* or to subsp. *africana* var. *africana*. Typical subsp. *cinerea* is confined to Asia.

Some or all leaflets 2 mm or more wide; leaves often large and up to 18 cm long, with pinnae up to 7,5 cm long; peduncles usually fascicled. . . . . .subsp.   *nyassana*

All leaflets less than 2 mm wide; leaves smaller than above, pinnae usually less than 4 cm long; peduncles single or sometimes fascicled:

Surfaces of leaflets (apart from the ciliate margins) glabrous or sometimes with few hairs on the lower surface only:

Glands on leaf-rhachis stipitate or columnar, 0,5–2 mm tall, present at the junction of each pinna pair or absent from some (very rarely the gland between the lowest pair stipitate and glands between the remainder ±sessile); leaflets 0,6–1,75(2) mm wide. . . . .
. . . . .subsp. *africana* var. *africana*

Glands on leaf-rhachis sessile or very shortly (to 0,3 mm) stipitate, present at junction of all pairs of pinnae, leaflets 0,5–0,8 mm wide. . . . .
. . . . .subsp. *africana* var. *setulosa*

Both surfaces of leaflets densely pubescent
. . . . .subsp. *africana* var. *pubescens*

(a) subsp. **nyassana** (*Taub.*) *Brenan* in Kew Bull. 12 : 358 (1958); in F.T.E.A. Legum.-Mimos. : 39 (1959); Brenan & Brummitt in Bol. Soc. Brot., Sér. 2, 39 : 96 (1965); in F.Z. 3,1 : 40, t. 9 (1970). Type: Malawi, *Buchanan* 195 (B, holo.†; K!).

Dichrostachys nyassana Taub. in Engl., Pflanzenw. Ost.–Afr. C : 195 (1895); Harms in Engl., Pflanzenw. Afr. 3,1 : 398 (1915); Bak. f., Leg. Trop. Afr. 3 : 807 (1930); Burtt Davy, Fl. Transv. 2 : 349 (1932); Steedman, Trees etc. S. Rhod. 16, t. 12 (1933); Brenan, Checklist Tang. Terr. 344 (1949); Codd, Trees & Shrubs Kruger Nat. Park 58, fig. 55 (1951); Gilbert & Boutique in F.C.B. 3 : 199 (1952); Torre in Mendonca, Contr. Conhec. Fl. Mocamb. 2 : 90 (1954); Torre in C.F.A. 2 : 265 (1956). Type as above. *D. major* Sim, For. Fl. P.E. Afr. 54, t. 36A (1909). Type: Mozambique, "Lourenzo Marques and

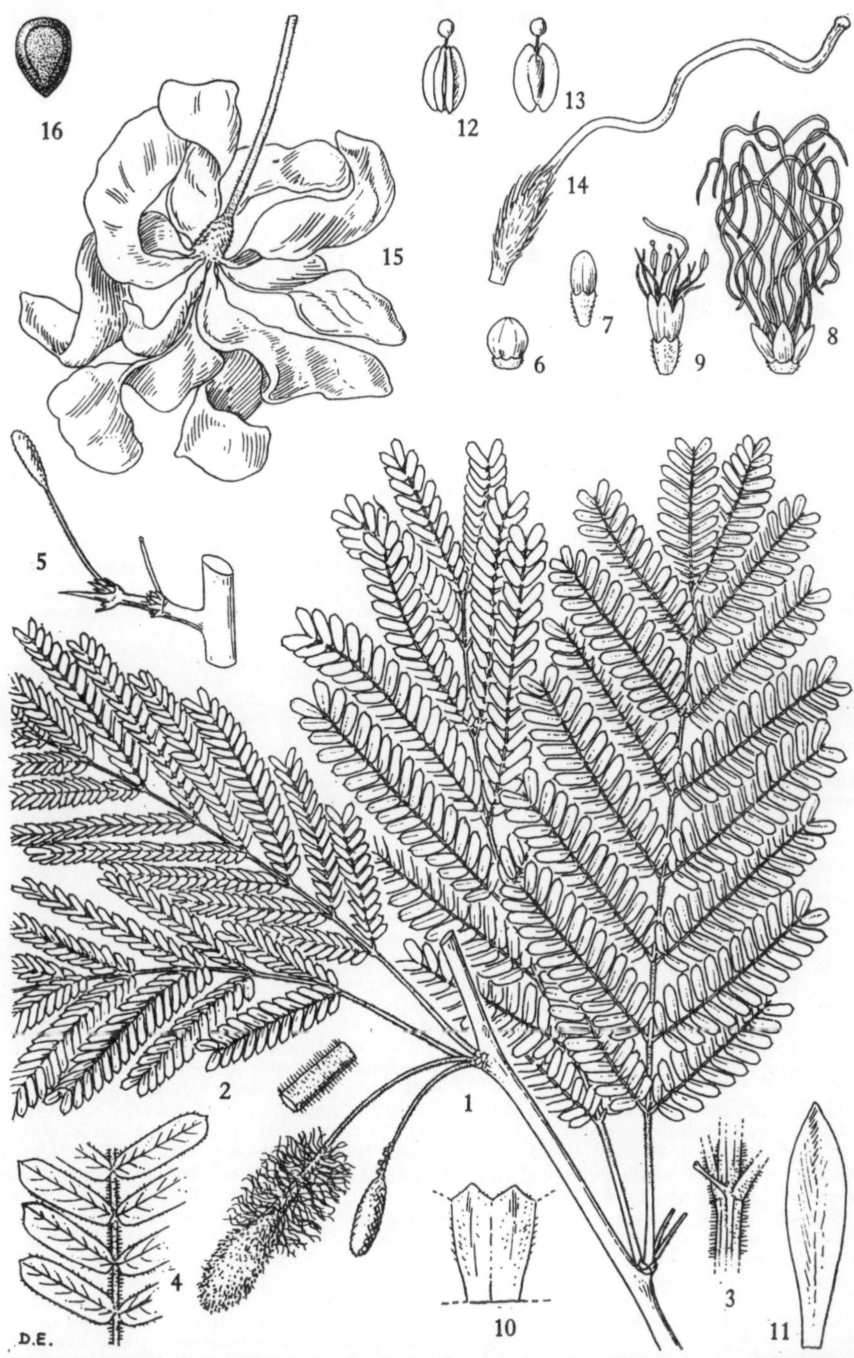

FIG. 15.—**Dichrostachys cinerea** subsp. **nyassana**. 1, flowering branch, × ⅔; 2, portion of petiole, showing indumentum, × 4; 3, part of rhachis of leaf, showing gland, × 4; 4, part of pinna, × 4, all from *Duff* 45; 5, spine, × ⅔, *Lusaka Natural History Club* 162; 6, neuter flower bud, × 4; 7, fertile flower bud, × 4; 8, neuter flower, × 4; 9, fertile flower, × 4; 10, part of calyx, × 12; 11, corolla-lobe, × 12; 12 and 13, two views of anther, × 12; 14, ovary, × 12, all from *Duff* 45; 15, cluster of pods, × ⅔, *White* 2486; 16, seed, × 2, *Boaler* 889. Reproduced by permission of the Editorial Board of Flora Zambesiaca.

Maputa up to the Lebombos", *Sim 6248* (whereabouts unknown). *D. glomerata* subsp. *nyassana* (Taub.) Brenan in Kew Bull. 11 : 188 (1956). Type as for *D. nyassana.*

Young branchlets densely spreading-pubescent. Leaves often large and up to 18 cm long, with 6–11 pairs of pinnae which are up to 7,5 cm long; leaflets 5–12 × 2–3 mm, margins shortly ascending- to appressed-ciliate or subglabrous. Peduncles ± densely spreading-pubescent, usually appearing fascicled. Pods 10–15 mm wide, loosely to tightly coiled.

Found in Zaire, Rwanda, Uganda, Tanzania, Angola, Zambia, Malawi, Rhodesia, Mozambique, the Transvaal, Swaziland and Natal. Appears to occur most commonly in woodland.

TRANSVAAL.—2231 (Pafuri): Kruger National Park, Mabasa, *Lang sub TRV 32192.* 2330 (Tzaneen): southern bank of Mtataspruit, Westfalia Estate (—CA), *Scheepers 50;* New Agatha (—CC), *McCallum 543.* 2430 (Pilgrim's Rest): Shiluvane, *Junod 691* (K). 2431 (Acornhoek): Bushbuck Ridge, *Pritchard 34.* 2528 (Pretoria): Wonderboompoort, *Leendertz 609.* 2530 (Lydenburg): Nelspruit (—BD), *Rogers 4769;* between White River and Nelspruit (—BD), *Burtt Davy 1505;* Waterval-Bo (—CB), *Rogers 14874.* 2531 (Komatipoort): Kruger National Park, 1,6 km N. of Pretorius Kop (—AB), *Codd 5192;* Barberton (—CC), *Pott 5306.* Grid ref. unknown: 32 km from Sibasa at Ivy Dell, *Rodin 4119.*

SWAZILAND.—2631 (Mbabane): Tulwane (—BC), *Compton 28800;* Sipofaneni (—DA), *Compton 28120.* 2731 (Louwsburg): 3,2 km E. of Goedgegun on Hlatikulu road, farm Buckwood, *Ross 1667* (NH, NU); *Ross 1668* (NH).

NATAL.—2731 (Louwsburg): Ngotshe, *Gerstner 2516* (NH). 2732 (Ubombo): Jozini, *Ross 1127* (K, NH, NU). 2832 (Mtubatuba): Hluhluwe Game Reserve (—AA), *Ross 942* (NH); Dukuduku (—AC), *Strey 7336.* 2930 (Pietermaritzburg): Shongweni Dam, *Morris 570.* 2931 (Stanger): New Guelderland (—AD), *Stewart 132* (NH); Umhlanga (—CA), *B. J. Huntley 75* (NH). Grid ref. unknown: Durban Flats, *Wood 1449* (NH).

In its typical form in tropical Africa subsp. *nyassana* is distinct and easily recognized by its broad leaflets, large leaves and usually fascicled peduncles. However, in our area leaflet width, leaf size and the arrangement of the peduncles provide no discontinuity between subsp. *nyassana* and subsp. *africana* var. *africana* and some specimens from the eastern Transvaal, Swaziland and Natal are extremely difficult to place with certainty. Indeed, it is sometimes a matter of opinion whether they should be assigned to subsp. *nyassana* or to subsp. *africana* var. *africana.* The problem of differentiating depauperate specimens of subsp. *nyassana* and robust specimens of subsp. *africana* var. *africana* in our area is therefore a very real and difficult one. The decision to uphold subsp. *nyassana* was taken because over most of its range in tropical Africa it is a ±distinct taxon.

There is some evidence to suggest that subsp. *nyassana* has slightly different ecological preferences to subsp. *africana* in our area, but detailed field studies are required to substantiate this.

(b) subsp. **africana** *Brenan & Brummitt* in Bol. Soc. Brot., Sér. 2, 39 : 77 (1965); Schreiber in F.S.W.A. 58 : 15 (1967); Brenan & Brummitt in F.Z. 3,1 : 42, t. 11 (1970). Type: Mozambique, Lourenzo Marques, Quinta do Umbeluzi, *Gomes e Sousa 3466* (K, holo.!).

Young branchlets ± densely pubescent to ± glabrous. Leaves usually smaller than in subsp. *nyassana,* with 4–19 pairs of pinnae which are up to ± 4 cm long; leaflets 2–7 × (0,3)0,5–1,75(2) mm wide, margins strongly ciliate to sparsely appressed-ciliate or glabrous. Pods 6–15 mm wide, loosely to tightly coiled.

This subspecies occurs throughout most of tropical Africa from the Cape Verde Is., Senegal, Ethiopia southwards to South West Africa, Botswana, the Transvaal, Swaziland, Natal and the northern Cape Province, but absent from rain-forest regions; three varieties occur in our area.

(i) var. **africana.**

Brenan & Brummitt in Bol. Soc. Brot. Sér. 2,39 : 78 (1965); in F.Z. 3,1 : 42, t. 10 fig. C (1970)*.

Mimosa nutans Pers., Syn. Pl. 2 : 266 (1807). Type: Senegal, *Adanson sub Herb. Jussieu* (P, holo.!).

Desmanthus nutans (Pers.) DC., Prodr. 2 : 446 (1825). Type as for *Mimosa nutans. D. trichostachys* DC., l.c. : 445 (1825); Mém. Leg. 444 (1827),t. 67 (1826). Type: Senegal, *Bacle & Perrottet* (G, syn.;

---

*Since Brenan & Brummitt's paper on *Dichrostachys* was published in 1965 important changes affecting autonyms (automatically established names) were introduced into Article 26 of the latest edition of the International Code of Botanical Nomenclature (1972). One of these changes is the rejection in certain circumstances of the previous ruling that autonyms must always be adopted for taxa which include the type of the correct name of the next higher taxon. In some instances this results in a name which was correct when published now being made retrospectively incorrect, and thus enforcing the adoption of another, often undesirable, name. *Dichrostachys cinerea* subsp. *africana* var. *africana* is such an example. Included in this variety was *Cailliea dichrostachys* Guill. & Perr. var. *leptostachys* (DC.) Guill. & Perr., so that the correct name for var. *africana,* which was itself correct under the Code when published, is now var. *leptostachys* under the new Code. This requires a new combination for var. *leptostachys.* However, as an attempt is to be made at the Leningrad Congress in 1975 to have the recent changes in the Code affecting autonyms reversed, it is considered undesirable to effect the new combination until the outcome of this attempt is known.

K, photo!). *D. leptostachys* DC., l.c. 445 (1825); Mém. Leg. 443 (1827), pro parte quoad lectotypum: Senegal, *Rousillon* (G, lecto.; K, photo!).

Cailliea dichrostachys Guill. & Perr. in Guill., Perr. & A. Rich., Fl. Sen. 240 (1832), nom. illegit. *C. dichrostachys* var. *leptostachys* (DC.) Guill & Perr., l.c. 239 (1832). Type as for *D. leptostachys*. *C. nutans* (Pers.) Skeels in U.S. Dept. Agric. Bur. Pl. Ind. Bull. 248 : 61 (1912). Type as for *Mimosa nutans*.

Acacia spinosa E. Mey., Comm. 170 (1836). Type: Natal, Durban [Port Natal], *Drege* (P, iso.!). "*A. cinerea* Spr.?" sensu Krauss in Flora 27 : 359 (1844) quoad specim. *Krauss 326*. *A. engleri* Schinz in Mém. Herb. Boiss. 1 : 107 (1900). Type: SouthWest Africa, between Ondonga and Uukuambi, *Rautanen* 211 (Z, lecto!).

Dichrostachys nutans (Pers.) Benth. in Hook., J. Bot. 4 : 353 (1841); Harv. in F.C. 2 : 278 (1862); Oliv. in F.T.A. 2 : 333 (1871); Benth. in Trans. Linn. Soc. Lond. 30 : 382 (1875); Harms in Warb., Kunene–Samb. Exped. 244 (1903); Sim, For. Fl. P.E. Afr. 53, t. 38A (1909); Harms in Engl., Pflanzenw. Afr. 3, 1 : 397, fig. 228 (1915). Type as for *Mimosa nutans*. *D. caffra* Meisn. ex Benth. in Hook., J. Bot. 4 : 354 (1841); Krauss in Flora 27 : 359 (1844), nomen nudum. *D. lugardiae* N.E. Br. in Kew Bull. 1909 : 106 (1909) ("lugardae"). Syntypes: Botswana, Ngamiland, Kwebe, *Lugard 42* (K!); *Mrs. Lugard 78* (K!). *D. arborea* N.E. Br. l.c. : 106 (1909); Burtt Davy, Fl. Transv. 2 : 350 (1932); O. B. Miller in J. S. Afr. Bot. 18 : 31 (1952). Type: Botswana, Totin, near Lake Ngami, *Lugard 27* (K, holo.!). *D. cinerea* subsp. *africana* var. *lugardiae* Brenan & Brummitt in Bol. Soc. Brot., Sér. 2, 39 : 91 (1965); in F.Z. 3, 1 : 44, t. 10 fig. G (1970). Types as for *D. lugardiae*. *D. cinerea* subsp. *africana* sensu Schreiber in F.S.W.A. 58 : 15 (1967) pro parte.

Glands on leaf-rhachis stipitate or columnar, 0,5–2 mm tall, present at the junction of each pinna pair or absent from some (very rarely the gland between the the lowest pair stipitate and glands between the remaining pinnae ± sessile); surfaces of leaflets (apart from ciliate margin) glabrous or sometimes with few hairs on the lower surface only; leaflets 0,6–1,75(2) mm wide.

Var. *africana* occurs more or less throughout the range of the subspecies. Occupies a diverse range of habitats including woodland, forest margins, dry thornveld, bushveld, grassland and scrub. Often forming dense thickets in disturbed areas.

S.W.A.—1715 (Ondangua): bordering Angola near Oshikango, *Rodin 2670*. 1724 (Katima Mulilo): 62,4 km from Katima Mulilo on road to Linyanti, *Killick & Leistner 3126*. 1917 (Tsumeb): Tsumeb (—BA), *Dinter 7519*; Otavi (—CB), *Dinter 5330*. 2216 (Otjimbingwe): Kuiseb, *Fleck 439a* (Z). 2217 (Windhoek): Windhoek, *Moss 17968* (BM). 2218 (Gobabis): Gobabis, *Liebenberg 4650*. Grid ref. unknown: Tsoachaub, *Fleck 488a* (Z); between Ondonga and

Uukuambi, *Rautanen 211* (Z); between Outjo and Etosha, *Werdermann & Oberdieck 2315*; farm Kumkauas, Grootfontein district, *Kinges 3022*.

TRANSVAAL.—2229 (Waterpoort): Dongola area, farm De Klundert 759, *Codd 4855*. 2230 (Messina): Makonde, *Van Warmelo 5115/5*. 2327 (Ellisras): between Ellisras and Villa Nora, *Acocks 8816*. 2329 (Pietersburg): 28,8 km E. of Pietersburg on road to Tzaneen, *Van Vuuren 1585*. 2330 (Tzaneen): Duiwelskloof, *Galpin 9651*. 2428 (Nylstroom): Warmbaths, *Irvine 114*. 2429 (Zebediela): Potgietersrust, *Thode A 1693*. 2430 (Pilgrim's Rest): Strydom Tunnel, *Strey 7894*. 2431 (Acornhoek): Klaserie, *Strey 7937*. 2527 (Rustenburg): Rustenburg, *Rogers 22359*. 2528 (Pretoria): Wonderboom, *Thode A417* (NH). 2530 (Lydenburg): 9,6 km from Nelspruit on Barberton road, *Wells 2010*. 2531 (Komatipoort): Kruger National Park, 1,6 km N. of Pretorius Kop, *Codd 5192*. 2731 (Louwsburg): 1,6 km N. of Pongola River on road to Gollel, *Ross 1705*.

SWAZILAND.—2631 (Mbabane): Malinda Hills, *Compton 27331*. 2731 (Louwsburg): Nsoko to Maloma, *Pole Evans 3406 (15)*.

NATAL.—2632 (Bela Vista): Ndumu Game Reserve, *Ross 2431*. 2731 (Louwsburg): 6,4 km from Mkuze on Nongoma road, *Ross 1638*. 2732 (Ubombo): between Jozini and Ingwavuma, foothills of Lebombo Mts., *Moll 4007*. 2828 (Bethlehem): Royal Natal National Park, *Ross 1606* (NH, NU). 2830 (Dundee): 1,6 km from Muden on Weenen road, *Ross 641* (NH, NU). 2831 (Nkandla): Umfolozi Game Reserve, Mpila, *Kluge 22* (NH). 2832 (Mtubatuba): Hluhluwe Game Reserve, *Ward 1592*. 2930 (Pietermaritzburg): Bisley, near Pietermaritzburg, *Ross 2097*. 2931 (Stanger): Lower Tugela valley, Essiena farm No. 2, *Edwards 3032*. 3030 (Port Shepstone): Gundrift, *Strey 8616*.

Subsp. *africana* var. *africana* grades into subsp. *nyassana* in our area and, as discussed under the latter, difficulty is sometimes experienced in deciding whether a specimen should be assigned to subsp. *africana* var. *africana* or to subsp. *nyassana*.

Subsp. *africana* var. *africana*, on the other hand, also grades into subsp. *africana* var. *lugardiae* (N.E. Br.) Brenan & Brummitt and into subsp. *argillicola* Brenan & Brummitt var. *hirtipes* Brenan & Brummitt in our area. There is continuous variation in leaflet width and other morphological characters between var. *africana* and var. *lugardiae* and depauperate specimens of var. *africana* tend to be confused with robust specimens of var. *lugardiae*. In view of this ± continuous variation, it is felt that little is to be gained by continuing to recognize both var. *africana* and var. *lugardiae*. Consequently var. *lugardiae* has been relegated to synonymy under var. *africana*.

In F.Z. 3, 1 : 38 (1970) subsp. *argillicola* var. *hirtipes* was distinguished from subsp. *africana* by having fewer pinnae pairs and narrower pods. In our area, however, specimens with few pinnae pairs which key out to subsp. *argillicola* var. *hirtipes* often have pods up to 1,1 cm wide so that pod width fails to provide a discontinuity between the two taxa. Although the pods on some specimens of subsp.

argillicola var. *hirtipes* are loosely coiled, on others the pods are strongly coiled and no distinction can be drawn between them and specimens of subsp. *africana* var. *africana* on the degree of coiling of the pods. The number of pinnae pairs likewise provide no discontinuity between the two taxa. As there do not appear to be any well-defined morphological, geographical or ecological discontinuities between var. *africana* and the specimens which key out to subsp. *argillicola* var. *hirtipes* in our area, the latter are also included in subsp. *africana* var. *africana*.

Two specimens collected near Ndumu in northern Tongaland on the border of Mozambique, namely, *Strey & Moll* 4014, 4020, fall within the limits of subsp. *africana* var. *plurijuga* Brenan & Brummitt. Although having slightly narrower leaflets than usual for var. *africana*, for the present these specimens are also included in var. *africana*.

Subsp. *forbesii* (Benth.) Brenan & Brummitt, with subglabrous to sparsely or densely puberulous peduncles and glabrous to sparsely appressed-puberulous young branchlets, was recorded from Natal by Brenan & Brummitt in Bol. Soc. Brot., Sér. 2,39 : 102 (1965) and in F.Z., l.c. : 40. The taxonomic significance of a ± glabrous peduncle in our area is difficult to assess. There is a suggestion that the peduncles of flowering specimens on some plants are pubescent, while peduncles on fruiting specimens from the same plant are glabrescent or glabrous. It is unknown whether plants with glabrous peduncles have distinct ecological preferences, or whether they are merely variants within populations of plants with predominantly pubescent peduncles. As little is known about subsp. *forbesii* at present, the specimens in our area with ± glabrous peduncles are, for the most part, included in var. *africana*.

Until the taxonomic significance of the differential characters employed to distinguish some of the taxa has been investigated in the field, the broad view of var. *africana* adopted here is preferred.

(ii) var. **pubescens** *Brenan & Brummitt* in Bol. Soc. Brot., Sér 2,39 : 86 (1965), in F.Z. 3,1 : 44, t. 10 fig. D (1970). Type: Mozambique, Gaza, Guijá, Aldeia da Barragem, *Barbasa & Lemos* 8149 (K, holo.!; COI; LISC; LMJ).

Glands on leaf-rhachis stipitate or columnar, 0,5–2 mm tall, present at the junction of the basal pinna pair and up to 5 of the distal pairs of pinnae; both surfaces of leaflets densely pubescent.

Found in Rhodesia, Mozambique, the Transvaal and Swaziland.

TRANSVAAL.—2328 (Baltimore): Villa Nora, *Acocks 8815*; farm Kaalhoek, 192 km N. of Potgietersrust, *Kinges 1335*. 2531 (Komatipoort): Kaapmuiden, *Rogers 25047*.

SWAZILAND.—2631 (Mbabane): Ranches Irrigation Settlement, *Compton 24669*.

Apparently infrequent and irregular in its occurrence in our area.

(iii) var. **setulosa** (*Welw. ex Oliv.*) *Brenan & Brummitt* in Bol. Soc. Brot., Sér. 2,39 : 93 (1965); in F.Z. 3,1 : 45, t. 10 fig. E (1970). Type: Angola, near Lopolo, *Welwitsch* 1800 (BM, iso.!).

Dichrostachys nutans var. *setulosa* Welw. ex Oliv. in F.T.A. 2 : 333 (1871); Hiern, Cat. Afr. Pl. Welw. 1 : 308 (1896). *D. glomerata* sensu Fl. Pl. Afr. 23 : t. 894 (1943), non (Forsk.) Chiov. sensu stricto. *D. cinerea* subsp. *africana* sensu Schreiber in F.S.W.A. 58 : 15 (1967) pro parte.

Acacia kalachariensis Schinz in Mém. Herb. Boiss. 1 : 114 (1900). Type: Kalahari, without precise locality, *Fleck* 408a (Z, holo.!).

Glands on leaf-rhachis sessile or very shortly (to 0,3 mm) stipitate, present at the junction of all pairs of pinnae; leaflets 0,5–0,8 mm wide, usually strongly and densely ciliate, hairs sometimes also present on the lower surface only.

Found in Tanzania, Angola, South West Africa, Botswana, Rhodesia, the Transvaal and northern Cape Province. Appears to favour sandy soils.

S.W.A.—1718 (Kuring-Kuru): 14,8 km S.W. of Nzinzi down Omuramba Mpungu, *De Winter 4003*. 1719 (Runtu): Runtu, *De Winter 3725*; 17,6 km W. of Sambusu Mission Station, *De Winter & Marais 4971*. 1721 (Mbambi): Mbambi, 56 km W. of Andara, *Le Roux 1067*. 1813 (Ohopoho): 19,2 km S. of Ohopoho, *De Winter & Leistner 5810*. 1917 (Tsumeb): 17,5 km S W. of Otavi on road to Otjiwarongo, *De Winter 2846*. 1920 (Tsumkwe): Tsumkwe, *Story 6443*; Samangeigei [Tsammagaigai], *Maguire 2091*. Grid ref. unknown : 32 km S. of Osiri, *Liebenberg 4684*.

TRANSVAAL.—2429 (Zebediela): Potgietersrust, *Hutchinson 1936*; *Leendertz 5968*. 2528 (Pretoria): Pienaars River, on road N.E. from Hammanskraal (—AD), *Story 1488*; Meintjies Kop, grounds of Botanical Research Institute (—CA), *Verdoorn sub PRE 27070*; Arcadia (—CA), *C. A. Smith 1651*.

CAPE.—2525 (Mafeking): 16 km S.E. of Pitsani, *Leistner 559*. 2624 (Vryburg): Palmyra (—AC), *Breuckner 1142*; "Moshesh", near Mosito (—BB), *Breuckner 259*. 2723 (Kuruman): 33,6 km N.W. of Kuruman, *Leistner 1062*. Grid ref. unknown: Armadillo Creek, Vryburg District, *Burtt Davy 13846*.

The ± sessile glands on the leaf-rhachis at the junction of each pinna pair appear to be a fairly reliable diagnostic character. Var. *setulosa*, which has a similar facies to var. *lugardiae* (now included in var. *africana*), differs from the latter almost solely on the type of glands on the leaf-rhachis. In general little difficulty has been experienced in referring specimens to this taxon except in the central Transvaal where var. *setulosa* almost grades into var. *africana*. These problematical specimens have a stalked gland at the junction of the lowest pinna pair and ± sessile glands at the junction of all of the other pinnae pairs.

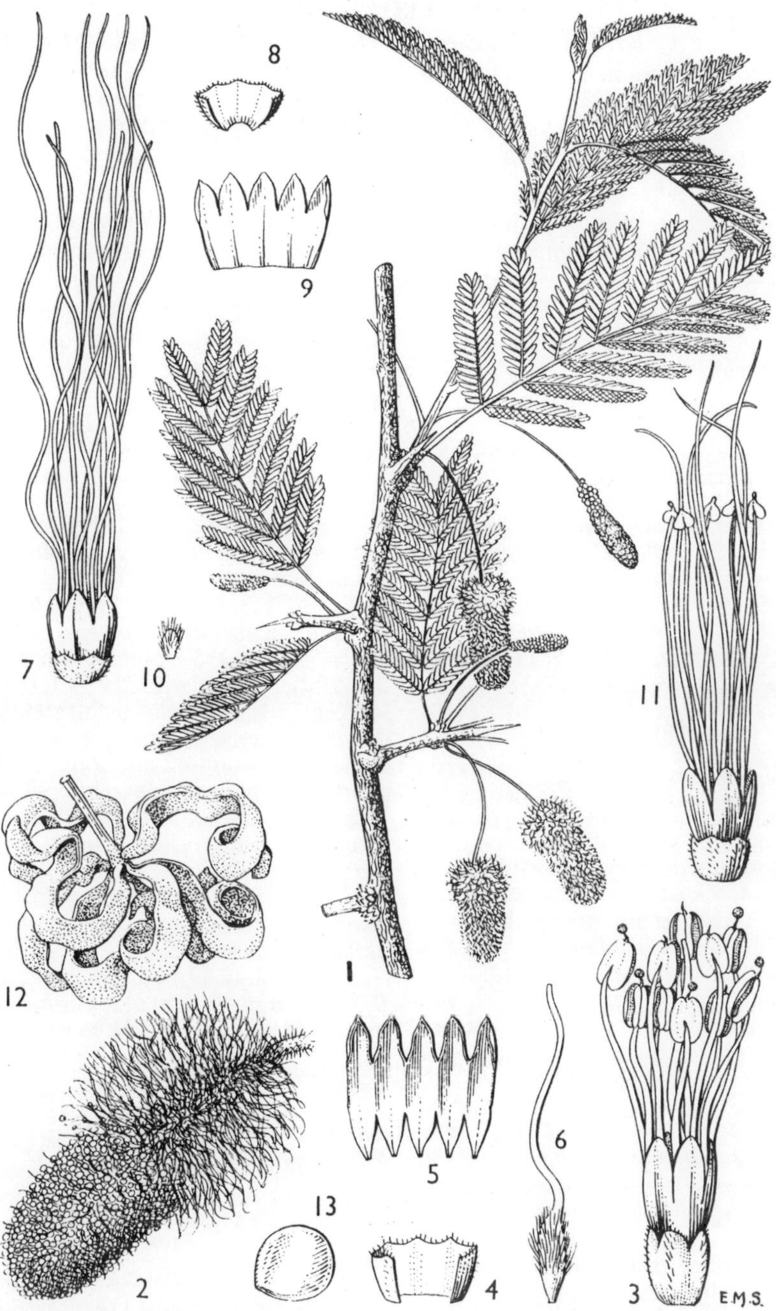

FIG. 16.—**Dichrostachys cinerea** subsp. **africana.** 1, flowering branch, × ⅔; 2, inflorescence, × 2; 3, fertile flower, × 12; 4, calyx, × 12; 5, corolla, × 12; 6, ovary, × 12; 7, neuter flower, × 12; 8, calyx of neuter flower, × 12; 9, corolla of neuter flower, × 12; 10, rudimentary ovary of neuter flower, × 12; 11, neuter flower showing intermediate stage in reduction of stamens, × 12; all from *Drummond & Hemsley* 1178; 12, cluster of pods, × ⅔; 13, seed, × 3, both from *Burtt* 1762. Reproduced by permission of the Editor of Flora of Tropical East Africa.

3453   8. **XEROCLADIA**

Xerocladia *Harv.* in F.C. 2 : 278 (1862); Benth. & Hook.f., Gen. Pl. 1 : 591 (1865); Harv., Gen. Pl. ed.2 : 92 (1868); Benth. in Trans. Linn. Soc. Lond. 30 : 381 (1875); Taub. in Pflanzenfam. 3, 3 : 118 (1892); Harms in Engl., Pflanzenw. Afr. 3, 1 : 392 (1915); Bak.f., Leg. Trop. Afr. 3 : 806 (1930); Phill., Gen. 392 (1951); Hutch., Gen. Fl. Pl. 1 : 292 (1964); Schreiber in F.S.W.A. 58 : 19 (1967). Type species: *X. viridiramis* (Burch.) Taub.

Small rigid much-branched shrubs. *Stipules* spinescent, in pairs, short and recurved. *Leaves* bipinnate, small, with 1 pinna pair; pinnae each with 6–12 pairs of linear-oblong leaflets. *Inflorescence* capitate, solitary in the axils of the leaves. *Flowers* in globose heads, hermaphrodite, 5-merous. *Calyx* divided almost to the base. *Petals* free, except basally. *Stamens* 10, filaments linear, free, the 5 opposite the petals short, the 5 alternating with the petals longer but scarcely exceeding the petals; anthers small, with a minute deciduous gland apically. *Ovary* shortly stipitate. *Pods* sessile, broadly falcate-ovate or semi-orbicular, plano-compressed, the lower suture arched and winged, 1-seeded, indehiscent. *Seeds* compressed, subcircular-elliptic, smooth.

An endemic monotypic genus recorded from Namaqualand and South West Africa.

The name *Xerocladia* is derived from the Greek words *xeros* and *kladion*, meaning dry and branch respectively.

Xerocladia viridiramis (*Burch.*) *Taub.* in Bot. Centr. 47 : 395 (1891); Dinter in Deutsch-Sudwest-Afrika 78 (1909); Bak.f., Leg. Trop. Afr. 3 : 806 (1930); Schreiber in F.S.W.A. 58 : 19 (1967). Type: Cape, Carnarvon Distr., Karel Krieger's Grave, *Burchell* 1586 (K, holo.!).

Acacia viridiramis Burch., Trav. 1 : 300 (1822); DC., Prodr. 2 : 457 (1825); Harv. in F.C. 2 : 284 (1862). Type as above.

Xerocladia zeyheri Harv. in F.C. 2 : 278 (1862); Benth. in Trans. Linn. Soc. Lond. 30 : 381 (1875). Type: Cape, Calvinia Distr., Springbokkeel, *Burke & Zeyher 558* (K, holo.!; P!; PRE!; TCD!).

Small rigid much-branched shrub up to 1 m high; branches often somewhat zigzag, pale green to olive, terete, cano-puberulous and substriate. *Stipules* spinescent, in pairs, recurved, reddish-brown, up to 3,5 mm long. *Leaves* bipinnate, small: petiole up to 3 mm long, cano-puberulous, with a stalked reddish-brown gland up to 1 mm high just below the point of attachment of the pinnae; rhachillae up to 1,5 cm long, glabrous to cano-puberulous; 1 pinna pair per leaf; leaflets in 6–12 subopposite pairs, up to 3(4) × 1(1,3) mm, oblong or linear-oblong, obtuse or acute apically, sparingly to densely pubescent on the margins only or sometimes on the lower surface also, seldom glabrous throughout, usually with a small reddish-brown gland at the base of each leaflet. *Inflorescences* capitate, pedunculate, solitary in the axils of the leaves. *Flowers* hermaphrodite, 5-merous, sessile, in heads up to 7 mm in diameter; peduncles up to 8 mm long, cano-puberulous. *Calyx* divided almost to the base, segments 1–2 mm long, up to 1,2 mm wide, densely villous externally. *Petals* free except basally, up to 2,5(3) mm long, ± 1 mm wide, linear-lanceolate to -oblong, glabrous. *Stamens* 10, filaments linear, free, the 5 opposite the petals short, the 5 alternating with the petals longer but scarcely exceeding the petals; anthers up to 0,8 mm long, with a minute deciduous apical gland. *Ovary* up to 2 mm long, 1,6 mm wide, shortly stipitate, densely villous; style glabrous; stigma truncate. *Pods* sessile, often clustered, chestnut- to reddish- or purplish-brown, up to 1,5 cm long and 1,5 cm broad, broadly falcate-ovate to semi-orbicular, compressed, the lower suture arched and winged, 1-seeded, indehiscent. *Seeds* chestnut-brown, compressed, up to 6,5 × 5 mm, subcircular-elliptic, smooth.

Found in South West Africa and Namaqualand. Occurs in sandy river beds, on river banks, alluvium and saline flats.

S.W.A.—2416 (Maltahöhe): Namseb (—DD), *Pearson 9345* (K); Christiana farm (—DD), *Steyn sub PRE 26458*. 2417 (Mariental): farm Orab, *Kinges 3428*. 2517 (Gibeon): 4,8 km S. of Asab (—DB), *Hardy 1946*; farm Gavetamas (—DB), *Giess, Volk & Bleissner 6854*. 2618 (Keetmanshoop): 54,4 km E. of Keetmanshoop on road to Aroab, *De Winter 3357*. 2619 (Aroab): farm Gross Aub, *Giess 8364*.

CAPE.—3019 (Loeriesfontein): Springbokkeel, *Burke & Zeyher 558*. 3021 (Vanwyksvlei): Vanwyksvlei, *Acocks 1749*. 3022 (Carnarvon): Kleinfontein (—BC), *Marloth 5069*; Karel Krieger's Grave (—CA), *Burchell 1586* (K). 3023 (Britstown): near Rosedale, *Reyneke sub PRE 31901*. Grid ref. unknown: Bushmanland, *Marloth 8072*.

X. viridiramis is a most distinctive and easily recognized plant.

FIG. 17.—**Xerocladia viridiramis**. 1, flowering and fruiting branch, × ⅔; 2, flower, with one anther removed, × 12; 3, ovary, × 12; 4, fruit, × 2, all from *Acocks* 1749.

3458   9. **AMBLYGONOCARPUS**

Amblygonocarpus *Harms* in Engl. & Prantl, Nat. Pflanzenfam., Nachtrag 1 : 191 (1897); Harms in Bot. Jahrb. 26 : 255 (1899); Harms in Engl., Pflanzenw. Afr. 3, 1 : 396 (1915); Bak. f., Leg. Trop. Afr. 3 : 803 (1930); Gilbert & Boutique in F.C.B. 3 : 217 (1952); Keay in F.W.T.A. ed. 2, 1 : 492 (1958); Brenan in F.T.E.A. Legum.–Mimos. : 32 (1959); Hutch., Gen. Fl. Pl. 1 : 290 (1964); Brenan & Brummitt in F.Z. 3, 1 : 35 (1970). Type species: *A. andongensis* (Welw. ex Oliv.) Exell & Torre (*A. schweinfurthii* Harms).

Unarmed tree, glabrous. *Leaves* bipinnate, eglandular; pinnae 2–6 pairs; pinnae each with several pairs of alternate or sometimes subopposite leaflets. *Inflorescences* of solitary or paired axillary racemes. *Flowers* hermaphrodite, pedicellate. *Calyx* gamosepalous, very small, with 5 (rarely 6) teeth. *Petals* 5 (rarely 6), free. *Stamens* 10 (rarely 12), free, fertile; anthers eglandular apically even in bud. *Pods* straight or nearly so, oblong, woody, indehiscent, bluntly tetragonal or subterete in section, internally septate between the seeds. *Seeds* brown, smooth, hard, unwinged.

A monotypic genus occurring in tropical Africa.

Closely related to *Tetrapleura* Benth. but differs from this genus in having eglandular anthers and pods which are bluntly tetragonal or subterete in section, while the pods of *Tetrapleura* are cruciform in section owing to the presence of a thick wing-like projection running longitudinally along each of the valves. Immature pods of *Amblygonocarpus* may have four rather prominent ribs, simulating the shape in section of those of the genus *Tetrapleura*.

The generic name *Amblygonocarpus* is a Greek compound meaning "blunt-angled fruit"; in allusion to the pods of *A. andongensis*.

Amblygonocarpus andongensis (*Welw. ex Oliv.*) *Exell & Torre* in Bol. Soc. Brot., Sér. 2, 29 : 42 (1955); Torre in C.F.A. 2 : 264 (1956); Keay in F.W.T.A. ed. 2, 1 : 492 (1958); Brenan in F.T.E.A. Legum.-Mimos. : 34, fig. 9 (1959); F. White, For. Fl. N. Rhod. 90 (1962); Brenan & Brummitt in F.Z. 3, 1 : 35, t. 8 (1970); Palmer & Pitman, Trees S. Afr. 2 : 823 (1973). Type: Angola, Cuanza Norte, Pungo Andongo, *Welwitsch* 618 (LISU, ? holo.; BM!; K!; P!).

Tetrapleura andongensis Welw. ex Oliv. in F.T.A. 2 : 331 (1871); Bak. f., Leg. Trop. Afr. 3 : 803 (1930). Type as above. *T. obtusangula* Welw. ex Oliv. in F.T.A. 2 : 331 (1871). Type: Angola, Cuanza Norte, Golungo Alto, *Welwitsch* 1751 (BM, drawing!).

Amblygonocarpus schweinfurthii Harms in Bot. Jahrb. 26 : 255 (1899); Harms in Engl. Pflanzenw. Afr. 3, 1 : 396, t.277 (1915); Eyles in Trans. Roy. Soc. S. Afr. 5 : 363 (1916); Bak. f., Leg. Trop. Afr. 3 : 804 (1930). Syntypes: Sudan, Seriba Siber Ruchama, *Schweinfurth ser.* 11, 92 (B†; BM!; K!); Seriba Agad, *Schweinfurth* 1692 (B †; BM!; K!); Angola, Malange, *Marques* 23 (B†). *A. obtusangulus* (Welw. ex Oliv.) Harms in Bot. Jahrb. 26 : 256 (1899); Bak. f., Leg. Trop. Afr. 3 : 804 (1930); Brenan, Checklist Tang. Terr. 343 (1949); Gilbert & Boutique in F.C.B. 3 : 217 (1952); O. B. Miller in J.S. Afr. Bot. 18 : 28 (1952); Gomes e Sousa, Dendrol. Moçamb. Estudo Geral. 1 : 226, t. 31 (1966). Type as for *Tetrapleura obtusangula*.

Unarmed tree up to 15 m high, glabrous throughout; bark greyish-brown to black, fissured, reticulate or scaly. *Leaves* bipinnate, eglandular: petiole 4–7,5 cm long; rhachis 2–15 cm long; pinnae 2–5 opposite or subopposite pairs; rhachillae 6,5–15 cm long; leaflets 5–9 on each side of the rhachilla, alternate or sometimes subopposite, 1,2–2,5 × 0,7–1,5 cm, elliptic to obovate-elliptic, usually emarginate apically, on petiolules 1,5–3 mm long. *Racemes* (3)6–12 cm long, on peduncles 1–3 cm long, axillary. *Flowers* yellowish-white, on pedicels 1,5–3,5 mm long. *Calyx* cupular, very small, 0,5–1 mm long, usually 5-toothed. *Petals* free, 3–4 × 0,8–1 mm. *Stamens* usually 10, filaments 5–6 mm long; anthers eglandular. *Ovary* up to 2,5 mm long, very shortly stipitate. *Pods* 8–17 × 1,8–3 cm, straight or nearly so, bluntly tetragonal or subterete in section, brown, glossy, woody, indehiscent, rounded or pointed apically. *Seeds* elliptic, 10–13 × 7–8 × 4–5 mm.

Widely distributed in the savanna regions of tropical Africa from Ghana to the Sudan, and southwards to the Caprivi Strip, Botswana, Rhodesia and Mozambique. Occurs in deciduous woodland.

S.W.A.—1724 (Katima Mulilo): Katima Mulilo, *West 3252*. Grid ref. uncertain: Caprivi Strip, E. of Cuando River, *Curson 974*.

More material of this species from our area is required.

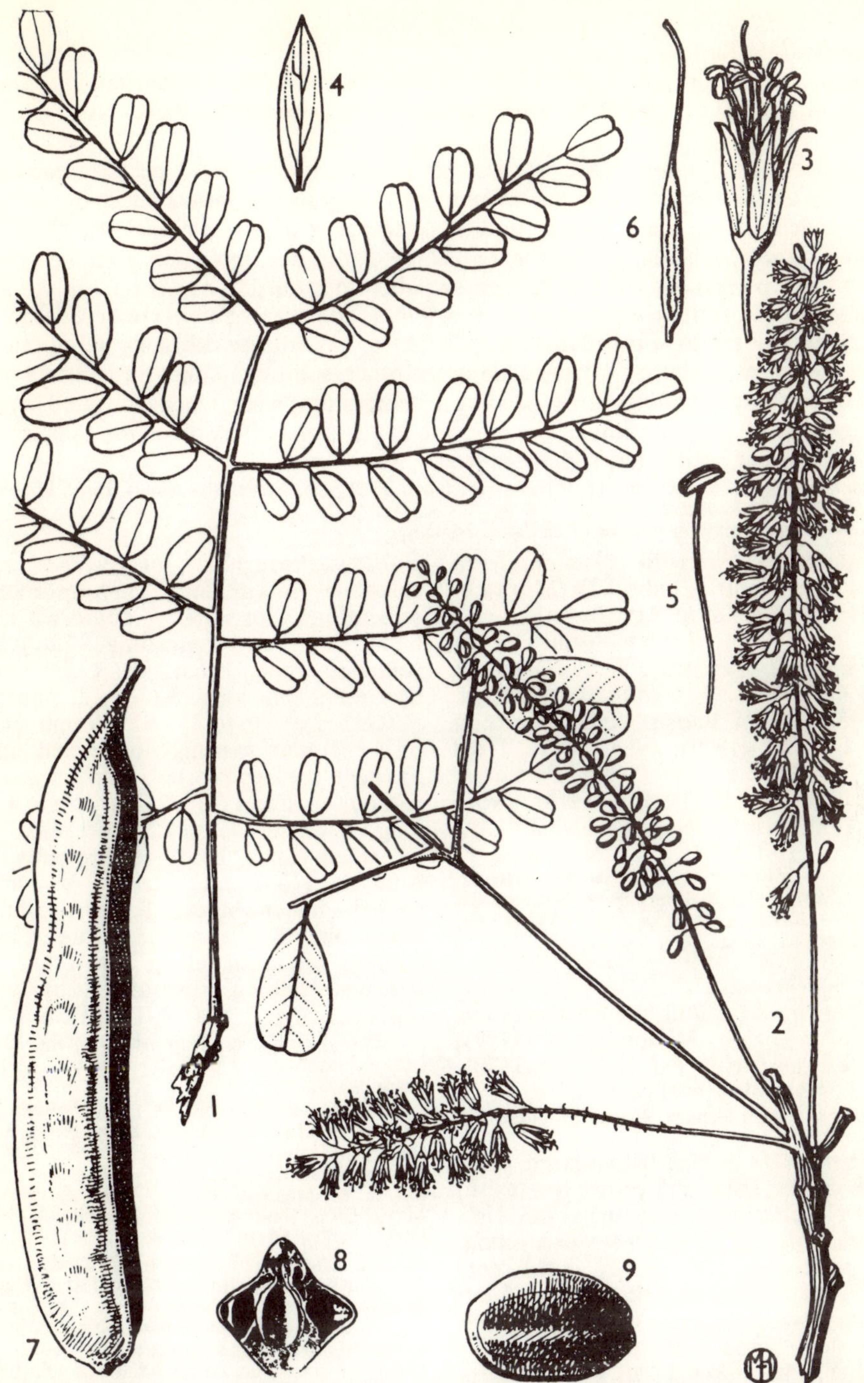

FIG. 18.—**Amblygonocarpus andongensis**. 1, leaf, × ½, *Eggeling* 6409; 2, part of flowering branch, × 1; 3, flower, × 4; 4, petal, × 6; 5, stamen, × 6; 6, ovary, × 6, all from *Eggeling* 3421; 7, pod, × ⅔; 8, cross-section of pod, × ⅔; 9, seed, × 2, all from *Dalziel* 26. Reproduced by permission of the Editor of Flora of Tropical East Africa.

3460   10. **NEWTONIA**

Newtonia *Baill.* in Bull. Soc. Linn. Par. 1 : 721 (1888); Brenan in Kew Bull. 10 : 180 (1955); in F.T.E.A. Legum.–Mimos. : 23 (1959); Hutch., Gen. Fl. Pl. 1 : 286 (1964); Brenan & Brummitt in F.Z. 3, 1 : 28 (1970). Type species: *N. duparquetiana* (Baill.) Keay (*N. insignis* Baill.).

Unarmed trees. *Leaves* bipinnate; rhachis usually with a gland between each pair of opposite pinnae; pinnae each with one to many pairs of leaflets. *Flowers* sessile or nearly so, in spikes or spiciform racemes, hermaphrodite, 5-merous. *Calyx* gamosepalous, pubescent or puberulous outside, sometimes on the margins only. *Petals* free, separated from the gynophore base by a short perigynous zone, pubescent or puberulous outside, sometimes on the margins only. *Stamens* 10, fertile; anthers with or without an apical gland. *Ovary* densely pilose outside. *Pods* straight or somewhat curved, flattened, at maturity dehiscing along one of the margins, the valves remaining attached along the other, splitting neither transversely nor into layers. *Seeds* flattened, oblong, surrounded by a membranous wing, much elongated lengthwise in the direction of the pod; cotyledons elongate in the same direction as the radicle; funicle slender, attached at or near one end of the seed.

A genus of 14 or more species, 11 of them over much of tropical Africa, the rest in tropical S. America. One species found in southern Africa.

The genus was named in honour of Sir Isaac Newton.

Newtonia hildebrandtii (*Vatke*) *Torre* in Mendonca, Contr. Conhec. Fl. Mocamb. 2 : 89 (1954); Brenan in Kew Bull. 10 : 181 (1955); in F.T.E.A. Legum.–Mimos. : 25 (1959); Dale & Greenway, Kenya Trees & Shrubs 305 (1961); F. White, For. Fl. N. Rhod. 433 (1962); Brenan & Brummitt in F.Z. 3,1 : 30 (1970); Van Wyk, Trees Kruger Nat. Park 1 : 176 (1972); Palmer & Pitman, Trees S. Afr. 2 : 819 (1973). Type: Kenya, Teita District, Ndi, *Hildebrandt 2492* (B, holo. †; BM!; K!).

Piptadenia hildebrandtii Vatke in Oest. Bot. Zeitschr. 30 : 273 (1880); Eyles in Trans. Roy. Soc. S. Afr. 5 : 364 (1916); Bak. f., Leg. Trop. Afr. 3 : 793 (1930); Brenan, Checklist Tang. Terr. 346 (1949). Type as above.

var. hildebrandtii.

Brenan in Kew Bull. 10 : 181 (1955); in F.T.E.A. Legum.–Mimos. : 25 (1959); Brenan & Brummitt in F.Z. 3,1 : 30 (1970); Ross, Fl. Natal 194 (1973).

Piptadenia sp. sensu Henkel, Woody Pl. Natal 237 (1934).

Tree up to 18 m high with a large, often rounded, crown; bark dark brown to greyish, rough or sometimes smooth; branchlets puberulous to shortly pubescent when young. *Leaves* puberulous to shortly pubescent: petiole 0,5–1,3 cm long; rhachis 2–4,5 cm long, with a sessile usually cylindrical gland between each pinna pair; pinnae (2)4–5(7) pairs; rhachillae 1,8–4,5 cm long; leaflets (8)10–19(22) pairs, 4–9,5 × 1–2 mm, ± linear-oblong or oblong, glabrous apart from the marginal cilia or sometimes sparingly pubescent on the lower surface, sometimes glabrous throughout. *Inflorescences* spicate; spikes 3–10 cm long; inflorescence axes puberulous with small ± appressed hairs or hairs short and spreading. *Flowers* pale yellowish-white, subsessile. *Calyx* cupular, up to 1,2 mm long, 5-toothed, puberulous. *Petals* 5, free, up to 3,2 × 0,9 mm, glabrous or sometimes sparingly pubescent apically. *Stamens* 10, free ,up to 5 mm long; anthers without an apical gland. *Ovary* up to 1,8 mm long, stipitate, densely pilose. *Pods* bright red when young, becoming brown with age, 8–20 × 1,3–2 cm, straight or somewhat curved, linear-oblong, flattened, dehiscing along one of the margins, the valves remaining attached along the other. *Seeds* flattened, oblong, surrounded by a membranous wing, 2,8–4 × 1,2–2 cm.

Found in Kenya, Tanzania, Zambia, Rhodesia, Mozambique and Natal (Tongaland). Occurs on sandy soils.

NATAL.—2632 (Bela Vista): Ndumu Game Reserve, *Moll 1754*; Ndumu Game Reserve, southern shore of Nyamiti Pan, *Ross 2433*. 2732 (Ubombo): 1,6 km E. of Makanes Pont (—AB), *Ross 1967*; 2,4 km E. of Makanes Pont (—AB), *Edwards 2986*; Mkuzi Game Reserve (—CB), *Ward 4559*; False Bay Park (—CD), *Strey 7344*.

Var. *hildebrandtii* occurs on the sandy soils of the Tongaland plain at altitudes below 130 m. One of the dominant species in the dry sand forest, more infrequently growing near water, for example, on the southern shore of the Nyamiti Pan in the Ndumu Game Reserve. The Tonga name for this species is umFomothi.

Var. *pubescens* Brenan, which is recorded from Tanzania, Rhodesia and Mozambique, differs from var. *hildebrandtii* in having the leaflets±densely pubescent or puberulous on both surfaces.

FIG. 19.—**Newtonia hildebrandtii**. 1, flowering branch, × ⅔, *Ross* 1967; 2, flower, × 6; 3, corolla, opened out, × 6; 4, gynoecium, × 12, all from *Ward* 2094; 5, pod, × ⅔; 6, dehisced pod showing the valves still attached along one margin, × ⅔; 7, seed, × 1, all from *Moll* 1754.

3462   **11. XYLIA**

Xylia *Benth.* in Hook., J. Bot. 4 : 417 (1842); Benth. & Hook.f., Gen. Pl. 1 : 594 (1865); Benth. in Trans. Linn. Soc. Lond. 30 : 373 (1875); Taub. in Pflanzenfam. 3, 3 : 121 (1892); Harms in Engl., Pflanzenw. Afr. 3, 1 : 404 (1915); Bak. f., Leg. Trop. Afr. 3 : 809 (1930); Gilbert & Boutique in F.C.B. 3 : 210 (1952); Keay in F.W.T.A. ed. 2, 1 : 495 (1958); Brenan in F.T.E.A. Legum.–Mimos. : 29 (1959); Hutch., Gen. Fl. Pl. 1 : 291 (1964); Brenan & Brummitt in F.Z. 3, 1 : 33 (1970). Type species: *X. xylocarpa* (Roxb.) Taub. (*X. dolabriformis* Benth.).

Unarmed trees. *Leaves* bipinnate; petiole bearing a gland at its apex at the junction of the solitary pair of pinnae; pinnae each with few to many pairs of leaflets. *Inflorescences* capitate, pedunculate, axillary or supra-axillary, solitary or paired or sometimes in threes, sometimes $\pm$ racemosely aggregated on short shoots. *Flowers* in round heads, male or hermaphrodite, 5-merous, sessile or pedicellate. *Calyx* gamosepalous, with 5-lobes. *Corolla* with 5 lobes $\pm$ united below, $\pm$ pubescent or puberulous outside. *Stamens* 10, fertile; anthers each with a caducous apical gland (rarely and only in extra-African species absent). *Ovary* pubescent. *Pods* usually obliquely obovate to oblanceolate or dolabriform, woody, compressed, dehiscing from the apex downwards into 2 recurving valves. *Seeds* lying transversely or obliquely in the pod, each sunk in a depression in the valve, usually brown, smooth, compressed, exendospermous.

A genus of $\pm$ 13 species in the tropics of the Old World, mostly in Africa and in Madagascar. One species occurs in southern Africa.

The generic name *Xylia* is derived from the Greek word for wood; in allusion to the hard wood or perhaps to the woody pods of the species of this genus.

Xylia torreana *Brenan* in Kew Bull. 12 : 359 (1958); Von Breitenbach, Indig. Trees S.Afr. 2 : 313 (1965); Gomes e Sousa, Dendrol. Mocamb. 1 : 228, t.33 (1966); Brenan & Brummitt in F.Z. 3,1 : 35, t.7 (1970); Van Wyk, Trees Kruger Nat. Park 1 : 178 (1972); Palmer & Pitman, Trees S.Afr. 2 : 823 (1973). Type: Mozambique, Maringua's village, 10 km N. of River Save, *Chase* 2244 (K, holo.!; BM!; LISC; SRGH).

Xylia africana sensu Torre in Mendonça, Contr. Conhec. Fl. Moçamb. 2 : 93 (1954).

Tree up to 15 m high, with rough dark brown to grey bark; branchlets, petioles, leaf-rhachillae and peduncles sparingly to densely brown-pubescent or tomentellous. *Leaves*: petiole 2-7 cm long, with a gland just below the junction of the solitary pair of pinnae; rhachillae 5-16 cm long, often with a gland just below the junction of some of the pairs of leaflets; leaflets 4-6 pairs, (3,5)4–7,5 (12) $\times$ 2–4,2(6) cm, narrowly ovate or rarely narrowly elliptic, rounded or slightly cordate basally, tomentose on both surfaces when young, the upper becoming glabrous, the lower remaining $\pm$ densely pubescent at maturity, especially on the midrib and veins. *Stipules* linear, up to 1 cm long, deciduous. *Flowers* yellow, in heads up to 1,8 cm in diameter, on axillary peduncles 2–3,5 cm long; pedicels up to 1,5 mm long, densely appressed-pubescent with the hairs longer and denser than on the calyx; interfloral bracts spathulate, 2–3 mm long. *Calyx* tubular-campanulate, up to 4,5 mm long, lobes up to 1,5 mm long, appressed-pubescent. *Corolla* up to 6 mm long, lobes up to 2,25 $\times$ 0,75 mm, appressed-pubescent. *Stamens* 10, filaments free, up to 1 cm long; anthers up to 0,6 mm long, each with a caducous apical gland. *Ovary* up to 3 mm long, 1,5 mm wide, pubescent; style glabrous, up to 8 mm long; stigma terminal. *Pods* obliquely obovate to oblanceolate, 9–12 $\times$ 3,2–5 cm, woody, compressed, brown-tomentellous at least in part, dehiscing from the apex downwards, the 2 valves recurving. *Seeds* lying transversely or obliquely in the pod, each sunk in a depression in the valve, compressed, brown, $\pm$ 1,2 $\times$ 0,9 cm, smooth; areole $\pm$ 6 $\times$ 4 mm.

Found in Rhodesia, Malawi, Mozambique and the north-eastern Transvaal. Occurs in deciduous woodland and, in our area, in sandveld.

TRANSVAAL.—2231 (Pafuri): Kruger National Park, Nwambiya, *Van der Schijff & Marais 3678*; *Van der Schijff 5689*; S.E. of Klopperfontein on Portuguese border, *Van der Schijff 2907*.

As no flowering specimens of *X. torreana* have been collected in our area, the description of the flowers has been drawn up from material collected outside of our area. More material of *X. torreana*, particularly flowering material, is required from our area.

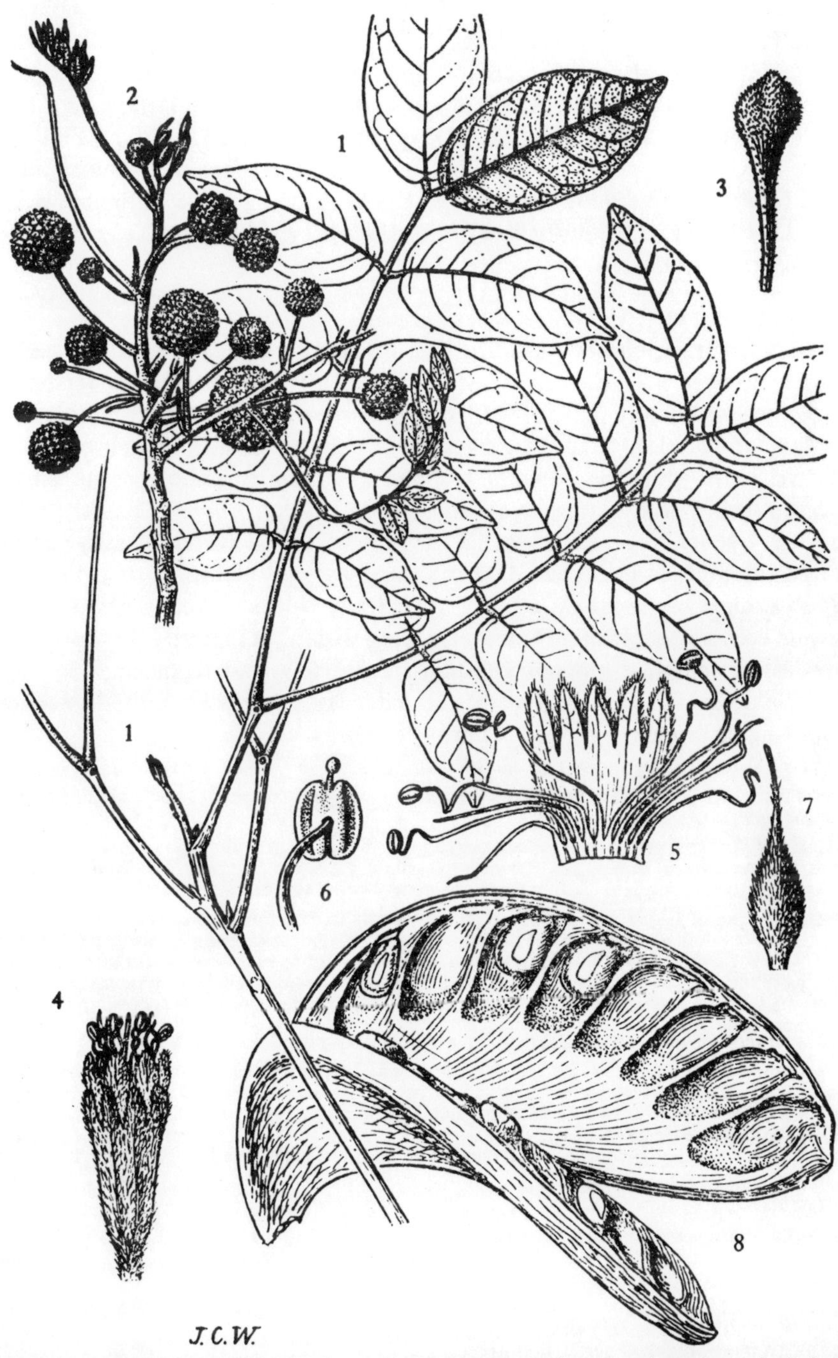

FIG. 20.—**Xylia torreana.** 1, sterile branch, × ⅔, *Torre* 2721; 2, flowering branch, × ⅔; 3, bract, × 8; 4, flower, × 4; 5, corolla, × 4; 6, anther, × 14; 7, ovary, × 4; 8, pod, × ⅔, all from *Dawe* 463. Reproduced by permission of the Editorial Board of Flora Zambesiaca.

3467   12. **ELEPHANTORRHIZA**

Elephantorrhiza *Benth.* in Hook., J. Bot. 4 : 344 (1841); Harv. in F.C. 2 : 277 (1862); Benth. &
Hook.f., Gen. Pl. 1 : 590 (1865); Harv., Gen. Pl., ed. 2 : 91 (1868); Benth. in Trans. Linn. Soc.
Lond. 30 : 365 (1875); Taub. in Pflanzenfam. 3, 3 : 122 (1892); Sim, For. Fl. Cape Col. 209
(1907); Harms in Engl., Pflanzenw. Afr. 3, 1 : 398 (1915); Phillips in Bothalia 1 : 187 (1923);
Bak.f., Leg. Trop. Afr. 3 : 800 (1930); Burtt Davy, Fl. Transv. 2 : 331 (1932); Phill., Gen. 392
(1951); Torre in C.F.A. 2 : 263 (1956); Brenan in F.T.E.A. Legum.–Mimos. : 19 (1959);
Hutch., Gen. Fl. Pl. 1 : 288 (1964); Schreiber in F.S.W.A. 58 : 16 (1967); Brenan & Brummitt
in F.Z. 3, 1 : 23 (1970); Ross in Bothalia 11 : 247 (1974). Type species: *E. elephantina* (Burch.)
Skeels (*E. burchellii* Benth.).

   Prosopis sensu E. Mey., Comm. 1 : 165 (1836); Eckl. & Zeyh., Enum. 259 (1836), non *Prosopis* L., Mant.
1 : 10 (1767).

   Unarmed small trees, shrubs or suffrutices, often with a greatly enlarged underground
rootstock or a number of rootstocks. *Leaves* bipinnate; petioles eglandular; pinnae 3–42
pairs; pinnae each with many pairs of leaflets. *Inflorescences* of spiciform racemes which are
axillary, solitary or fascicled, often ± aggregated. *Flowers* normally hermaphrodite, 5-merous,
usually pale yellowish-white, on pedicels 1–2 mm long. *Calyx* gamosepalous, small, 1–2,5
mm long, 5-toothed. *Petals* 5, free or shortly united basally. *Stamens* 10, fertile; filaments
4–7,5 mm long, free among themselves, slightly adnate to the corolla basally; anthers with
a usually rapidly deciduous apical gland. *Ovary* usually sessile, glabrous; style filiform; stigma
terminal. *Pods* straight or somewhat curved, not spirally twisted, often large and up to 45 cm
long, somewhat compressed, without transverse septa within; at maturity the valves separating
from the persistent margins, but not splitting transversely into segments; the outer layer
(exocarp) of the pod-wall often peeling off the inner layer (endocarp), the layers remaining
intact or breaking up irregularly. *Seeds* often ± compressed.

   A genus of 9 species restricted to Africa south of the equator.

   The generic name *Elephantorrhiza* is a Greek compound meaning "elephant root"; in allusion to the large
roots of *E. elephantina*.

   The underground root systems of each species need to be investigated as they may provide useful additional
means of distinguishing some of the species. For example, there is a suggestion that *E. obliqua* and *E. elephantina*
have different root systems, but field observations are necessary to substantiate this.

   In keying out species of *Elephantorrhiza*, emphasis is usually laid on the habit of the plants, that is, whether
the plants are suffrutices with unbranched aerial stems or whether they are shrubs or small trees with branched
aerial stems. This is the character employed in the first dichotomy of the key and, as far as it is known, it is a
fairly reliable character. However, *E. elephantina*, which typically has unbranched aerial stems, may prove an
exception when the growing apex has been damaged because then the stems sometimes develop lateral branches.

Suffrutex with unbranched (unless damaged) annual aerial stems up to 0,75(1) m high:

    Pinnae (1)2–6 pairs per leaf; leaflets 4–13(21) pairs per pinna, 2–6,5 mm wide, ovate to ovate-oblong, very
        oblique basally, midrib starting in the distal corner of the leaflet-base and gradually becoming ±
        central in the leaflet, lateral nerves and veins usually prominent; confined to the Transvaal..1. *E. obliqua*
    Pinnae (2)8–17 pairs per leaf; leaflets (7)12–45(55) pairs per pinna, 0,5–2(2,5) mm wide, linear to linear-
        oblong, rarely narrowly oblanceolate, asymmetric basally; widespread............2. *E. elephantina*

Shrub or small tree with woody branched aerial stems up to 7 m high, very rarely a suffrutex with branched
    procumbent stems:

    Leaflets with the midrib marginal throughout, in (17)27–40(50) pairs per pinna, 3–7,5 × 0,5–1,2 mm
    ....9. *E. suffruticosa*
    Leaflets with the midrib central or nearly so, at least towards the apex:

        Suffrutex; branched aerial stems procumbent, longitudinally striate, glabrous to densely puberulous;
            leaves with (2)5–10 pinnae pairs...3. *E. woodii*
        Shrub or small tree; branched aerial stems erect, glabrous:

            Leaflets only slightly asymmetric basally, with the proximal side cuneate to slightly rounded,
               1,5–3,5(4,5) mm wide; leaves with (1)4–8(9) pairs of pinnae....................4. *E. burkei*

Leaflets ± strongly asymmetric basally, with the proximal side broadly rounded-truncate to almost auriculate and the distal side cuneate; leaves with 2–41 pairs of pinnae:

Leaves with (9)12–30(41) pinnae pairs; leaflets 0,7–2,5 mm wide, usually on very short petiolules; pods 1,3–2,2 cm wide, very narrow in proportion to their length, when mature the position of the seeds usually marked by distinct raised bumps...........................6. *E. goetzei*
Leaves with 2–12(14) pinnae pairs; leaflets 0,9–3,5 mm wide, sessile; pods 2–3,9 cm wide, ± compressed, position of seeds not marked by distinct raised bumps:

Leaflets 0,9–1,5 mm wide; racemes short, 4–5,5 cm long, usually aggregated on short lateral branchlets, less frequently solitary or fascicled; pods 12–18 × 2–3,2 cm; restricted to the eastern Transvaal..5. *E. praetermissa*
Leaflets (1)1,5–3,5 mm wide; racemes 5,5–9,5 cm long, axillary, solitary or paired; pods (15)18,5–40,5 × 2–3,9 cm; restricted to South West Africa:

Pinnae (2)6–14 pairs per leaf; calyx up to 1,5 mm long; pods 3–3,9 cm wide....8. *E. schinziana*

Pinnae 3–7(9) pairs per leaf; calyx 2–2,25 mm long; pods 2–2,5 cm wide........7. *E. rangei*

1. Elephantorrhiza obliqua *Burtt Davy*

in Kew Bull. 1921: 191 (1921); Phillips in Bothalia 1 : 189 (1923); Burtt Davy, Fl. Transv. 2 : 332 (1932) pro parte excl. specim. *Rogers* 22011; Ross in Bothalia 11 : 248 (1974). Type: Transvaal, between Carolina and Oshoek, at an outspan ± 1,6 km from Robinson's, *Burtt Davy* 2976 (BM, holo.!; FHO!; K!).

Suffrutex producing at ground level annual herbaceous stems up to 30 cm high from a number of underground rhizomes; aerial stems unbranched (rarely branched after damage to the main apex), longitudinally striate, pubescent or glabrous. *Leaves* pubescent or glabrous: petiole 2–6 cm long; rhachis (0)1,5–9 cm long; pinnae (1)2–6 pairs; rhachillae 2–11 cm long; leaflets 4–13(21) pairs per pinna, 5,5–15 × 2–6,5 mm, very oblique, ovate to ovate-oblong, broadly truncate basally, asymmetric and attached by one corner, midrib starting in the distal corner of the leaflet-base and gradually becoming almost central in the leaflet, usually with 2–3 other prominent veins arising from the leaflet-base, midrib and lateral nerves prominent above and below, acute or distinctly mucronate apically, glabrous or sparingly pubescent on the margins. *Racemes* axillary, often solitary, on the lower or the apical part of the stem, 3,5–6 cm long (including the peduncle), glabrous or very sparingly pubescent. *Flowers* yellowish-white, on pedicels up to 1,5 mm long, with minute reddish glands at the base of the pedicels. *Calyx* campanulate, up to 2 mm long, shortly 5-toothed, glabrous. *Petals* shortly united below, up to 4,5 mm long, 1 mm wide, linear-oblong, inflexed apically, glabrous. *Stamens* free among themselves, slightly adnate to the corolla

basally; filaments up to 7,5 mm long; anthers up to 0,8 mm long, with a deciduous apical gland. *Ovary* up to 2 mm long, glabrous, sessile. *Pod* (only one collected) dark purplish-brown, 11 × 4 cm, straight, compressed, prominently transversely venose.

Apparently confined to the Transvaal. Occurs in grassland.

E. obliqua is readily distinguished from all other species by its large ovate leaflets with prominent venation. *E. obliqua* appears to have a different underground root system to *E. elephantina*, but field observations are required to substantiate this.

Stems pubescent; petioles, rhachides and rhachillae sparingly pubescent........(a) var. *obliqua*

Stems glabrous; petioles, rhachides and rhachillae glabrous..................(b) var. *glabra*

(a) var. **obliqua.**

Phillips in Bothalia 1 : 189 (1923); Burtt Davy, Fl. Transv. 2 : 332 (1932) pro parte excl. specim. *Rogers* 22011; Ross in Bothalia l.c. : 248. Type as above.

TRANSVAAL.—2630 (Carolina): between Carolina and Oshoek, at an outspan ± 1,6 km from Robinson's farm, *Burtt Davy* 2976 (BM, FHO, K); Billy's Vlei, Lake Chrissie, *Pole Evans sub PRE 13185*.

More material of *E. obliqua*, particularly fruiting material, is required.

Rogers 22011 from the Pietersburg District of the Transvaal, cited under *E. obliqua* var. *obliqua* by Burtt Davy l.c. : 332 (1932), is in fact *Dichrostachys cinerea* (L.) Wight & Arn. subsp. *nyassana* (Taub.) Brenan.

(b) var. **glabra** *Phillips* in Bothalia 1 : 189, t.5 fig. 1 (1923); Burtt Davy, Fl. Transv. 2 : 332 (1932); Ross in Bothalia l.c. : 249. Syntypes: Transvaal, Middelburg Distr., Botsabelo, *Eiselen sub PRE 1229* (GRA!; K!; PRE!); Middelburg, *Jenkins sub TRV 9128* (PRE!).

E. transvaalensis Phillips ined.

TRANSVAAL.—2529 (Witbank): 8 km N. of Middelburg, Aasvoegelskop (—CB), *Dyer 3934* (PRE); Doornkop (—CB), *Du Plessis 1074* (PRE); near Middelburg (—CD), *Gower s.n.* (PRE).

Codd 10119 from Bellevue farm near Twenty-four Rivers in the Waterberg District of the Transvaal resembles *E. obliqua*. The stem, petioles, rhachides and rhachillae are glabrous or almost so, and the leaves have up to 6 pinnae pairs and up to 19 pairs of leaflets per pinna. The leaflets are 9–18 × 3–4 mm, ± oblong, oblique basally, with an excentric midrib and two other prominent veins arising from the leaflet-base, conspicuously venose and distinctly mucronate apically. Although leaflet shape differs somewhat from the leaflet shape of the syntypes, *Codd* 10119 is closer to *E. obliqua* than to any of the other species and, for the present, is referred to *E. obliqua* var. *glabra*. Unfortunately *Codd* 10119 is sterile. Further collections are required to indicate whether or not *Codd* 10119 falls within the range of variation of *E. obliqua*.

2. Elephantorrhiza elephantina (*Burch.*) *Skeels*

2. **Elephantorrhiza elephantina** (*Burch.*) *Skeels* in U.S. Dept. Agric. Bur. Pl. Ind. Bull. 176 : 29 (1910); Bak.f., Leg. Trop. Afr. 3 : 800 (1930); Burtt Davy, Fl. Transv. 2 : 332 (1932); O.B. Miller in J. S. Afr. Bot. 18 : 31 (1952); Leistner, Mem. Bot. Surv. S. Afr. 38 : 123, t.14 (1967); Schreiber in F.S.W.A. 58 : 16 (1967); Van der Schijff & Snyman in J. Arn. Arb. 51 : 114 (1970); Brenan & Brummitt in F.Z. 3,1 : 27 (1970); Ross, Fl. Natal 194 (1973); in Bothalia 11 : 249 (1974). Type: Cape Province, Kuruman Distr., between Matlowing [Mashowing] River and Kuru, *Burchell* 2410 (K, holo.!; P!).

Acacia elephantina Burch., Trav. 2 : 236 (1824). Type as above. *A. elephantorhiza* DC., Prodr. 2 : 457 (1825) nom. illegit. Type as above.

Prosopis elephantorrhiza (DC.) Spreng., Syst. Cur. Post. iv : 165 (1827); Eckl. & Zeyh., Enum. 259 (1836). Type as above. *P. elephantina* (Burch.) E. Mey., Comm. 165 (1836). Type as above.

Elephantorrhiza burchellii Benth. in Hook., J. Bot. 4 : 344 (1841) nom. illegit.; Harv. in F.C. 2 : 277 (1862); Benth. in Trans. Linn. Soc. Lond. 30 : 365 (1875); MacOwan in Agric. J. Cape G.H. 10 : 29 (1897); Schinz in Mém. Herb. Boiss. 1 : 116 (1900); Sim., For. Fl. Cape Col. 209, t.16, viii (1907); Dinter, Deutsch-Sudwest-Afrika Fl. Forst und landwirtschaft. Frag. 78 (1909); Harms in Engl., Pflanzenw. Afr. 3,1 : 400, t.229 (1915); Dinter in Feddes Repert. 17 : 190 (1921); Hofmeyer in S. Afr. J. Nat. Hist. 3 : 215 (1921); Phillips in Bothalia 1 : 189, t.5 fig. 2 (1923); Marloth, Fl. S. Afr. 2 : fig. 26 (1925). Type as above. *E. rangei* ("*rangeri*") sensu Phillips in Bothalia 1 : 192, t.5 fig. 5 (1923) pro parte quoad specim. *Dinter* 2264 et *Herb. Mus. Austro-Afr.* 4485, non Harms sensu stricto. *E. dinteri* Phillips ined.

Suffrutex producing at ground level annual stems 20–90 cm high from the woody end of a ± elongate rhizome; aerial stems unbranched except for inflorescences (rarely branched after damage to the main apex); young stems glabrous or rarely pubescent. *Leaves* glabrous or sparingly pubescent: petiole 1,3–3,6(8) cm long; rhachis 3,5–13,5(17,5) cm long; pinnae 2–4 pairs in lower leaves, increasing to 7–17 opposite or subopposite pairs in upper leaves; rhachillae 3–9(10,5) cm long; leaflets (7)12–45(55) pairs per pinna, (4)5–10(15) × (0,3)0,5–2(2,5) mm, linear to linear-oblong, rarely narrowly oblanceolate, glabrous or almost so, base nearly always asymmetric, with the proximal side rounded to cuneate, apex symmetric to asymmetric, acute, usually mucronate or rarely obtuse, lateral nerves and veins prominent or not. *Racemes* usually confined to the lower part of the stem, axillary, solitary or clustered, (2)3,5–8(12) cm long (including the peduncle), glabrous or very rarely pubescent. *Flowers* yellowish-white, on pedicels up to 1,5 mm long, pedicels articulated near the middle, with minute reddish or reddish-brown glands at the base of the pedicels. *Calyx* shortly campanulate, up to 1,75 mm long, 5-toothed, glabrous. *Petals* free or slightly connate basally, 2,75–3,75 mm long, up to 1 mm wide, linear-oblong, inflexed apically, glabrous. *Stamens* free among themselves, slightly adnate to the corolla basally; filaments up to 6,5 mm long; anthers up to 1 mm long, with a deciduous apical gland. *Ovary* up to 1,75 mm long, linear, shortly stipitate, glabrous. *Pods* dark brown or reddish-brown, (5)9,5–15(21) × 3–5,7 cm, straight or slightly curved, oblong, compressed, usually prominently transversely venose, often umbonate over the seeds, at maturity the valves separating from the persistent margins, the outer layer of the pod-wall peeling off the inner layer, the layers usually breaking up irregularly. *Seeds* 18–26 × 13–18 × 6–13 mm, ± ellipsoid.

Found in South West Africa, Botswana, Rhodesia, Mozambique, the Transvaal, Orange Free State, Swaziland, Natal, Lesotho and Cape Province. Occurs in grassland and open scrub; often gregarious.

S.W.A.—2217 (Windhoek): Bodenhausen, *Seydel 2388*. 2219 (Sandfontein): Oas, *Seydel 3761*. Grid ref. unknown: Lichtenstein, *Dinter s.n.* (Z); Kalahari, Nosob, *Fleck 399a* (Z); *Fleck 398a* (Z).

TRANSVAAL.—2330 (Tzaneen): Houtbosch, *Rehmann 6280* (Z). 2429 (Zebediela): Percy Fyfe Nature Reserve, *Huntley 1492*. 2430 (Pilgrim's Rest): Pilgrim's Rest, *Rogers 23066*. 2431 (Acornhoek): Kruger National Park, 29 km from Satara on Rabelais road, *Van der Schijff 3291*. 2527 (Rustenburg): Rustenburg, *Nation 225* (K). 2528 (Pretoria): Groenkloof, *Phillips 3051*. 2529 (Witbank): ± 11 km from Middelburg on road to Hendrina, *Marsh 115*. 2531 (Komatipoort): Barberton, *Galpin 562*. 2627 (Potchefstroom): Vereeniging, *Burtt Davy 15084*. 2628 (Johannesburg): 12,8 km from Heidelberg on Brakpan road, *Marsh 57*. 2629 (Bethal): 11 km from Ermelo on road to Hendrina, *Marsh 96*.

O.F.S.—2627 (Potchefstroom): Sasolburg, *Theron 569*. 2727 (Kroonstad): near Kroonstad, *Pont 454* (Z). 2825 (Boshof): 24 km from Kimberley along Boshof road, *Badenhorst 86*. 2827 (Senekal): Doornkop, *Goossens 901*. 2828 (Bethlehem): Bethlehem, *Phillips 3186*. 2829 (Harrismith): Harrismith, *Sankey 35* (K). 2926 (Bloemfontein): P.O. De Burg, Bloemfontein, *Cyrus sub PRE 8794*. 2927 (Maseru): Ladybrand, *Rogers sub TRV 5057*. Grid ref. unknown: Olifantsfontein, *Rehmann 3512* (K, Z).

SWAZILAND.—2631 (Mbabane): Black Mbuluzi Falls, *Compton 28175*; Evelyn Baring Bridge, *Compton 29160*.

NATAL.—2729 (Volksrus): Laingsnek, 18,4 km from Newcastle, *Marsh 65*. 2732 (Ubombo): 4,8 km W. of Jozini, Lebombo Mts., *Edwards 2914*. 2829 (Harrismith): Harts Hill, near Colenso, *Strey 9942*. 2831 (Nkandla): Mtunzini, *Mogg 5803*. 2832 (Mtubatuba): Hluhluwe Game Reserve, *Ward 2704*. 2930 (Pietermaritzburg): Inanda, *Wood 634* (K, Z). 2931 (Stanger): Mount Moreland, *Wood 2607* (NH).

LESOTHO.—2828 (Bethlehem): Leribe, *Dieterlen 46* (P).

CAPE.—2623 (Morokweng): bank of Matlowing River, between Takoon [Takun] and Molito, *Burchell 2310* (K). 2624 (Vryburg): farm Palmyra, 96 km N.W. of Vryburg, *Rodin 3532*. 2722 (Olifantshoek): between Matlowing River and Kuru, *Burchell 2410* (K, P). 2723 (Kuruman): between source of Kuruman River and Kosi Fontein, *Burchell 2537* (K.). 2824 (Kimberley): Kimberley, *Marloth 852* (GRA). 3126 (Queenstown): "Prospect", Queenstown, *Galpin 1917* (GRA). 3128 (Umtata): Umtata aerodrome. *Strey 11164*. 3226 (Fort Beaufort): Shiloh, *Baur 379* (GRA, K); between Klipplaatrivier and Swart Kei, *Drege* (BM, K).

E. elephantina, commonly known as "Elands Boontjie", is the commonest and most widespread species. *E. elephantina* shows considerable variation in the number of pinnae pairs and in the number, size and shape of the leaflets. This variation appears to be to some extent geographical. There is a tendency for specimens from South West Africa, Botswana, the western portion of Rhodesia, the western Transvaal, Orange Free State and northern Cape to have leaves with fewer than 10 pinnae pairs, fewer than 26 leaflet pairs per pinna and leaflets more than 8 × 1 mm. The leaflets in these areas are frequently glaucous and the midrib is close to the distal margin basally but gradually becomes ± centric so that the leaflets are ± symmetric apically. In the eastern areas of Rhodesia, Mozambique, the eastern Transvaal, Swaziland and

Natal there is a tendency for specimens to have leaves with more than 10 pinnae pairs, more than 26 leaflet pairs per pinna and leaflets less than 8 mm long and 1 mm wide. The midrib in these specimens is very close to the distal margin of the leaflets throughout their length as in *E. suffruticosa* and the leaflets are asymmetric apically. The Pretoria District of the Transvaal appears to be a critical area for to the west the leaves tend to have fewer than 10 pinnae pairs and fewer pairs of large leaflets, while to the east the leaves tend to have more than 10 pinnae pairs and more numerous pairs of smaller leaflets. The extremes, for example *Seydel 3761* from Oas in South West Africa and *Wood 634* from Inanda in Natal, look very different but as there is ± continuous variation throughout and the individual characters often vary independently no means has been found of satisfactorily delimiting the two groups.

The thick red underground rootstocks were at one time used for tanning and dyeing. Burtt Davy, Fl. Transv. 2 : 332 (1932), reports having dug up a rootstock ± 8 metres long at Vereeniging.

Although the exocarp of the ripe pod is fairly hard, it readily absorbs water and soon starts to disintegrate. Seeds often germinate within the moist disintegrating pods on the surface of the soil. The interesting and unusual type of germination of the seeds of *E. elephantina* is discussed by Hofmeyer in S. Afr. J. Nat. Hist. 3 : 215 (1921) and by Van der Schijff and Snyman in J. Arn. Arb. 51 : 114 (1970).

3. **Elephantorrhiza woodii** *Phillips* in Bothalia 1 : 193, t.5 fig. 6 (1923); Ross, Fl. Natal 194 (1973); in Bothalia 11 : 250 (1974). Type: Natal, Klip River Distr., Pieters, near Colenso, *Wood 7958* (NH, holo.!; PRE!).

Suffrutex (but see below) producing at ground level annual procumbent branched stems up to 60 cm long from an elongate rhizome; aerial stems longitudinally striate, glabrous to densely puberulous. *Leaves* glabrous to densely puberulous: petiole 0,8–1,6 cm long; rhachis (1)3,5–8,5(13) cm long, distinctly sulcate above; pinnae (2)5–10 opposite or subopposite pairs; rhachillae 1,8–6 cm long; leaflets 12–28 pairs per pinna, 2,5–6(9) × 1–1,8(2,25) mm, linear to linear-oblong, glabrous, asymmetric basally, midrib starting in the distal corner of the leaflet-base and gradually becoming almost central in the leaflet, proximal side of base rounded, apex symmetric or asymmetric, acute or obtuse, mucronate, midrib prominent or not, lateral veins indistinct. *Racemes* axillary, usually solitary, 4,5–9,5 cm long (including the peduncle), glabrous to densely puberulous. *Flowers* yellowish-white, on pedicels up to 1,25 mm long, pedicels articulated near the middle, with minute glands at the base of the pedicels. *Calyx* up to 1,5 mm long, shortly 5-toothed, glabrous.

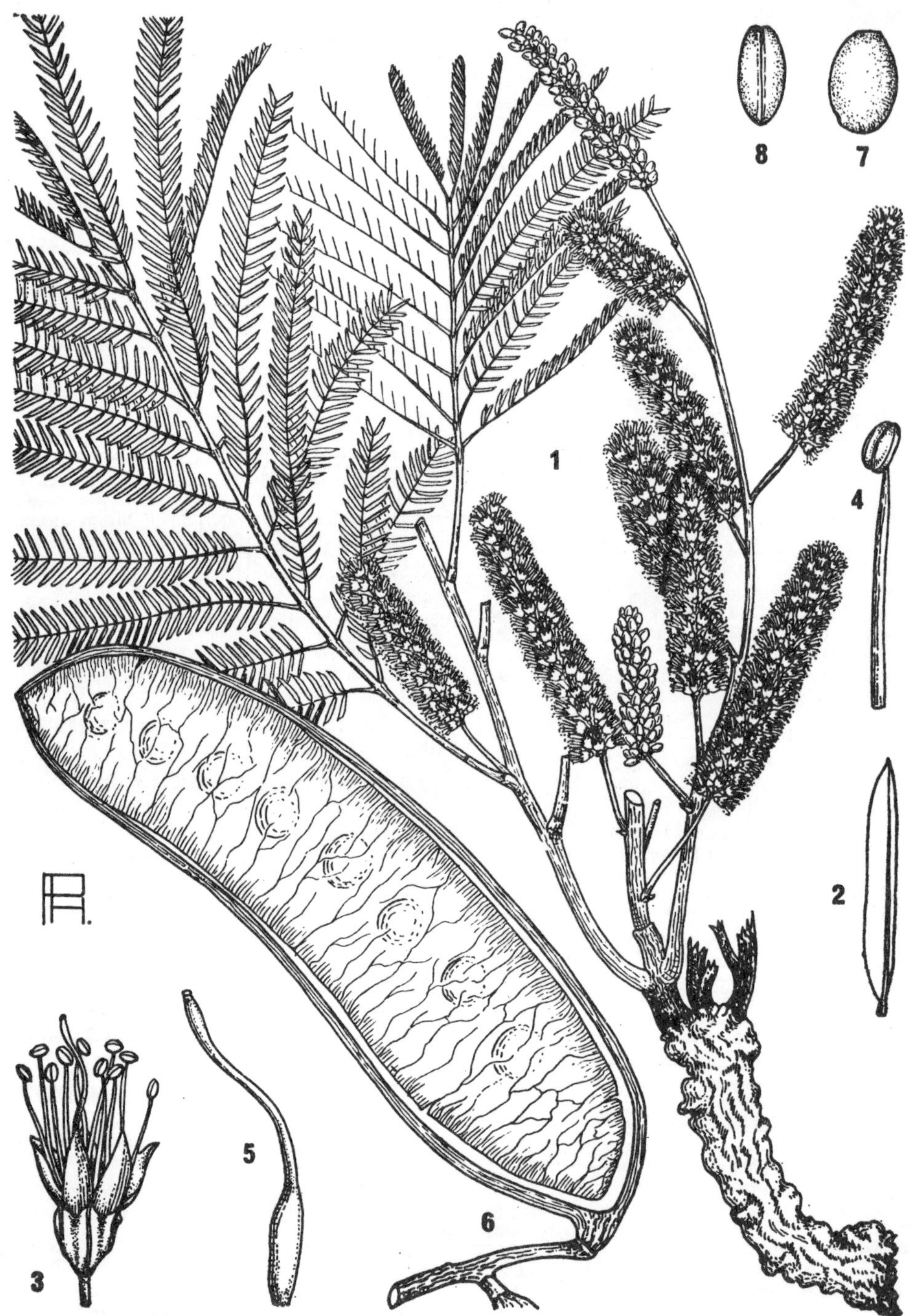

FIG. 21.—**Elephantorrhiza elephantina.** 1, flowering stems arising from part of the underground rhizome, × ⅔, *Seydel* 2388; 2, leaflet, × 4; 3, flower, × 4; 4, stamen, × 6; 5, gynoecium, × 6, all from *Marsh* 96; 6, pod, × ⅔; 7, seed, × ⅔; 8, seed, in profile, × ⅔, all from *Nation* 225.

Petals free or slightly connate basally, up to 3,25 mm long, 1,25 mm wide, lanceolate, slightly inflexed apically, glabrous. *Stamens* free among themselves, slightly adnate to the corolla basally; filaments up to 6 mm long; anthers up to 1 mm long, with a deciduous apical gland. *Ovary* up to 2 mm long, linear, subsessile, glabrous. *Pod* (only one ever collected) dark reddish-brown, 9 × 3,2 cm, falcate, compressed, prominently transversely venose, umbonate over the seeds. *Seeds* immature.

Found in Natal and Lesotho. Occurs in grassland.

Phillips, in Bothalia 1 : 189 (1923), keyed out *E. woodii* under the species which are "shrubs or small trees with a distinct aerial stem." As there is no information about the habit of the plants in Medley Wood's collectors book or on the herbarium sheets, it is thought that Phillips assumed *E. woodii* was a shrub or small tree because the aerial stems are branched. During recent years several attempts have been made to re-collect *E. woodii* in the type locality and eventually, in 1970, Mr R. G. Strey succeeded in finding a few plants. However, the plants were procumbent and not shrubby or arborescent as indicated by Phillips. The habit notes in the above description are based on the specimens collected by Mr. Strey and subsequently cultivated at the Natal Herbarium, Durban. Each year the branched procumbent aerial stems die back and are replaced by a new set of stems the following season. The stems are relatively weak and show no signs of assuming an erect posture. More field observations are required to establish whether *E. woodii* ever does grow as an erect shrub. At the time of collecting the living plants, concern was expressed for the future of the species in the type locality, at least, because the plants only survive in the narrow strips of uncultivated land which may themselves be cultivated at any time.

E. woodii is closely related to *E. elephantina* but differs in having branched procumbent aerial stems and leaflets with a slightly different venation. It differs from all of the other species with branched stems in that the stems are procumbent and longitudinally striate.

Stems glabrous or almost so; petioles, rhachides
    and rhachillae glabrous or almost so; inflore-
    scence axes ± glabrous......(a) var. *woodii*
Stems pubescent; petioles, rhachides and rhachil-
    lae pubescent; inflorescence axes pubescent
    ....(b) var. *pubescens*

Unfortunately not enough material is available to evaluate the taxonomic significance of the degree of pubescence of the stems, leaves and inflorescence axes as a means of distinguishing varieties within this species.

(a) var. **woodii.**

Phillips in Bothalia 1 : 193 (1923); Ross, Fl. Natal 194 (1973); in Bothalia 1.c. : 251 (1974). Type as above.

NATAL.—2829 (Harrismith): Pieters (—DB), *Strey 9760*; Harts Hill, near Colenso (—DB), *Strey 10000*; Ladysmith (—DB), *Geekie 46* (NU).

More material of var. *woodii* is required.

The following note written by Phillips on *Wood 7958* (NH) indicates that he had initially intended calling this specimen *E. wahlbergii* (Harv.) Phillips: "This is undoubtedly the plant described by Harvey as *Entada ?wahlbergii* in Fl. Cap. 2 : 277. I propose the new combination *Elephantorrhiza wahlbergii*, Phill. 8/6/19."

(b) var. **pubescens** *Phillips* in Bothalia 1 : 193 (1923); Ross, Fl. Natal 194 (1973); in Bothalia 1.c. : 251 (1974). Type: Natal, Estcourt District, near Little Tugela, 1219 metres, *Wood 2867* (NH, holo.).

E. pubescens Phillips ined.

NATAL.—probably 2929 (Underberg): near Little Tugela, *Wood 2867* (NH). Grid ref. unknown: between Pietermaritzburg and Newcastle, Nov. 1883, *Wilms 1973* (BM).

?LESOTHO.—no precise locality, *Cooper 2279* (K).

Wilms 1973 (BM), collected between Pietermaritzburg and Newcastle in 1883, appears to be referable to *E. woodii* var. *pubescens*. However, the specimen has wider leaflets and a somewhat different facies to typical var. *pubescens*. More material is desirable but, as the specimen was not well localised, there seems little likelihood of further material being collected.

Cooper 2279, a flowering specimen with immature leaves, also appears to be referable to var. *pubescens*. Although Cooper gave the locality of collection as "Basutoland", it is very doubtful whether the specimen was collected in present-day Lesotho.

Mogg sub PRE 9644, an immature and rather fragmentary specimen from Charlestown in northern Natal, is extremely difficult to place with certainty. The specimen in the Kew Herbarium consists of a single stem bearing axillary racemes and very young foliage, the details of which are scarcely discernible. The stem is 17,5 cm high, longitudinally striate and pubescent. The leaves have 2 pinnae pairs and up to 16 leaflets per pinna. The immature leaflets are up to 4,5 × 1,75 mm, the midrib is excentric basally and the leaflet-apex is distinctly mucronate. This specimen is hesitantly referred to *E. woodii* var. *pubescens*, but additional and better material from this area is required to establish the identity of the plants.

More material of var. *pubescens*, and from definite localities, is required. Pods would be of particular interest.

4. **Elephantorrhiza burkei** *Benth*. in Hook., Lond. J. Bot. 5 : 81 (1846); Harv. in F.C. 2 : 278 (1862); Benth. in Trans. Linn. Soc. Lond. 30 : 365 (1875); Phillips in Bothalia 1 : 192, t.5 fig. 4 (1923); Bak. f., Leg. Trop. Afr. 3 : 801 (1930); Burtt Davy, Fl. Transv. 2 : 332 (1932); O. B. Miller in J. S. Afr. Bot. 18 : 31 (1952); Wild, Guide Fl. Vict. Falls 149 (1953); Brenan & Brummitt in

F.Z. 3,1 : 27 (1970); Palmer & Pitman, Trees S. Afr. 2 : 825 (1973); Ross in Bothalia 11 : 251 (1974). Type: Transvaal, Magaliesberg, *Burke & Zeyher* (K, holo.!; BM!; TCD!; Z!, ? iso.).

E. elephantina (Burch.) Skeels var. *burkei* (Benth.) Merr. in Contr. Gray Herb. 59 : 18 (1919). Type as above.

A branched shrub or small tree 1–3(6) m high, occasionally as small as 0,3 m, but then the stems distinctly woody and branched and the inflorescences normally borne on lateral shoots of the current season's growth; bark dark-grey to reddish; young branchlets glabrous. *Leaves* glabrous or almost so: petiole 2,6–6,5 cm long; rhachis 3,6–14,5 cm long; pinnae (1)4–8(9) pairs; rhachillae 3,5–12,5 cm long; leaflets (9)12–23(32) pairs per pinna, 7–17 × 1,5–3,5(5) mm, narrowly oblanceolate to very narrowly elliptic or linear-oblong, usually glaucous, glabrous, base slightly asymmetric (less so than in *E. elephantina*), with the proximal side rounded to cuneate, apex symmetric, obtuse to rounded, generally mucronate, lateral nerves and veins prominent or not. *Racemes* axillary, solitary or fascicled, often on lateral shoots, 5–10(12,5) cm long (including the peduncle), glabrous. *Flowers* yellowish-white, on pedicels up to 2 mm long, pedicels articulated near the middle, with minute reddish glands at the base of the pedicels. *Calyx* campanulate, up to 2,5 mm long, 5-toothed, the teeth up to 0,75 mm long, glabrous. *Petals* shortly united basally, up to 4,5 mm long, 1 mm wide, linear-oblong, inflexed apically, glabrous. *Stamens* free among themselves, slightly adnate to the corolla basally; filaments up to 5 mm long; anthers up to 0,75 mm long, with a deciduous apical gland. *Ovary* up to 2 mm long, glabrous. *Pods* dark brown to reddish-brown, 10–19(28) × 2,5–4 cm, straight or slightly curved, oblong, compressed, sometimes prominently transversely venose, at maturity the valves separating from the persistent margins, the outer layer of the pod-wall peeling off the inner layer, the layers remaining intact or breaking up irregularly. *Seeds* ± 9–13 × 8–12 mm.

Found in Botswana, Rhodesia, Mozambique and the Transvaal. Favours rocky situations, in woodland, grassland and scrub.

TRANSVAAL.—2329 (Pietersburg): University College of the North, 28,8 km from Pietersburg on road to Tzaneen, *Van Vuuren 1293.* 2425 (Gaberones): Lekkerlach, *Louw 597* (NH). 2428 (Nylstroom): 15,2 km N. of Warmbaths, *Marais 1236.* 2429 (Zebediela): Chuniespoort, *Pole Evans sub PRE 19452.* 2431 (Acornhoek): Kruger National Park, 25,6 km N.E. of Skukuza, *Codd 5723.* 2527 (Rustenburg): Wolhuterskop, *Schweickerdt 1642.* 2528 (Pretoria): end of Daspoort range about 9,6 km E. of Pretoria, *Phillips 3040.* 2529 (Witbank): Loskop Dam Nature Reserve, *Mogg 30609.*

E. burkei differs from *E. elephantina* primarily in being a shrub or tree with branched perennial aerial stems and not a suffrutex with annual unbranched aerial stems (unless damaged). The leaflet-base in *E. burkei* is less asymmetric than in *E. elephantina*, and the midrib soon becomes ± centric. The leaflets of *E. burkei* are typically larger than those of *E. elephantina*. *E. burkei* appears to have smaller seeds than *E. elephantina*, but more fruiting material is required to confirm this.

5. Elephantorrhiza praetermissa *J. H. Ross* in Bothalia 11 : 252 (1974). Type: Transvaal, Lydenburg Distr., Steelpoort valley, near Sarahshof, *Codd 9830* (PRE, holo.!; BM!; K!).

Shrub 1–2 m high; young branchlets grey- or reddish-brown, glabrous. *Leaves* glabrous: petiole 2,2–4 cm long; rhachis 4–9 cm long, sulcate above, sometimes with minute scattered dark glands; pinnae (3)5–10(12) opposite or subopposite pairs; rhachillae (2,8)3,5–6(7) cm long; leaflets 20–40 pairs, 5–10 × 0,9–1,5 mm, linear or linear-oblong, sessile, glabrous, asymmetric basally, midrib starting in the distal corner of the leaflet-base and gradually becoming almost central in the leaflet, proximal side of base rounded, apex rounded or acute, nearly symmetric, lateral nerves not visible or inconspicuous beneath, sometimes with minute dark purplish glands at the base of the leaflets. *Inflorescences* racemose, racemes solitary, fascicled or aggregated on abbreviated lateral branchlets, 4–5,5 cm long (including the peduncle), glabrous. *Flowers* yellowish-white, pedicellate, pedicels 1,5–2 mm long, articulated near or below the middle, with minute dark reddish glands at the base of the pedicels. *Calyx* 0,75–1,25 mm long, 5-toothed, glabrous. *Petals* shortly united basally, 2–3 mm long, linear-oblong, glabrous. *Stamens* free among themselves, slightly adnate to the corolla basally; filaments 4–5 mm long; anthers with a deciduous apical gland. *Ovary* ± 2 mm long, linear, glabrous. *Pods* dark brown or reddish-brown, 12–18 × 2–3,2 cm, oblong, straight or slightly curved, compressed, obscurely or prominently venose, at maturity the valves separating from the persistent margins. *Seeds* ± 15 × 13 × 3,5 mm.

E. praetermissa appears to have a rather restricted distribution in the eastern Transvaal. Occurs on dry wooded hillsides.

TRANSVAAL.—2429 (Zebediela): 25 km N.N.W. of Schoonoord (—DB), *Acocks 20969*. 2430 (Pilgrim's Rest): 59,2 km from Lydenburg on road to Steelpoort via Tweefontein (—CC), *Vorster 2129*; 74,4 km from Lydenburg on road to Steelpoort via Tweefontein (—CC), *Vorster 2128*.

E. praetermissa is most closely related to *E. goetzei* and to *E. elephantina*. Both *E. goetzei* and *E. elephantina* occur in the eastern Transvaal, but the specimens in question differ from the material of both of these species.

E. praetermissa differs from typical *E. goetzei* in having consistently fewer pinnae pairs. The leaflets of *E. praetermissa* differ from those of *E. goetzei* in having a somewhat thicker texture and in being ± sessile; the leaflets of typical *E. goetzei* usually have distinct petiolules. In *E. goetzei* the pods are long and narrow in proportion to their length (15–44 × 1,3–2,2 cm) and, when mature, the position of the seeds is marked by distinct raised bumps. In *E. praetermissa* the pods are shorter and broader (12–18 × 2–3,2 cm), ± compressed, and lack distinct raised bumps over the seeds. The seeds of *E. praetermissa* are ± compressed in contrast to the ellipsoid or lenticular seeds of *E. goetzei*, and they are smaller than those of the latter. Although the length of the racemes provides no discontinuity between the two species, the racemes of *E. praetermissa* are consistently short and are much shorter than is usual in *E. goetzei*.

E. praetermissa differs from *E. elephantina* in being a robust shrub 1–2 m high and in having branched aerial stems. The leaflets of *E. praetermissa* differ slightly in texture and lack the ± conspicuous venation of typical *E. elephantina*, while the pods tend to be slightly narrower than is usual in *E. elephantina*.

Although *E. praetermissa* is described as locally common by collectors, very few specimens have been collected. More material is required.

6. **Elephantorrhiza goetzei** (*Harms*) Harms in Engl., Pflanzenw. Afr. 3,1 : 400 (1915); Bak.f., Leg. Trop. Afr. 3 : 802 (1930); Brenan, Checklist Tang. Terr. 344 (1949); Wild, Guide Fl. Vict. Falls 149 (1953); Williamson, Useful Pl. Nyasal. 52 (1955); Brenan in F.T.E.A. Legum.-Mimos. : 19, fig. 4 (1959); White, For. Fl. N. Rhod. 91 (1962); Brenan & Brummitt in F.Z. 3,1 : 24, t.4 (1970); Palmer & Pitman, Trees S. Afr. 2 : 827 (1973) Ross in Bothalia 11 : 253 (1974). Type: Tanzania, Rufiji District, *Goetze 82* (B, holo.† , BM, drawing!; K, iso.!).

Piptadenia goetzei Harms in Bot. Jahrb. 28 : 397 (1900). Type as above.

subsp. **goetzei.**

Brenan & Brummitt in Bol. Soc. Brot., Sér. 2,39 : 189 (1965); in F.Z. 3,1 : 24 (1970); Ross in Bothalia 11 : 253 (1974).

E. rubescens Gibbs in J. Linn. Soc. Bot. 37 : 441 (1906); Eyles in Trans. Roy. Soc. S. Afr. 5 : 364 (1916). Type: Rhodesia, Matopo Hills, *Gibbs 184* (BM, holo.!). *E.* cf. *petersiana* sensu Gomes e Sousa, Pl. Menyharth. 70 (1936). *E.* cf. *goetzei* (Harms) Harms, Torre in C.F.A. 2 : 263 (1956). *E.* sp. sensu Torre l.c. : 264.

Acacia rehmanniana sensu M. A. Exell in Bol. Soc. Brot. Sér. 2,12 : 16 (1937), non Schinz.

Shrub or small tree 1–4(7) m high; bark grey-brown to dark brown or reddish-brown to purplish, often becoming blackish; young branchlets glabrous. *Leaves* glabrous or nearly so: petiole 1–5(7,5) cm long; rhachis 6–20(45,5) cm long, sulcate above; pinnae 3–30(41) opposite or subopposite pairs; rhachillae 1,8–9 cm long; leaflets 9–40(48) pairs, 3,5–12 × 0,7–2,75 mm (in our area), linear-oblong to narrowly oblong, midrib starting in the distal corner of the leaflet-base, gradually becoming almost central in the leaflet, proximal side of the base rounded and almost auriculate, apex acute to rounded and mucronate, nearly symmetric, glabrous, lateral nerves and veins not or scarcely visible. *Racemes* solitary, fascicled or borne on short lateral shoots, (2)5–20 cm long (including the peduncle), glabrous. *Flowers* yellowish-white, sometimes tinged with pink or purple, on pedicels up to 1 mm long, pedicels articulated near the middle, with minute pale yellowish-white glands at the base of the pedicels. *Calyx* 1,5–1,75 mm long, with 5 acute teeth, glabrous. *Petals* shortly united below, becoming almost free in open flowers, 2,5–3 mm long, linear-oblong, inflexed apically, glabrous. *Stamens* free among themselves, slightly adnate to the corolla basally; filaments up to 4,5 mm long; anthers up to 1 mm long, with a deciduous apical gland. *Ovary* up to 2 mm long, linear, glabrous. *Pods* dark brown or reddish- or purplish-brown, 15–30(44) × 1,3–2,2 cm, linear, straight or curved, raised over the seeds, at maturity the valves separating from the persistent margins, the outer layer of the pod-wall peeling off the inner layer, the layers remaining intact or breaking up irregularly. *Seeds* 11–20 × 9–18 × 7–12 mm, ellipsoid to lenticular.

Found in Tanzania, Angola, Botswana, Zambia, Rhodesia, Malawi, Mozambique and the Transvaal. Occurs in woodland of various types and scrub; favours rocky places.

TRANSVAAL.—2331 (Phalaborwa): Kruger National Park, Shingwedzi, in Lebombo Mts., *Van der Schijff 3848*. 2429 (Zebediela): Wolkberg, 14,4 km S. of Boyne on road to Welcome Gold Mine, *Codd 10393*. 2430 (Pilgrim's Rest): 28 km S.E. of Gravelotte (—BB), *Codd 9477*; Abel Erasmus Pass, c. 3 km S. of Strydom Tunnel (—BC), *Vorster & Coetzer 2099*. 2431 (Acornhoek): Kruger National Park, ± 29 km from Satara on Rabelais road (—BC), *Van der Schijff 3290*; Kruger National Park, 10,4 km from Nwanedzi [Ngwanetsi] on Satara road (—BD), *Story 3967*.

In the area delimited for Flora Zambesiaca *E. goetzei* frequently produces its flowers when the plant is leafless. In the Transvaal, however, *E. goetzei* usually produces flowers together with the leaves.

It is possible that *E. petersiana* Bolle in Peters, Reise Mossamb. Bot. 1 : 9 (1861) is an earlier name for *E. goetzei*. If this were ever confirmed, then *E. petersiana* would be the correct name for this species. The holotype of *E. petersiana*, now destroyed, was a flowering specimen (without leaves) collected by Peters at Sena in Mozambique. Unfortunately the type description is too imperfect to enable the species to be positively identified. Bak.f., Leg. Trop. Afr. 3 : 802 (1930) shed no light on the identity of *E. petersiana*.

Burtt Davy, Fl. Transv. 2 : 332 (1932), based his *E.* (?) sp. nov.? [=E. (?) elongata Burtt Davy ined.] on *Burtt Davy* H2304 collected at Potgietersrust in the Transvaal. This specimen, which is quite leafless, has ±straight immature pods up to 21 × 1,7 cm. The pods of *Burtt Davy* H2304 resemble those of *E. goetzei* fairly closely and, although *E. goetzei* has not been recorded from Potgietersrust subsequently, it seems likely that Burtt Davy's specimen is referable to *E. goetzei*. E. (?) sp. nov. ? is therefore a probable synonym of *E. goetzei*.

Subsp. *lata* Brenan & Brummitt is recorded from Zambia and Rhodesia. It differs from subsp. *goetzei* in having fewer pinnae pairs and fewer pairs of larger leaflets.

7. **Elephantorrhiza rangei** *Harms* in Bot. Jahrb. 49 : 420 (1913); Dinter in Feddes Repert. 17 : 190 (1921); Bak.f., Leg. Trop. Afr. 3 : 802 (1930); Range in Feddes Repert. 30 : 148 (1932); Schreiber in F.S.W.A. 58 : 17 (1967) pro parte quoad specim. *Range 455*; Ross in Bothalia 11 : 254 (1974). Type: South West Africa, Keetmanshoop Distr., Naute, near Keetmanshoop, *Range 455* (B, holo. †, BM, drawing!; BOL, SAM, iso.!).

A branched shrub (? or small tree) to 4 m high; young branchlets reddish-brown to purplish, glabrous. *Leaves* glabrous: petiole 2–4,5 cm long; rhachis (1,5) 3–7,5 cm long [petiole and rhachis together described as 2–15 cm long by Harms]; pinnae 3–7[9] opposite or subopposite pairs, sometimes 1–3 pairs on immature leaves; rhachillae 5–8,5[9] cm long; leaflets 24–36 pairs per pinna, 6–9[12] × 1–2,75[4] mm, linear-oblong to oblong, sometimes slightly falcate, midrib starting in the distal corner of the leaflet-base, gradually becoming almost central in the leaflet, proximal side of the base rounded to almost auriculate, apex rounded to acute, mucronate, almost symmetric, lateral nerves inconspicuous, glabrous. *Racemes* axillary, solitary or paired, 5,5–8 cm long (including the peduncle), glabrous. *Flowers* greenish-yellow, on pedicels 1–1,75 mm long, pedicels articulated just below the middle, with minute yellowish glands at the base of the pedicels. *Calyx* campanulate, 2–2,25 mm long, glabrous, 5-toothed. *Petals* shortly united basally, 3–4 mm long, 1 mm wide, oblong, inflexed apically, glabrous. *Stamens* free among themselves, slightly adnate to the corolla basally; filaments up to 5,5 mm long; anthers up to 0,9 mm long, with a deciduous apical gland. *Ovary* up to 4 mm long, shortly stipitate, linear, glabrous. *Pods* dark brown or reddish-brown, 18,5–20[22] × 2–2,5 cm, straight or almost so, oblong, compressed, slightly umbonate over the seeds, transverse venation relatively inconspicuous, at maturity the valves separating from the persistent margins. *Seeds* unknown.

Known only from the type locality in South West Africa. Ecology unknown.

S.W.A.—2617 (Bethanie): Naute, near Keetmanshoop (—DD), *Range 455* (BOL, SAM).

The above description was drawn up from two isotypes in the SAM Herbarium. The extreme dimensions given in square brackets were recorded by Harms in his type description.

There is considerable variation in leaflet size even on a single branch; the upper leaves often have small leaflets and the lower leaves larger leaflets.

E. rangei bears a superficial resemblance to *E. suffruticosa* but differs in having larger and broader leaflets in which the midrib is ± centric apically, slightly longer pedicels, and larger flowers.

The specimens cited by Phillips in Bothalia 1 : 192 (1923) under *E. rangei* ("*rangeri*"), and on which t.5 fig. 5 was based, are in fact referable to *E. elephantina*.

E. rangei is known only from the type collection. It has never been re-collected since Jan. 1908 and the possibility exists that it is now extinct. A thorough search for this plant in the type locality is most desirable in an attempt to evaluate its present conservation status.

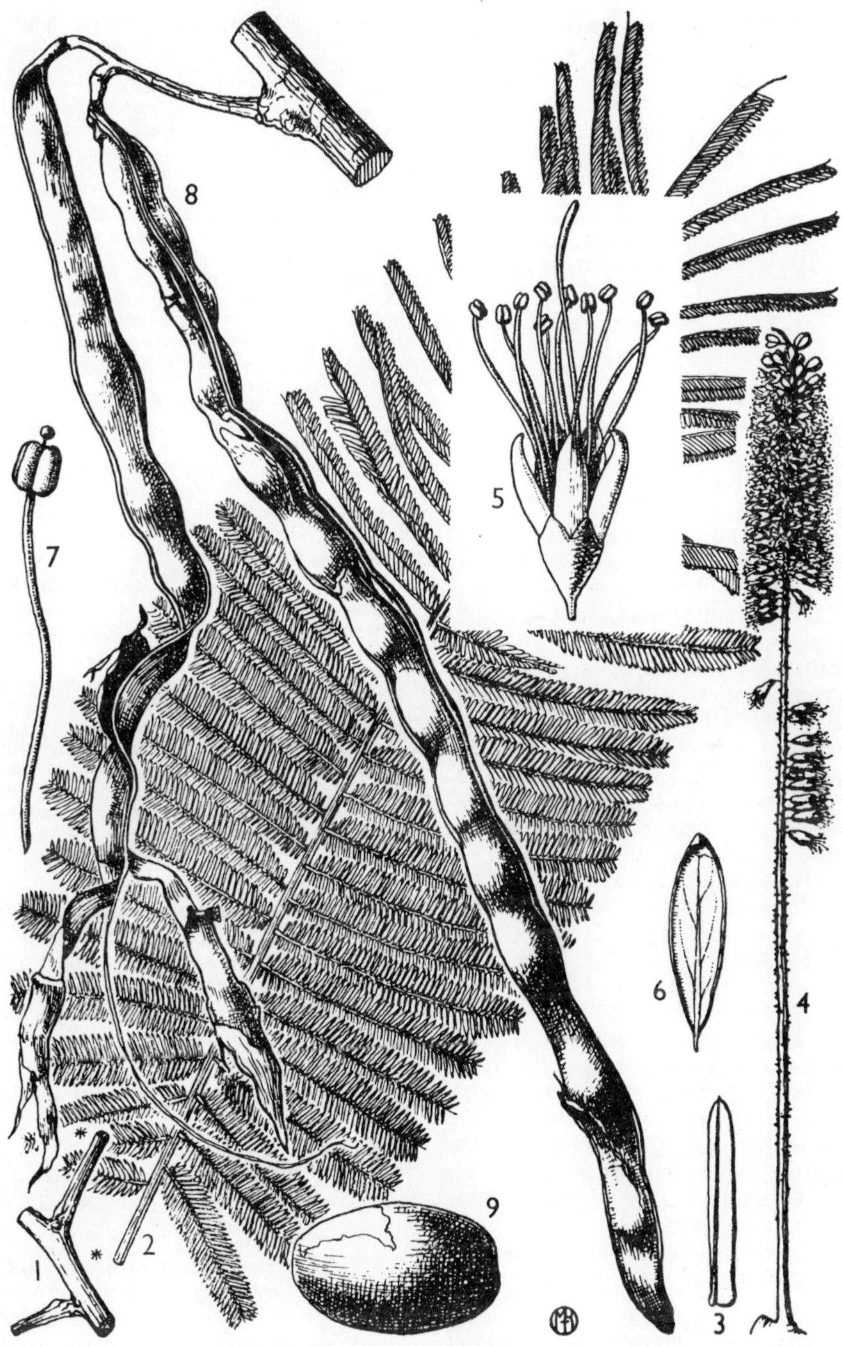

FIG. 22.—**Elephantorrhiza goetzei** subsp. **goetzei**. 1, part of branch with petiole-bases, × ¼; 2, leaf, detached from petiole-base of 1, × ¼; 3, leaflet, × 6, all from *Milne-Redhead & Taylor* 9549; 4, flowering raceme, × 1½; 5, flower, × 9; 6, petal, × 15; 7, anther, × 25, all from *Andrada* 1452; 8, pods, × 1; 9, seed, × 3, both from *Jackson* 1418. Reproduced by permission of the Editor of Flora of Tropical East Africa.

8. Elephantorrhiza schinziana *Dinter* in Feddes Repert. 17 : 190 (1921); Bak.f., Leg. Trop. Afr. 3 : 802 (1930); Schreiber in F.S.W.A. 58 : 17 (1967); Ross in Bothalia 11 : 255 (1974). Type: South West Africa, Grootfontein Distr., Otavi, *Dinter 745* (SAM, lecto.!).

A branched shrub (? or small tree) up to 2,5 m high; bark grey- to dark- or reddish-brown; young branchlets grey- or reddish- to purplish-brown, glabrous. *Leaves* glabrous: petiole 2,2–3,5(5,2) cm long; rhachis (4,5) 7,5–14,5(20,5) cm long; pinnae (2)6–11(14) opposite or subopposite pairs; rhachillae 5,5–10(14) cm long; leaflets (14)21–40 pairs per pinna, (5)7–14 × 1,5–3,5 mm, linear-oblong to oblong, midrib starting in the distal corner of the leaflet-base, gradually becoming almost central in the leaflet, proximal side of base rounded, apex rounded and sometimes distinctly mucronate, nearly symmetric, lateral nerves scarcely visible, glabrous, somewhat glaucous. *Racemes* axillary, solitary or paired, 7–9,5 cm long (including the peduncle), glabrous. *Flowers* yellowish-white, on pedicels up to 0,75 mm long, pedicels articulated towards the apex, with minute yellowish glands at the base of the pedicels. *Calyx* cupular, up to 1,5 mm long, glabrous, shortly 5-toothed. *Petals* shortly united basally, 3–3,75 mm long, 1 mm wide, linear-oblong, inflexed apically, glabrous. *Stamens* free among themselves, slightly adnate to the corolla basally; filaments up to 5 mm long; anthers up to 0,8 mm long, with a deciduous apical gland. *Ovary* up to 2,25 mm long, glabrous. *Pods* dark brown or reddish-brown, (15)19–30(40,5) × 3–3,9 cm, straight or slightly curved, oblong, compressed, umbonate over the seeds, prominently transversely venose, at maturity the valves separating from the persistent margins, the outer layer of the pod-wall peeling off the inner layer, the layers tending to break up irregularly. *Seeds* immature, mature seeds unknown.

Known only from the Grootfontein District in South West Africa. Ecology unknown.

S.W.A.—1917 (Tsumeb): farm Heidelberg near Tsumeb (—BB), *Dinter 1689* (SAM); Otavi (—CB), *Dinter 745* (SAM), *Dinter 5300* (BOL, PRE, Z). Grid ref. unknown: Grootfontein District, farm Asis, *Volk 767* (M).

The original specimens on which Dinter based his description are no longer available for study. Fortunately, however, both syntypes, namely *Dinter 745*

and 1689, are represented in the South African Museum collections. There is one sheet of *Dinter 745* and three sheets of *Dinter 1689*. One of the sheets of *Dinter 1689* is a mixed gathering consisting of a vegetative shoot of *E. suffruticosa* and a pod of *E. schinziana*, while the flowers in the capsule could belong to either species. On the second sheet of *Dinter 1689* there is a vegetative shoot of *E. schinziana* with a mature pod attached and, in addition, there is a flowering specimen which is leafless apart from an extremely young leaf or shoot on which no details are discernible. In view of the mixed gathering of *E. schinziana* and *E. suffruticosa* on the first sheet of *Dinter 1689*, the possibility exists that the flowering specimen on the second sheet belongs to *E. suffruticosa* and not to *E. schinziana*. It will be recalled that in South West Africa *E. suffruticosa* usually flowers when leafless. As none of the other specimens of *E. schinziana* examined were in flower, the details of the flowers in the above description were taken from this second specimen of *Dinter 1689*. It is possible therefore that the flowers described are those of *E. suffruticosa* and not of *E. schinziana*. The third sheet of *Dinter 1689* consists of a single leaf and two mature pods (the valves of one pod are mounted separately which gives the impression that there are three pods).

There is considerable variation in leaflet size even on a single branch; some of the upper leaves often have distinctly smaller leaflets than the leaflets on the lower leaves.

E. schinziana has not been re-collected since Jan. 1939 and there is a possibility that it is now extinct. A thorough search for this plant in the type localities is most desirable in an attempt to evaluate its present conservation status. If *E. schinziana* is re-discovered, it is important that an effort be made to collect both flowering and fruiting material. Flowering material is essential to establish whether or not the flowers on which the above description was based are in fact those of *E. schinziana*.

9. Elephantorrhiza suffruticosa *Schinz* in Mém. Herb. Boiss. 1 : 117 (1900); Dinter, Deutsch-Sudwest-Afrika Fl. Forst und land-wirtschaft. Frag. 78 (1909); Harms in Engl., Pflanzenw. Afr. 3,1 : 400 (1915); Dinter in Feddes Repert. 17 : 190 (1921); Phillips in Bothalia 1 : 193, t.5 fig. 7 (1923); Bak. f., Leg. Trop. Afr. 3 : 801 (1930); Schreiber in F.S.W.A. 58 : 17 (1967); Brenan & Brummitt in F.Z. 3,1 : 26 (1970); Ross in Bothalia 11 : 255 (1974). Type: Angola, Huila district, "Kilevi am Kunene" (south of Humbe), *Schinz 2071* (Z, lecto.!).

A branched shrub or small tree 1–5 m high; bark grey-brown to dark- or reddish-brown; young branchlets glabrous or sometimes puberulous to shortly pubescent. *Leaves* glabrous to puberulous or shortly pubescent: petiole (0,6)1,5–3,5 cm long; rhachis (0,5)10–17(25,4) cm long; pinnae (2)15–27(42) opposite or subopposite pairs;

rhachillae (1,4)2–3,5(6,8) cm long; leaflets (17)27–40(50) pairs per pinna, 3–7,5 × 0,5–1,2 mm, linear-oblong to linear, rarely almost falcate, midrib marginal throughout, proximal side rounded basally, apex asymmetric, obtuse to acute, often mucronate, lateral nerves and veins not or scarcely visible, glabrous or sometimes sparingly pubescent on the margins. *Racemes* axillary, solitary or 2–3 together, or borne on lateral shoots, (4)6–14(18) cm long (including the peduncle), pubescent or sometimes glabrous. *Flowers* yellowish-white, on pedicels up to 1 mm long, pedicels articulated near the middle, with minute reddish or reddish-brown glands at the base of the pedicels. *Calyx* cupular, up to 1 mm long, shortly 5-toothed, glabrous or sometimes very sparingly pubescent. *Petals* shortly united basally, 3–3,75 mm long, 1 mm wide, linear-oblong, inflexed apically, glabrous. *Stamens* free among themselves, slightly adnate to the corolla basally; filaments up to 5 mm long; anthers up to 0,8 mm long, with a deciduous apical gland. *Ovary* up to 2 mm long, linear, glabrous. *Pods* dark brown or reddish-brown, 8,5–30,5 × 1,8–2,25 cm, straight or slightly curved, linear-oblong to oblong, compressed, usually prominently transversely venose, umbonate over the seeds, at maturity the valves separating from the persistent margins, the outer layer of the pod-wall peeling off the inner layer, the layers remaining intact or breaking up irregularly. *Seeds* 13–15 × 9–12 mm, roughly ellipsoid.

Found in Angola, South West Africa, Rhodesia and Mozambique. Occurs in woodland, grassland and in broken country; often among rocks. The ecological preferences of *E. suffruticosa* are not clear and more information is required.

S.W.A.—1814 (Otjitundua): Otusemba, *Story 5920*. 1815 (Okahakana): 84,4 km N. of Okakeujo on road to Ondongua, *De Winter 3617*. 1816 (Namutoni): Amutele, *Schinz 2070* (Z). 1817 (Tsintsabis): 48 km N. of Tsumeb, *Rodin 2605*. 1916 (Gobaub): farm Zukov 337, *De Winter 3019*. 2016 (Otjiwarongo): Otjiwarongo, *Liebenberg 4909*. 2116 (Okahandja): Okahandja, *Dinter 314* (BM, E, GRA, Z). 2117 (Otjosondu): Omupanda, *Wulfhorst s.n.* (Z). 2214 (Swakopmund): Swakopmund, *Seydel 1146* (Z). 2215 (Trekkopje): Ebony Mine, *Schenck 457* (Z). 2216 (Otjimbingwe): Auas Mts., pass between Haris and Aub, *Pearson 9658* (K). 2217 (Windhoek): Windhoek, *Rogers 29755*. 2317 (Rehoboth): Rehoboth, *Fleck 499a* (Z). 2416 (Maltahöhe): Bull's Mouth Pass, *Pearson 8919*; Bullsport, Naukluft Mts., *Hardy 1982*. 2417 (Mariental): Voigtsgrund, *Keet 1662*. Grid ref. unknown: "sandige stellen bei Ombalambuenge", *Rautanen 242* (Z).

The narrow leaflets of *E. suffruticosa*, with the midrib marginal throughout the length of the leaflet, are most diagnostic and enable this species to be readily distinguished from all of the other *Elephantorrhiza* species.

There appears to be a discontinuity in the distribution of *E. suffruticosa* between eastern and central Rhodesia and South West Africa and Angola. No significant morphological differences have been so far noted between specimens from these two areas of distribution except for an inconsistent tendency for some leaflets to be more acute and mucronate in Rhodesia than in South West Africa and Angola. In addition, in specimens from Rhodesia there are usually ± conspicuous reddish or orange glands at the base of the leaflets, while in Angola and South West Africa the glands are inconspicuous and pale yellow or even absent.

In the Flora Zambesiaca area *E. suffruticosa* usually produces flowers together with the leaves, while *E. goetzei* usually flowers when leafless. In South West Africa, however, *E. suffruticosa* frequently flowers when leafless, while in the Transvaal *E. goetzei* usually produces flowers together with the leaves. In the area delimited for the Flora of Southern Africa, therefore, the reverse situation tends to prevail.

The reason for selecting *Schinz 2071* from Kilevi am Kunene in Angola as the lectotype of *E. suffruticosa* is discussed in Bothalia 11 : 256 (1974). As mentioned in Bothalia l.c., having selected *Schinz 2071* as the lectotype of *E. suffruticosa*, it was most disconcerting to find that Schinz, in Mém. Herb. Boiss. 1 : 105 (1900), cited *Schinz 2071* from Olukonda–Oshiheke in Amboland, South West Africa, as one of the syntypes of *Acacia arenaria* Schinz. Fortunately I have examined this syntype of *Acacia arenaria* and can vouch for its identity. In any event, it seems unlikely that *Schinz 2071* from Olukonda–Oshiheke in South West Africa, a syntype of *Acacia arenaria*, would ever be confused with *Schinz 2071* from Kilevi am Kunene in Angola, the lectotype of *Elephantorrhiza suffruticosa*. It is as well, however, to draw attention to the existence of these two *Schinz* specimens each numbered 2071.

Insufficiently known species

10. **Elephantorrhiza sp.** Ross in Bothalia 11 : 257 (1974).

Suffrutex producing at ground level unbranched, longitudinally striate, glabrous stems 60–80 cm high. *Leaves* glabrous or almost so: petiole 3–4,5 cm long; rhachis 7,5–18 cm long; pinnae 3–8 opposite or subopposite pairs; rhachillae 7,5–10 cm long; leaflets 13–22 pairs per pinna, 9–11 × 3–5 mm, very oblique, broadly truncate basally or sometimes slightly auricled on the proximal side, asymmetric and attached by one corner, the midrib starting in the distal corner of the leaflet-base and gradually becoming almost central in the leaflet, rounded to acute or distinctly mucronate apically, glabrous throughout or with few minute marginal cilia. *Flowers* and pods unknown.

Known from two gatherings from the eastern Cape.

CAPE.—3128 (Umtata): Umtata aerodrome (—DA), *Strey 11073*; *Strey 11165*.

The two Strey gatherings do not appear to match material of any of the existing species. Like *E. elephantina* and *E. obliqua*, the specimens have unbranched aerial stems. The Strey specimens, which were growing in association with *E. elephantina*, appear to differ from this species in having larger leaflets, and from *E. obliqua* in having larger leaves with more numerous pinnae and leaflet pairs, and leaflets without a conspicuous venation. More material, particularly fertile material, is required to enable a positive identification to be made. The specimens seem to be most closely related to *E. elephantina* and the possibility exists that they are only a variant of this species.

3468   13. **ENTADA**

Entada *Adans.*, Fam. Pl. 2 : 318 (1763), nom. conserv.; DC., Prodr. 2 : 424 (1825); Mém. Leg. 419 (1826); Benth. in Hook., J. Bot. 4 : 332 (1841); Harv. in F.C. 2 : 276 (1862); Benth. & Hook. f., Gen. Pl. 1 : 589 (1865); Harv., Gen. Pl., ed. 2 : 91 (1868); Oliv. in F.T.A. 2 : 325 (1871); Benth. in Trans. Linn. Soc. Lond. 30 : 363 (1875); Sim, For. Fl. Cape Col. 209 (1907); Harms in Engl., Pflanzenw. Afr. 3, 1 : 401 (1915); Bak. f., Leg. Trop. Afr. 3 : 784 (1930); Burtt Davy, Fl. Transv. 2 : 333 (1932); Phill., Gen. 392 (1951); Gilbert & Boutique in F.C.B. 3 : 220 (1952); Torre in C.F.A. 2 : 257 (1956); Brenan in F.T.E.A. Legum.–Mimos. : 9 (1959); Hutch., Gen. Fl. Pl. 1 : 288 (1964); Brenan in Kew Bull. 20 : 361 (1966); Schreiber in F.S.W.A. 58 : 17 (1967); Brenan in F.Z. 3, 1 : 13 (1970). Type species: *E. pursaetha* DC. (*E. monostachya* DC.).

Gigalobium P. Br., Hist. Jamaic. 362 (1789).

Adenopodia C. Presl, Epimel. Bot. 206 (1849).

Pusaetha L. ex Kuntze, Rev. Gen. Pl. 1 : 204 (1891); Taub. in Pflanzenfam. 3, 3 : 122 (1892).

Entadopsis Britton in N.Amer. Fl. 23 : 191 (1928); Gilbert & Boutique in F.C.B. 3 : 203 (1952).

Trees (not in our species), scandent shrubs, suffrutices or lianes; prickles present or absent. *Leaves* bipinnate; petioles eglandular or sometimes glandular; rhachis sometimes ending in a tendril; pinnae each with one to many pairs of leaflets. *Inflorescences* of spiciform racemes or spikes which are axillary or supra-axillary, solitary or clustered and often ± aggregated. *Flowers* hermaphrodite or male, 5-merous. *Calyx* gamosepalous, with 5 short teeth. *Petals* 5, free or nearly so or ± connate, separated from the ovary-base by a very short perigynous zone composed of stamens adnate to an apparent corolla-tube. *Stamens* 10, fertile, free or united basally; anthers with a usually very caducous apical gland. *Ovary* sessile or shortly stipitate; stigma terminal. *Pods* straight or curved and almost falcate, flat or rarely spirally twisted, sometimes enormous; at maturity the valves (but not the margins) splitting transversely into 1-seeded segments from which the outer layer (exocarp) of the pod-wall often peels off, the inner layer (endocarp) persisting as a closed envelope around the seed; the segments usually falling away from the margins which persist as a continuous but empty frame. *Seeds* (in the African species at least) ± compressed, elliptic or subcircular in outline, deep brown, smooth.

A genus of ± 30 species, widespread and mainly tropical; about 20 species in Africa and Madagascar, about four species in America. Four species occur in our area.

Entada is a southern Indian (Malabar) name used by van Rheede for *E. monostachya* (i.e. *E. pursaetha*).

Leaf-rhachis ending in a forked tendril; large unarmed woody liane with 1–2 pinnae pairs per leaf; leaflets 3–5 pairs, 25–75 × 11–35 mm; pods gigantic, up to 2 m long, 7–15 cm wide, woody......1. *E. pursaetha*

Leaf-rhachis not ending in a forked tendril; suffrutex, scandent shrub or woody climber, unarmed or armed with scattered recurved prickles, with 1–7 pinnae pairs per leaf, leaflets 6–13 pairs, smaller than above; pods at most 0,25 m long, 6 cm wide, coriaceous:

Suffrutex with erect annual stems, unarmed; pinnae 2–4 pairs per leaf; leaflets 15–40 × 7–20 mm; flowers yellowish-white..2. *E. arenaria*

Woody climber or scandent shrub, unarmed or armed; pinnae 1–7 pairs per leaf; leaflets 7,5–20 ×
   1,75–7(9) mm; flowers yellowish-white or dark purple or red:

   Unarmed slender woody climber; pinnae (1) 2 (3) pairs per leaf; petioles eglandular; flowers dark purple
     or red; ovary glabrous, sessile..3. *E. wahlbergii*

   Woody climber or scandent shrub, usually armed with scattered recurved prickles, sometimes unarmed;
     pinnae 5–7 pairs per leaf; petioles with a gland a short distance above the pulvinus; flowers
     yellowish-white; ovary densely villous, on a distinct stipe..........................4. *E. spicata*

1. Entada pursaetha *DC.*, Prodr. 2 : 425 (1825); Mém. Leg. 421 (1826); Brenan in Kew Bull. 10 : 161 (1955); Keay in F.W.T.A. ed. 2,1 : 490 (1958); Brenan in F.T.E.A. Legum.-Mimos. : 12 (1959); in F.Z. 3,1 : 15, t.3D (1970); Ross Fl. Natal 194 (1973). Type: A plant cultivated in Mauritius, *Delessert* (G-DC, lecto.; K, photo.!).

Entada monostachya DC., Prodr. 2 : 425 (1825); Mém. Leg. 422 (1826). Type: Rheede, Hort. Malabar. 9 : 151, t.77 (1689). *E. scandens* sensu Harv. in F.C. 2 : 276 (1862); Sim, For. Fl. P.E. Afr. 52 (1909), non (L.) Benth. *E. gigas* sensu Bak. f., Leg. Trop. Afr. 3 : 785 (1930) pro parte; Henkel, Woody Pl. Natal 237 (1934); Gilbert & Boutique in F.C.B. 3 : 220 (1952), non (L.) Fawc. & Rendle. *E. phaseoloides* sensu Brenan, Checklist Tang. Terr. 344 (1949); Torre in Mendonca, Contr. Conhec. Fl. Mocamb. 2 : 87 (1954), non (L.) Merr. *E. gogo* (Blanco) I. M. Johnston in Sargentia 8 : 137 (1949). Type as for *Adenanthera gogo* Blanco.

Adenanthera gogo Blanco, Fl. Filip. 353 (1837). No type specimen extant.

Large liane up to 75 m long, climbing to the tops of the tallest trees, unarmed; young branchlets glabrous or rarely (but not in our area) pubescent. *Leaves*: petiole up to 5 cm long, eglandular; rhachis ending in a forked tendril; pinnae 1–2 pairs; rhachillae 5,5–15 cm long; leaflets 3–5 pairs, 25–75 × 11–35 mm, elliptic to obovate-elliptic, obtuse or rounded apically, emarginate, glabrous or nearly so except for puberulence on the midrib above and near the base of the leaflet beneath. *Stipules* up to 3 mm long, deciduous. *Inflorescences* spicate, spikes axillary on lateral branches which are sometimes leafless and abbreviated, the spikes thus aggregated; spikes 6,5–15 cm long, axes pubescent; peduncles 1–6 cm long. *Flowers* yellowish-white or greenish-yellow, sessile or on pedicels up to 0,5 mm long. *Calyx* cupular, ± 1,25 mm long, shortly 5-toothed, glabrous or rarely (but not in our area) pubescent. *Petals* free, ± 2,5 mm long, glabrous. *Stamens* linear, ± 6 mm long; anthers with an apical caducous gland. *Ovary* up to 2,25 mm long, glabrous. *Pods* gigantic, 0,5–2 m long, 7–15 cm wide, woody, straight or curved, but not spirally twisted, outer leathery layer of the pod-wall falling away to expose a reddish-brown woody rigid inner layer, ultimately breaking up transversely into 1-seeded segments. *Seeds* deep brown, ± 5 × 3,5–5 cm, hard, smooth.

Widely distributed in tropical Africa; also from India to China, the Philippines, Guam and N. Australia. In our area it is found at low altitudes on the Zululand coast, particularly in the vicinity of Port Durnford. Occurs in riverine fringing vegetation and in swamp forest, the plants climbing to the canopy of the tallest trees.

NATAL.—2831 (Nkandla): Port Durnford (—DD), *Lawn 1849* (NH); *Lawn 1849a* (NH); *Strey 9919*.

Some years ago a solitary plant, *Ward 5315* (GRA, NU), was found just south of Durban at Jeffels Hill-South, Isipingo. It is not known whether or not this plant was introduced. The plant has since disappeared and the species is now not known to occur south of the Mtunzini District.

Despite the size of the pods, they are seldom easily visible owing to the density of the supporting canopy. In falling, the pods often break up partially into transverse segments. It is indeed remarkable that these enormous pods develop from such small ovaries.

Few specimens of *E. pursaetha* have been collected in our area and more material is desired.

The seeds of *E. pursaetha*, and probably also those of *E. gigas* (L.) Fawc. & Rendle, are frequently washed up on the Natal beaches and sometimes are also found on the Transkei coast. *E. gigas* differs from *E. pursaetha* in that the gigantic pods are spirally twisted and the flowers are on distinct slender pedicels 1–2 mm long.

2. Entada arenaria *Schinz* in Mém. Herb. Boiss. 1 : 118 (1900); Schreiber in F.S.W.A. 58 : 17 (1967); Ross in Bothalia 11 : 126 (1973). Type: South West Africa, Grootfontein Distr., mittelauf des Omuramba Omatako, *Schinz 277* (Z, holo.!).

subsp. **arenaria**.
Ross in Bothalia 11 : 126 (1973)

Entada nana Harms in Warb., Kunene–Samb. Exped. 244 (1903); Harms in Engl., Pflanzenw. Afr. 3, 1 : 403 (1915); Bak. f., Leg. Trop. Afr. 3 : 787 (1930); Torre in C.F.A. 2 : 258, t.51 (1956); F. White, For. Fl. N. Rhod. 92 (1962); Brenan in F.Z. 3, 1 : 19 (1970). Type: Angola, Habungu, *Baum 471* (B, holo. †; E, iso.!).

Suffrutex with erect annual stems up to 1,2 m high, unarmed, stems densely pubescent to subglabrous when young, longitudinally striate. *Leaves* densely puberulous or pubescent: petiole 6–12 cm long, sulcate above; rhachis 4–17 cm long, sulcate above; pinnae 2–4 pairs; rhachillae 7,5–14 cm long; leaflets 7–13 pairs, (12)20–35(40) × 7–20 mm, narrowly oblong, oblong or obovate-oblong, very asymmetric basally, the proximal side rounded to cordate and the distal side cuneate to cuneate-rounded, rounded to emarginate apically, lateral nerves conspicuous, pubescent throughout on lower surface or pubescence confined to the midrib, sometimes glabrous throughout. *Inflorescences* spicate, axillary, solitary or 1–3 together; spikes 4–10 cm long, axes glabrous to sparingly pubescent; peduncles up to 1,5 cm long. *Flowers* pale yellowish-white, on pedicels up to 1,75 mm long. *Calyx* campanulate, 1–2 mm long, 5-toothed, glabrous. *Petals* free or very shortly united basally, ± 3 mm long, glabrous. *Stamens* up to 6,5 mm long; anthers with an apical caducous gland. *Ovary* up to 2 mm long, sessile, glabrous. *Pods* 11–22 × 3,5–5(6) cm with a stipe up to 3 cm long, compressed, curved or distinctly falcate, subcoriaceous, the valves splitting transversely into 1-seeded segments. *Seeds* deep brown, ± 12,5 × 9 mm, smooth.

Found in Angola, South West Africa, Zambia and Rhodesia. Occurs in deep sand, often in woodland.

S.W.A.—1718 (Kuring-Kuru): 4,8 km S. of Omuramba Mpungu on road to Tsintsabis, *De Winter 3911*. 1819 (Karakuwisa): 32 km S. of Runtu, *De Winter 3808*. 1820 (Tarikora): 16 km N. of Tamso on road to Kapupahedi, *De Winter & Marais 4718*. 1821 (Andara): near Shamvura Camp, *De Winter & Wiss 4448*. 1918 (Grootfontein): Omuramba Omatako, *Schinz 277* (Z). 1920 (Tsumkwe): 48 km N. of Gautscha Pan, *Story 6435*.

Subsp. *microcarpa* (Brenan) J. H. Ross, which occurs in Zaire and in Zambia, differs from subsp. *arenaria* in having smaller and narrower pods.

3. Entada wahlbergii *Harv.* in F.C. 2 : 277 (1862); Benth. in Trans. Linn. Soc. Lond. 30 : 364 (1875); Torre in Mendonca, Contr. Conhec. Fl. Mocamb. 2 : 87 (1954); Brenan in Kew Bull. 10 : 169 (1955) pro parte excl. specim. *Michelmore & Bullock*; Keay in F.W.T.A. ed. 2,1 : 492 (1958); Brenan in F.T.E.A. Legum.-Mimos.: 18 (1959) pro parte excl. specim. Tanganyika; Brenan in F.Z. 3,1 : 22, t.3C (1970); Ross, Fl. Natal 194 (1973); in Bothalia 11 : 125 (1973). Type: Natal, probably Zululand, *Wahlberg s.n.* (S, holo.; K, PRE, photo).

Pusaetha wahlbergii (Harv.) Kuntze, Rev. Gen. Pl. 1 : 204 (1891). Type as above.

Entada flexuosa Hutch. & Dalz. in F.W.T.A. 1 : 356 (1928); in Kew Bull. 1928 : 401 (1928). Type: Nigeria, Northern Prov., Nupe, *Barter 991* (K, holo.!).

Entadopsis flexuosa (Hutch. & Dalz) Gilbert & Boutique in F.C.B. 3 : 206 (1952). Type as for *Entada flexuosa*. *Entadopsis wahlbergii* (Harv.) Pedro in Bol. Soc. Est. Moçamb. 92 : 10 (1955). Type as for *Entada wahlbergii*.

Slender unarmed woody climber up to 4 m high; young branchlets glabrous, flexuous. *Leaves:* petiole 1,8–4,8 cm long, glabrous; rhachis 0–4 cm long; pinnae (1) 2 (3) pairs; rhachillae 2,5–6,5 cm long, one or more of the rhachillae, usually the terminal ones, sometimes modified into a tendril or spirally twisted basally and bearing leaflets above; leaflets 6–12 pairs, 8–18 × 1,75–4,5 mm, obliquely-oblong, asymmetric basally, rounded apically and usually slightly mucronate, glabrous. *Stipules* inconspicuous. *Inflorescences* spicate, axillary, solitary, often aggregated on short leafless shoots or occupying terminal parts of the shoots; spikes 3–5,5 cm long, axes glabrous; peduncles 4–10 mm long. *Flowers* dark purple or red, on pedicels 1–1,5 mm long. *Calyx* green, cupular, up to 1,5 mm long, 5-toothed, glabrous. *Petals* green, up to 3,5 mm long, 1,4 mm wide, united with the stamens and ovary into a basal tube ± 1 mm long, glabrous, apex of each petal with an inwardly deflexed appendage. *Stamens* dark purple or red, 4–6,5 mm long; anthers yellowish, with an apical caducous gland. *Ovary* up to 2,5 mm long, sessile, glabrous. *Pods* 11–28 × 2,8–4 cm, with a stipe 1–2 cm long, compressed, curved or often falcate, subcoriaceous, the valves splitting transversely into 1-seeded segments. *Seeds* 9–11 × 7–10 mm, deep brown, smooth; areole 5,5–6,5 × 4–5 mm.

Widespread in Africa from Portuguese Guinea and Mali to Nigeria, the Sudan, southwards to Zaire, Mozambique and Natal. Occurs in bushveld, valley scrub and on the banks of dry watercourses and streams, usually on dry sandy soil.

NATAL.—2831 (Nkandla): 22,4 km S.S.W. of Nongoma (—BA), *Acocks 13012*; Umfolozi Game Reserve, near Tobothi (—BD), *Ross 2059*; Umhlatuzi valley, Melmoth road, not far from Empelengeni bridge (—DA), *Lawn 2198* (NH). 2832 (Mtubatuba): Hluhluwe Game Reserve, *Ward 1892* (NH); *Ward 2230* (GRA, NH). 2931 (Stanger): Tugela Valley, below San Souci, *Edwards 1659*; 4,8 km from Mandini on Tugela mouth road, *Edwards 1630*. Grid ref. unknown: without locality, *Gerrard 1706* (BM, K).

FIG. 23.—**Entada spicata.** 1, flowering branch, × ⅔; 2, flower, × 8; 3, anther, × 12; 4, ovary, × 10, all from *Ward* 1898; 5, pod, × ⅔, *Ross* 962.

E. wahlbergii is a variable species. In our area it is found almost entirely in Zululand at altitudes below 350 m. There is but a single gathering south of the Tugela River.

A number of tropical species with dark purple or red flowers are closely related to *E. wahlbergii*. *E. stuhlmannii* (Taub.) Harms from Tanzania and Mozambique differs from *E. wahlbergii* in having comparatively large leaflets with readily visible lateral nerves and nervation. This species has often been confused with *E. wahlbergii* in the past; Bak. f., Leg. Trop. Afr. 3 : 788 (1930) based his account of *E. wahlbergii* on a specimen of *E. stuhlmannii*.

4. **Entada spicata** (*E. Mey.*) *Druce* in Rep. Bot. Soc. Exch. Club Br. Isl. 1916 : 621 (1917); Merr. in Contr. Gray Herb. 59: 19 (1919); Burtt Davy, Fl. Transv. 2 : 333 (1932); Henkel, Woody Pl. Natal 235 (1934); Ross, Fl. Natal 194 (1973); in Bothalia 11 : 125 (1973). Syntypes: Natal, Durban [Port Natal], *Drege* (not traced); Cape Province, Bashee River, *Drege* (BM!; E!; K!; OXF!; P!; PRE!; TCD!).

Mimosa spicata E. Mey., Comm. 164 (1836); Meisn. in Hook., Lond. J. Bot. 2 : 101 (1843). Syntypes as above.

Entada ? natalensis Benth. in Hook., J. Bot. 4 : 333 (1841); Benth. in Hook., Lond. J. Bot. 5 : 78 (1846); Harv. in F.C. 2 : 276 (1862); Benth. in Trans. Linn. Soc. Lond. 30 : 364 (1875); Wood & Evans, Natal Plants 1 : 33, t.39 (1898); Sim, For. Fl. Cape Col. 209 (1907). Syntypes as above. *E. natalensis* Benth. var. *aculeata* Harv. in F.C. 2 : 276 (1862). Syntypes: Plant raised at Cape Town from Natal seeds, *J. D. Watt* (TCD!).

Adenopodia spicata (E. Mey.) Presl., Epimel. Bot. : 207 (1849). Syntypes as for *Entada spicata*.

Woody climber or scandent shrub up to 6 m high, usually armed with numerous scattered recurved prickles up to 5 mm long, sometimes unarmed; young branchlets densely puberulous or pubescent. *Leaves* sparingly to densely puberulous: petiole 3,5–9 cm long, unarmed or armed with recurved prickles, adaxial gland 0,7–1,5 mm long, often slightly raised or columnar, usually a short distance above the pulvinus, sometimes more than one gland present; rhachis 3,5–10,5(13,5) cm long, sulcate above, unarmed or armed with recurved prickles; pinnae 5–7 pairs; rhachillae 4–9 cm long; leaflets 7–12 pairs, 7,5–20(24) × 2,5–7(9) mm, obliquely-oblong to obovate-oblong, asymmetric basally, rounded or obtuse apically, sometimes slightly mucronate, sparingly to densely puberulous on both surfaces or sometimes ± glabrous, sometimes with a conspicuous basal tuft of hairs to one side of the midrib on the lower surface. *Stipules* linear, up to 6 mm long, soon deciduous. *Inflorescences* spicate, axillary, solitary or 2–3 together, usually forming a terminal panicle; spikes 1,5–4,5 cm long, axes sparingly to densely puberulous; peduncles 0,8–3 cm long. *Flowers* pale yellowish-white, sessile. *Calyx* cupular, up to 1 mm long, very shortly 5-toothed or almost truncate, subglabrous to sparingly puberulous. *Petals* up to 2,75 mm long, united basally for ± 1 mm, glabrous. *Stamens* up to 6 mm long; anthers with an apical caducous gland. *Ovary* up to 1,5 mm long, densely villous, on a long stipe that elevates it above the corolla. *Pods* (5)8–13(16) × (1,3) 1,8–3(3,7) cm, with a stipe up to 1,5 cm long, compressed, slightly curved to falcate, subcoriaceous, the valves splitting transversely into 1-seeded segments, sometimes one margin of the pod armed with recurved prickles. *Seeds* deep brown, ± 7–10 × 5–7 mm, smooth; areole ± 5 × 3 mm.

Found in the Transvaal, Swaziland, Natal and eastern Cape Province. Often forms dense impenetrable thickets on stream banks or on forest margins and in forest clearings.

TRANSVAAL.—2330 (Tzaneen): 14,4 km E. of Louis Trichardt, farm Rustfontein (—AA), *Schlieben 7553*; Duiwelskloof, Westfalia, eastern flank of Piesang Kop (—CA), *Scheepers 958*. 2430 (Pilgrim's Rest): Mariepskop, Blyde River picnic spot, *Van der Schijff 7315*. 2531 (Komatipoort): Barberton, *Rogers 14012*. 2730 (Vryheid): "Groothoek", Madhlangampisiberg, Wakkerstroom, *Killick 3918*. Grid ref. unknown: Blouberg, *Smuts & Pole Evans 882*.

SWAZILAND.—2631 (Mbabane): Grand Valley Hills, *Compton 27444*.

NATAL.—2732 (Ubombo): 3,2 km from Ingwavuma on Ndumu road, *Ross 2348*. 2830 (Dundee): road to Middledrift from Kranskop, *Edwards 2730*. 2831 (Nkandla): Nkandla forest, *Edwards 2332*. 2832 (Mtubatuba): Hluhluwe Game Reserve, *Ward 1898*. 2930 (Pietermaritzburg): Umgeni Valley, Nagle Dam road, *Ross 962* (K, NU). 2931 (Stanger): Umdloti, *Ross 485* (K, NU). 3030 (Port Shepstone): Uvongo, *Strey 9439*.

CAPE.—3129 (Port St. Johns): Ntafufu, *Strey 8523*. 3226 (Fort Beaufort): Douglasdale, near Seymour, *Wells 3860*. 3227 (Stutterheim): near Kei Road (—DA), *Ranger 232*; East London (—DD), *Rogers 28098*. 3228 (Butterworth): The Haven, *J. L. Gordon-Gray 962* (GRA, NU), Grid ref. unknown: Pondoland, *Bachmann 642, 645, 647a* (Z).

E. spicata is the most widespread species of *Entada* in our area. There appears to be a tendency for an increase in the degree of pubescence of the lower leaflet surface towards the southern limit of distribution of the species.

Van der Schijff 3471, a sterile specimen with exceptionally large leaflets from the Pretorius Kop area of the Kruger National Park, is probably referable to *E. spicata*.

E. spicata is sometimes confused with *Acacia kraussiana* Meisn. ex Benth. but the two species may be readily distinguished. In *A. kraussiana* the flowers are in round heads, the pods do not split transversely into segments, and the lowest pair of leaflets on each pinna are greatly reduced and bract-like.

The record of *E. natalensis* Benth. (i.e. *E. spicata*) by Sim, For. Fl. P.E. Afr. 53 (1909), needs confirmation. Although the description indicates *E. spicata*, the occurrence of this species in Mozambique has never been confirmed and it is thought that Sim was probably referring to *E. schlechteri* (Harms.) Harms. *E. schlechteri*, which is closely related to *E. spicata*, differs in having fewer pinnae pairs, broader leaflets and dark red flowers.

INDEX

Page   *Page*

ACACIA Mill. 24
accola Maiden & Betche* 114
adunca *A. Cunn. ex G. Don** 114
albida *Del.* . . . . *fig.* 7, 44
amboensis Schinz 99
arabica (Lam.) Willd. 82
    var. *kraussiana* Benth. 82
arabica sensu E. Mey. 82
arenaria *Schinz* 97
armata *R. Br.** 109
ataxacantha *DC.* 46
    var. *australis* Burtt Davy . . . 46
atomiphylla Burch. 80
baileyana *F. Muell** 109
barbertonensis Schweick. 77
benthamii Rochebr., non Meisn. . . . 82
borleae *Burtt Davy* 77
brevispica *Harms* 67
    subsp. dregeana (*Benth.*) Brenan . . . 67
    var. *dregeana* (Benth.) Ross & Gordon-
      Gray 67
    var. *schweinfurthii* (Brenan & Exell) Ross
      & Gordon-Gray 68
burkei *Benth.* 56
caffra (*Thunb.*) Willd. 60
    var. *campylacantha* (Hochst. ex A. Rich.)
      Aubrev. 60
    var. *longa* Glover 61
    var. *namaquensis* Eckl. & Zeyh . . . 61
    var. *pechuelii* Kuntze 65
    var. *rupestris* Sim 46
    var. *tomentosa* Glover 61
    var. *tomentosa* sensu Bak.f. . . . 64
    var. *transvaalensis* Glover . . . 61
caffra sensu N.E. Br 64
caffra sensu Oliv. pro parte 55
caffra sensu Oliv. pro parte 57
caffra sensu Oliv. pro parte 60
caffra sensu Schinz 62
callicoma *Meisn.* 106
campylacantha Hochst. ex A. Rich. . . 58
capensis (Burm.f.) Burch. 70
catechu sensu Harms 64
catechu sensu E. Mey 61
catechu sensu Schweinf. 58
    subsp. *suma* (Roxb.) Roberty
    var. *campylacantha* (Hochst. ex A.
      Rich.) Roberty 60
    var. *baumii* Roberty 64
cinerea Schinz, non Spreng 64
circummarginata Chiov. 50
clavigera E. Mey. 95
    subsp. *clavigera* 96
cultriformis *A. Cunn. ex G. Don** . . . 113
cyanophylla Lindl.* 110
cyclopis A. Cunn. ex Loudon, nomen nudum* 112
cyclops *A. Cunn. ex G. Don** . . . 111
davyi *N.E. Br.* 77
dealbata *Link** 107
decurrens Willd.* 108
    var. *mollis* Lindl.* 108
decurrens sensu Bak. f. pro parte* . . . 107
delagoensis Harms 54

detinens Burch. 52
dulcis Marloth & Engl. 65
dulcis sensu Henkel 57
elata *A. Cunn. ex Benth.** 113
elephantina Burch. 140
elephantorhiza DC. 140
engleri Schinz 127
eriadenia Benth. 46
erioloba *E. Mey.* 78
erioloba *E. Mey.* × Acacia haematoxylon
    Willd. 81
erubescens *Welw. ex Oliv.* 65
etbaica sensu Torre 92
exuvialis *Verdoorn* 73
fallax E. Mey.. 61
farnesiana (*L.*) *Willd.** 106
ferox Benth. pro parte 52
ferox Benth. pro parte 56
fimbriata *A. Cunn. ex G. Don** . . . 114
fleckii *Schinz* 63
galpinii *Burtt Davy* 57
gansbergensis Schinz 62
gerrardii *Benth.* 93
    var. gerrardii. 93
    subsp. negevensis *Zohary* . . . 94
gillettiae Burtt Davy 91
giraffae Willd.. 81
    var. *espinosa* Kuntze 79
giraffae sensu auct. mult., non Willd. . . 78
giraffae Willd. × *Acacia haematoxylon*
    Willd. 81
glandulifera Schinz 75
glandulifera sensu Burtt Davy pro parte . . 74
glandulifera sensu Burtt Davy pro parte . . 76
goeringii Schinz 91
grandicornuta Gerstn. 96
haematoxylon *Willd.* 80
harmsiana Dinter 87
hebeclada *DC.* 100
    subsp. chobiensis (*O. B. Miller*) Schreiber 102
    subsp. hebeclada. 101
    subsp. tristis *Schreiber* . . . 102
    var. *stolonifera* Dinter . . . 102
hebeclada sensu Bews. 99
hebeclada sensu Harms 102
hebeclada sensu F. White pro parte . . 102
hebecladoides Harms 93
hereroensis *Engl.* 62
hermannii Bak. f. 97
heteracantha Burch. 89
hirtella E. Mey. 70
    var. *inermis* Walp. 70
hirtella sensu Sim 96
horrida Willd. pro parte 70
    var. *transvaalensis* Burtt Davy . . 70
horrida sensu auct. mult.. 70
inconflagrabilis Gerstn. 70
inermis Marloth 18
kalachariensis Schinz 128
karroo *Hayne* 69
    var. *transvaalensis* (Burtt Davy) Burtt
      Davy 70

---

* An asterisk signifies exotic species or genera; synonyms are in italics.

Page

kirkii *Oliv.*. 87
   subsp. kirkii 87
      var. *intermedia* Brenan 87
      var. *kirkii* 87
kraussiana *Meisn. ex Benth.*. 66
kwebensis N.E. Br. 65
lasiopetala sensu Burtt Davy 99
lebbeck (L.) Willd.* 22
litakunensis Burch. 89
longifolia (*Andr.*) *Willd.** 111
   var. longifolia 111
   var. sophorae (*Labill.*) *F. Muell. ex Benth.* 111
longipetiolata Schinz 65
lophantha Willd* 23
luederitzii *Engl.* 90
   var. luederitzii 91
   var. retinens (*Sim*) *Ross & Brenan* . 91
luederitzii Engl. pro parte 92
lugardiae N.E. Br. 46
maidenii *F. Muell.** 114
maras Engl. 89
marlothii Engl. 18
mearnsii De Wild.* 107
melanoxylon *R. Br.** 112
mellei Verdoorn 62
mellei sensu O. B. Miller. 61
mellifera (*Vahl*) *Benth.* 50
   subsp. detinens (*Burch.*) *Brenan* . . 52
   subsp. mellifera 51
mellifera sensu Henkel 55
mollissima sensu auct. mult.* . . . 107
montis-usti *Merxm. & Schreiber* . . . 53
mossambicensis Bolle 44
mossambicensis sensu Henkel . . . 56
multijuga Meisn. 61
natalitia E. Mey 70
nebrownii *Burtt Davy* 75
nebrownii sensu Burtt Davy pro parte, non
  sensu stricto 74
nebrownii sensu Burtt Davy pro parte, non
  sensu stricto 76
nigrescens *Oliv.* 54
   var. *pallens* Benth. 55
nilotica (*L.*) *Willd. ex Del.* 82
   subsp. *adstringens* (Schumach. & Thonn.)
     Roberty var. *kirkii* (Oliv.) Roberty. . 87
   subsp. kraussiana (*Benth.*) *Brenan*. *fig. 9,* 82
     var. *kraussiana* (Benth.) A. F. Hill . 84
   subsp. *subalata* sensu Boughey. . . 84
pallens (Benth.) Rolfe 55
paradoxa *DC.** 109
passargei Harms 55
pendula *A. Cunn. ex G. Don** . . . 114
pennata sensu Harv. pro parte . . . 68
pennata sensu E. Mey. 67
   var. *dregeana* Benth. 67
permixta *Burtt Davy* 76
   var. *glabra* Burtt Davy 73
podalyriifolia *A. Cunn. ex G. Don** . . 109
polyacantha *Willd.* 58
   subsp. campylacantha (*Hochst. ex A.*
     Rich.) *Brenan* . . . . . *fig. 8,* 58
   subsp. polyacantha 60
procera (Roxb.) Willd.* 23
pycnantha *Benth.** 110
redacta *J. H. Ross* 104

Page

reficiens *Wawra* 92
   subsp. reficiens 92
reficiens sensu Schreiber pro parte . . . 91
rehmanniana *Schinz* 98
rehmanniana sensu M.A. Exell 145
reticulata (L.) Willd. pro parte 70
retinens Sim 91
retinens sensu O. B. Miller 91
retinodes *Schlechtend.** 113
robusta *Burch.* 94
   subsp. clavigera (*E. Mey*) *Brenan* . . 95
   subsp. robusta 95
   subsp. usambarensis (*Taub.*) *Brenan* . 96
robusta sensu Codd 96
robynsiana *Merxm. & Schreiber* . . . 52
rogersii Burtt Davy 75
rostrata Sim 50
rufobrunnea N.E. Br. 97
saligna (*Labill.*) *Wendl.** 110
sambesiaca Schinz 96
schlechteri *Harms* 106
schweinfurthii *Brenan & Exell* 68
   var. schweinfurthii 68
senegal (*L.*) *Willd.* 48
   var. leiorhachis Brenan 50
   subsp. *mellifera* (Vahl) Roberty . . 51
   var. rostrata *Brenan*. 49
   subsp. *trispinosa* (Stokes) Roberty pro
     parte 50
senegal sensu O. B. Miller 57
sericocephala Fenzl 11
seyal sensu Sim 70
   var. *multijuga* sensu O. B. Miller . . 97
sieberana *DC.*. 98
   var. *vermoesenii* (De Wild.) Keay &
     Brenan 99
   subsp. *vermoesenii* (De Wild.) Troupin . 99
   var. woodii (*Burtt Davy*) *Keay & Brenan* 99
spinosa Marloth & Engl.. 49
spinosa E. Mey. 127
spirocarpa Hochst. ex A. Rich 90
spirocarpoides Engl. 89
sp. 77
sp. 105
sp. 1 50
sp. sensu Schinz 65
sp. nov. sensu Schinz. 70
sp. cf. hebeclada sensu Wood 99
sp. cf. stolonifera sensu Torre pro parte . 99
sp. aff. trispinosa sensu Schinz . . . 65
sp. cf. uncinata sensu Torre 92
stolonifera Burch. 101
   var. *chobiensis* O. B. Miller . . . 102
stuhlmannii *Taub.*. . . . . . *fig.* 11, 104
subalata sensu Brenan 84
suma sensu Benth. pro parte 60
swazica *Burtt Davy* 74
tenax Marloth. 52
tenuispina *Verdoorn* 73
terminalis sensu Court* 113
tortilis (*Forsk.*) *Hayne* 88
   subsp. heteracantha (*Burch.*) *Brenan* . 89
   subsp. spirocarpa (*Hochst. ex A. Rich.*)
     Brenan 90
     var. crinita *Chiov.* 90
   subsp. tortilis 89
trispinosa Marloth & Engl. ex Schinz . . 49

Page

tristis Welw. ex Oliv. 102
uncinata Engl. 92
uncinata sensu O. B. Miller 91
vermoesenii De Wild.. 99
verrucifera Harms. 87
viridiramis Burch.. 130
viscidula *A. Cunn. ex Benth.** 114
visite *Griseb.** 113
volkii Suesseng. 50
walteri Suesseng 75
welwitschii *Oliv.* 54
    subsp. delagoensis (*Harms*) *Ross & Brenan* 54
    subsp. welwitschii 54
welwitschii Oliv. pro parte 54
woodii Burtt Davy 99
xanthophloea *Benth.* . . . . . *fig.* 10, 85
Adenanthera gogo Blanco 151
Adenopodia C. Presl. 150
spicata (E. Mey.) Presl. 154
ALBIZIA Durazz.. 7
    adianthifolia (*Schumach.*) *W. F. Wight.* .
    . . . . *fig.* 2, 20
    amara (*Roxb.*) *Boiv.* 11
        subsp. sericocephala (*Benth.*) *Brenan* . . 11
    amara sensu Oliv.. 11
    anthelminitica (*A. Rich.*) *Brongn.* . . . 18
        var. *australis* Bak. f.. 18
        var. *pubescens* Burtt Davy 18
    antunesiana *Harms* 15
    brevifolia *Schinz* 10
    chinensis (*Osbeck*) *Merr.** 8
    distachya (Vent.) Macbride*. 23
    evansii Burtt Davy 19
    fastigiata (E. Mey.) Oliv. 20
    forbesii *Benth.*. 14
    gummifera sensu C.A. Sm. pro parte . . 20
    harveyi *Fourn.*. . . . . . . *fig.* 1, 12
    hypoleuca Oliv. 12
    julibrissin *Durazz.** 7
    lebbeck (*L.*) *Benth.** 22
        var. *australis* Burtt Davy 16
    lophantha (*Willd.*) *Benth.** 23
    lugardii N.E. Br. 55
    mossambicensis Sim 15
    pallida Fourn. 12
    pallida Harv. 12
    parvifolia Burtt Davy. 10
    petersiana (*Bolle*) *Oliv.* 19
        subsp. evansii (*Burtt Davy*) *Brenan* . . 19
    procera (*Roxb.*) *Benth.** 23
    rhodesica Burtt Davy. 16
    rogersii Burtt Davy 10
    sericocephala Benth. 11
    struthiophylla Milne-Redhead 11
    suluensis *Gerstn.* 17
    tanganyicensis *Bak.f.* 16
        subsp. tanganyicensis 16
    umbalusiana Sim 18
    versicolor *Welw. ex Benth* 14
        var. *mossambicensis* Schinz. 15
AMBLYGONOCARPUS Harms. 132
    andongensis (*Welw. ex Oliv.*) Exell & Torre.
    . . . . *fig.* 18, 132
    obtusangulus (Welw. ex Oliv.) Harms . . 132
    schweinfurthii Harms. 132

Page

Besenna A. Rich. 7
    anthelmintica A. Rich. 18
Cailliea Guill. & Perr. 123
    dichrostachys Guill. & Perr. 127
        var. *leptostachys* (DC.) Guill & Perr . . 127
    nutans (Pers.) Skeels 127
DESMANTHUS Willd.. 120
    leptostachys DC. 127
    nutans (Pers.) DC. 126
    sect. *Dichrostachys* DC. 123
    stolonifer DC. 121
    trichostachys DC. 126
    virgatus (*L.*) *Willd.* 120
DICHROSTACHYS (DC.) Wight & Arn.. . . 123
    arborea N.E. Br. 127
    caffra Meisn. ex Benth. 127
    cinerea (*L.*) *Wight & Arn.* 123
        subsp. africana *Brenan & Brummitt* . . 126
            var. africana *Brenan & Brummitt*
            . . . . *fig.* 16, 126
            var. *lugardiae* Brenan & Brummitt . 127
            var. plurijuga *Brenan & Brummitt* . 128
            var. pubescens *Brenan & Brummitt* . 128
            var. setulosa (*Welw. ex Oliv.*) *Brenan*
              & Brummitt 128
        subsp. argillicola *Brenan & Brummitt*. . 127
            var. hirtipes *Brenan & Brummitt* . 127
        subsp. forbesii (*Benth.*) *Brenan & Brum-*
           mitt. 128
        subsp. nyassana (*Taub.*) *Brenan* .*fig.* 15, 124
    glomerata (Forsk.) Chiov. 123
        subsp. *nyassana* (Taub.) Brenan . . 126
    glomerata sensu Fl. Pl. Afr. 128
    lugardiae N.E. Br. 127
    major Sim 124
    nutans (Pers.) Benth. 127
        var. *setulosa* Welw. ex Oliv. . . . 128
    nyassana Taub. 124
ELEPHANTORRHIZA Benth. 138
    burchellii Benth. 140
    burkei *Benth.* 143
    dinteri Phillips ined. 140
    elephantina (*Burch.*) *Skeels* . . . *fig.* 21, 140
        var. *burkei* (Benth.) Merr. 144
    goetzei (*Harms*) *Harms* 145
        subsp. goetzei . . . . . *fig.* 22, 145
        subsp. lata *Brenan & Brummitt* . . 146
    obliqua *Burtt Davy* 139
        var. glabra *Phillips* 139
        var. obliqua 139
    petersiana *Bolle* 146
    praetermissa *J. H. Ross* 144
    pubescens Phillips ined. 143
    rangei *Harms* 146
    rangei sensu Phillips 140
    rubescens Gibbs 145
    schinziana *Dinter.* 148
    suffruticosa *Schinz* 148
    sp. 149
    sp. cf. goetzei (Harms) Harms 145
    sp. cf. petersiana sensu Gomes e Sousa . 145
    transvaalensis Phillips ined. 140
    woodii *Phillips* 141
        var. pubescens *Phillips* 143
        var. woodii 143

<table>
<tr><td colspan="2" align="center">Page</td><td colspan="2" align="center">Page</td></tr>
</table>

ENTADA Adans.	150
arenaria *Schinz*	151
subsp. arenaria	151
subsp. microcarpa (*Brenan*) *J. H. Ross*	152
flexuosa Hutch. & Dalz	152
gigas sensu auct.	151
gogo (Blanco) I. M. Johnston	151
monostachya DC.	151
nana Harms	151
? *natalensis* Benth.	154
var. *aculeata* Harv.	154
phaseoloides sensu Brenan	151
pursaetha *DC.*	151
scandens sensu Harv.	151
schlechteri (*Harms*) *Harms*	155
spicata (*E. Mey.*) *Druce* .fig. 23,	154
stuhlmannii (*Taub.*) *Harms*	154
wahlbergii *Harv.*	152
Entadopsis Britton	150
flexuosa (Hutch. & Dalz.) Gilbert & Boutique	152
wahlbergii (Harv.) Pedro.	152
Faidherbia A. Chev.	24
albida (Del.) A. Chev.	45
Farnesia Gasparr.	24
Gigalobium P. Br.	150
Inga fastigiata (E. Mey.) Steud.	20
mellifera (Vahl) Willd.	51
sericocephala A. Rich.	11
LEUCAENA Benth.	115
glauca sensu auct. mult.,	115
leucocephala (*Lam.*) *De Wit** .fig. 12,	115
MIMOSA L.	117
adianthifolia Schumach.	20
amara Roxb.	11
asperata L.	117
caffra Thunb.	61
capensis Burm. f. pro parte	70
cinerea L.	123
decurrens Donn, nomen nudum*	108
decurrens Wendl.*	108
distachya Vent.*	23
eburnea sensu Boj.	70
farnesiana L.*	106
glauca sensu L. pro parte	115
glomerata Forsk.	123
lebbeck L.*	22
leucacantha Jacq.	70
leucocephala Lam.*	115
longifolia Andr.*	111
mellifera Vahl	51
natans L.f.	121
nilotica L.	82
nilotica sensu Burm. f.	70
nilotica sensu Thunb.	82
nilotica sensu Thunb.	70
nutans Pers.	126
pigra *L.* .fig. 13,	117
procera Roxb.*	23
prostrata Lam.	121
pudica *L.*	118
var. hispida *Brenan**	118
saligna Labill.*	110
senegal L.	49
spicata E. Mey.	154
sp. sensu Paterson	79
tortilis Forsk.	88
virgata L.*	120
Mimosa L. pro parte.	24
NEPTUNIA Lour.	120
natans (L.f.) Druce	121
oleracea *Lour.* .fig. 14,	121
prostrata (Lam.) Baill.	121
stolonifera (DC.) Guill. & Perr.	121
NEWTONIA Baill.	134
hildebrandtii (*Vatke*) *Torre*	134
var. hildebrandtii .fig. 19,	134
var. pubescens *Brenan*	134
Phyllodoce Link, non Salisb.	24
Piptadenia goetzei Harms	145
hildebrandtii Vatke	134
sp. sensu Henkel	134
Prosopis sensu E. Mey.	138
chilensis auct., non (Mol.) Stuntz	7
elephantina (Burch.) E. Mey.	140
elephantorrhiza (DC.) Spreng.	140
glandulosa *Torrey*.	7
juliflora auct., non (Swartz) DC.	7
? *kirkii* Oliv.	45
pubescens *Benth.*	6
velutina *Wooton*	7
Pusaetha L. ex Kuntze	150
wahlbergii (Harv.) Kuntze	152
Tetrapleura andongensis Welw. ex Oliv.	132
obtusangula Welw. ex Oliv.	132
Vachellia Wight & Arn.	24
XEROCLADIA Harv.	130
viridiramis (*Burch.*) *Taub.* .fig. 17,	130
zeyheri Harv.	130
XYLIA Benth.	136
africana sensu Torre	136
torreana *Brenan* .fig. 20,	136
Zygia sensu E. Mey.	7
fastigiata E. Mey.	20
petersiana Bolle	19